The Ethnobotany

of the

California Indians

Revised, Updated, Expanded

George R. Mead, Ph. D.

E-Cat Worlds Press

Comments and questions? –> gmead01@gmail.com

The Ethnobotany of the California Indians

LCCN 2014937740

Mead, George R.
The Ethnobotany of the California Indians /George R. Mead.
p. cm.
Includes bibliographical references
ISBN-13 978-0-9890927-9-1
1. Indians of North America - Ethnobotany - California.
2. Ethnobotany-California. I. Title.

E-Cat Worlds established its publishing program as a reaction to the large commercial publishing houses currently dominating the book industry and the smaller intellectual clones. It is interested in publishing works of fiction and non-fiction that are often deemed insufficiently profitable or commercial or that are not necessarily reflective of current literary trends and fads.

E-Cat Worlds, 57744 Foothill Road, La Grande OR 97850
www.ecatworldspress.com
SAN 255-6383
In the middle of nowhere - Creativity.

Second Edition - Updated, Revised, Expanded
Printed in the United States of America

Nonfiction

A History of Union County
The Ethnobotany of the California Indians, 2nd Edition
A History of The Chinese in The West: 1848-1880
Yachats. The Town Called "Dark Water at the Foot of the Mountains."

Fiction

From Grandeville.

Portal
Lair
Search
Not Again
And Again.
Magiwitch
Rebirth
Offspring
Holiday
Treasure
E'Nilt
Braidna

A Tale of The Feyra

Jonathon and Dee
Dee Of The Fontala
Dee and The People
Dee and The Golden Cartouche

The Seven Lands
(With Zakke L. Zacog)

Seventeen Siblings

CONTENTS

Preface (to the First Edition).

This work was written primarily for an audience of anthropologists rather than for botanists. Thus, this work is organized in a manner that the author felt would be easiest to use for his chosen target group. The plants are listed in alphabetical order by scientific name rather than in the usual botanical reference structure of family, genus, and species.

As the scientific nomenclature has a habit of changing through time, all the scientific names have been brought into accord with a single reference manual (Munz and Keck 1965; Munz 1968) for ease of use and for future checking as well as to keep a standardized naming convention through out the work. This "standardization" was accomplished by cross-checking the plant scientific names found in the various sources utilized against Jepson (1925), Bailey (1950), and Abrams (1940-1960). The marine algaes were checked against Smith (1944) and Gulberlet (1956); the lichens in Tuckerman (1882), Schneider (1898), and Duncan (1959); the fungi in Krieger (1936); and the mosses in Grout (1928-1940). All the tribal names found in the literature were brought into a single naming source, Heizer (1966).

The work is intended to answer, for the most part, a single question: How was this plant used/utilized by the various tribes of California? In addition answers (as provided in the literature) to other questions are listed, i.e.: What did this tribe call this plant? What is the food value of this plant? Are there known hazards associated with this plant?

In most cases, including all the fungi, plants (in the inclusive sense we are using the term) were selected for inclusion if, and only if, a scientific name was provided. If only a common name was provide, it was not included. Common names tend to be frequently so generic and cross-cutting across species as to be most often not precise enough. In those few cases where the common name was applicable to only a single genus and species, it was included.

Within this work, the synonyms and common names are those provided within the sources searched. Many of these could not be reconciled with Munz and Keck (1965), Munz (1968). This appears to be due to mis-spellings, typographic errors, etc., within the literature (an all to easy thing to do).

As you can see, plants usage has been handled under four categories: Food; Material; Medicine; and, Miscellaneous (a catch-all for everything else).

A Note(s) heading lists information such as: potentially poisonous properties, known season when ripens, etc.

Certain listings are by genus only, such as: Quercus spp. This stems, no pun intended, from ambiguity in the literature, especially when more than one species is growing in the area and the author did not more closely identity which species was being discussed.

Note.

While the title of this work is "The Ethnobotany of the California Indians," using the word "the" does not mean that there is a complete listing of every tribe that once lived within the political boundary, California. The gaps reflect the interest area(s) of those who wrote the literature as well as any data which I may not have found. To me, this work is a beginning, not an end. It is but one brick in the foundation of our knowledge of the ethnobotany of the original inhabitants of the land we call California.

This is as well, a work of literary archaeology. The original manuscript fell into a publishing black hole. This, then, is a re-creation based upon an older manuscript I found while cleaning out the attic. Also found were the original 3x5 cards and copies of research materials. The original research was done in the years 1966-1968. Additional research was done in the years 2000-2002.

The Revision (to the Second Edition).

While in a conversation with Gary Breschini of Coyote Press, California, he brought up the topic of doing a revision with newer information. I mumbled something about it taking at least a year to do as my computer managed to eat everything during its nervous breakdown, and that, I would do some searching, here and there, just to see. After some hours doing just that, it appeared to me that there was enough "new" stuff (new to me) to begin a revision, incorporating these new pieces of information. So then, here it is. In the process, started in the middle of 2013, I also "fixed" some strange formatting that somehow crept into the First Edition during the transmission, electronically, of the ms to the printer. In addition, I changed the overall format of the data presented as I thought that it might be more useful.

But, as seemingly easy projects have a tendency of doing, this one turned into a major exercise. In 2012 a major revision occurred, the release of "The Jepson Manual, Vascular Plants of California, Thoroughly Revised and

Expanded, 1600 pages." This meant that every entry herein had to be checked against the "correct" nomenclature as it was currently viewed by the botanists (Baldwin et al 2012).

In addition, new data was available for the "seaweeds:" Kyle et al 2009, Christopher et al 2006, Miller 2012, Hwan et al 2001; and the lichens and fungus: Jordan 1973; Blanco et al 2006; Nash et al 2001, Trudell and Ammirati 2009, Crespo et al 2007, Kyung et al 2005, Justo 2011.

The nomenclature listed under the heading Synonym(s) is/are all the labels put on the plant by the various authors utilizing whatever materials they had at hand to identify them with. These names are not necessarily what is currently recognized by the botanists but are listed because they were what was used at that moment.

The Plants

1   *Abies* spp.

   Family: *Pinacea*

Usage: Material: **Bear River**: used bark cut in two foot lengths for meat platters. **Yuki**: made fish spears 8-12 long. **Chimariko**: used bark to form house walls. **Maidu, Modoc (Lutuami)**: made dye from bark. **Chumash, Gabrielino**: made sea-worthy plank canoes when redoowd (*Sequoia sempervirens*) driftwood unavailable. **Shasta**: used wood to burn in sweathouse; inhaling smoke brought good fortune; made house posts.

          Medicine: **Shasta**: used for fire sweating and inhaling wood occasionally used to smoke, particularly for some special luck. **Tubatulabal**: dropped a live coal in the resin and inhaled smoke to relieve sinus problems, asthma, and other breathing difficulties. **Maidu**: used branches to prepare decoction as a diuretic. **Maidu, Modoc (Lutuami)**: used resin to cure wounds and sores. **Modoc (Lutuami)**: chewed resin to relieve sore throats. **Karok**: boiled needles, drank infusion as a tonic or cure for

various ailments. **Miwok:** decoction of needles used externally and internally to relieve colds and rheumatic pain; men and women used leaves as a deodorizer.
Citation(s): Nomland 1938; Foster 1944; Holt 1946; Dixon 1910; Strike 1994.

2   *Abies bracteata*
Common Name(s): Santa Lucia Fir.

Usage: Material: **Esselen:** source of firewood.
Citation(s): Breschini and Haversat 2004.

3   *Abies concolor*
Synonym(s): *Abies concolor lowiana.*

Usage: Food: **Kawaiisu (Tehachapi):** seeds generally considered inedible.

Material: **Miwok:** utilized mats made of the boughs to be used in the process of leaching acorns. **Kawaiisu (Tehachapi):** branches used to make windbreaks.

Medicine: **Washo:** soft resin from the bark was eaten to cure tuberculous; a teaspoon daily, or a little each day until cured; boiled bark infusion is drunk freely for tuberculous.
Citation(s): Merriam 1955; Train et 1941; Zigmond 1981.
Terminology: **Washo:** mah-<u>hah</u>-wa; shaw-<u>wa</u>-eh.
    Kawaiisu (Tehachapi): puugusivi.
Citation(s): Train et al 1941; Zigmund 1981.

4   *Abies grandis*
        Common Name(s): Lowland Fir.

        Usage: Medicine: **Karok**: the needles are boiled in a basket with hot stones and drunk; the tonic may be taken for illness. **Miwok**: applied externally and taken internally in case of colds and rheumatism.

        Citation(s): Schenck and Gifford: 1952; Barret and Gifford 1933.

        Terminlogy: **Karok**: maxaisarip.

        Citation(s): Schenck and Gifford 1952.

5   *Abies magnifica*
        Synonym(s): *Abies nobilis* var. *magnifica*
        Common Name(s): Red Fir.

        Usage: Food: **Maidu:** made tea from needles; ate gum and cambium.

        Citation(s): Strike 1994.

6   *Abronia* spp.
        Family: *Nyctaginaceae*
        Common Name(s): Sand Verbena, Wild Sand Verbena.

        Usage: Medicine: **Diegueño:** used as a diuretic.

        Citation(s): Strike 1994.

7   *Abronia villosa*
        Common Name(s): Wild Sand Verbena.

        Usage: Material: **Cahuilla:** childrens game used plants as a cover to play guess who was

hiding.
Citation(s): Bean and Saubel 1972.
Terminology: **Cahuilla:** nyuku.
Citation(s): Bean and Saubel 1972.

8   *Acacia gregorii*

Family: *Fabaceae*
Common Name(s): Catclaw.

Usage: Food: **Cahuilla:** pods eaten fresh, or, dried
                      and ground into flour to make mush
                      or cakes.

         Material: **Cahuilla:** considered outstanding
                      construction material and fine
                      firewood.

Citation(s): Bean and Saubel 1972.
Terminology: **Cahuilla:** amul.
Citation(s): Bean and Saubel 1972.
Note(s): the food – seeds:

protein   33.8%,
oil content 25.4%,
trace of alkaloids.

Citation(s): Bean and Saubel 1972.

9   *Acamptopappus sphaerocephalus*

Family: *Asteraceae*

Usage: Medicine: **Kawaiisu (Tehachapi):** plant cut off
                      at ground level, washed and mixed
                      with enough water to form a salve,
                      applied externally for relief of pain.

Citation(s): Zigmond 1981.

10   *Acamptopappus sphaerocephalus* var. *hirtellus*

Usage: Medicine: **Kawaiisu (Tehachapi)**: plant cut off at ground level, mashed them, washed and mixed with water to form a salve, applied externally for relief of pain.
Citation(s): Strike 1994.

11   *Acer* spp.
           Family: *Sapindaceae*
           Common Name(s): Maple.

Usage: Material: **Shasta**: used leaves to line food bins.
Citation(s): Holt 1946.

12   *Acer circinatum*
           Common Name(s): Vine Maple

Usage: Material: **Achomawi**: branches used for the frames of snowshoes.
         Misc: **Karok**: used as a "love medicine."
Citation(s): Merriam 1967; Schenck and Gifford 1952.
Terminology: **Karok**: mahsaan, "mountain saan;" sharis.
Citation(s): Schenck and Gifford 1951.

13   *Acer macrophyllum*
           Common Name(s): Big Leaf Maple, Oregon Maple, Broad-Leaved Maple, Tree Maple.

Usage: Material: **Maidu**: used sapwood (stems) as warp in basketry. **Miwok**: used sapwood as woof and wrapping; bark in rim binding in baskets. **Monache**: made musical bows about 4 feet long.

Yana: the bark was dried, pounded, and separated into filaments and used for pendant strands of women's skirts; mats of leaves used to cover layers of dried salmon stored for winter in baskets; leaves placed under and between layers of bulbs when they are cooked; made into paddle for stirring. **Wailaki**: made deer catching corral from bark. **Pomo**: man's skirt made of strips of inner bark. **Pomo (Southwestern)**: made dice and staves for gambling game. **Tolowa**: inner bark used for ordinary dress for women. **Cahuilla**: limbs used in house construction and considered good firewood.

Citation(s): Baumhoff 1958; Chesnut 1902; Gifford and Merriam 1955; Merril 1923; Sapir and Spier 1943;Schenck and Gifford 1952; Steward 1933; Goodrich et al 1980; Bean and Saubel 1972.

Terminology: **Karok**: saan, sanpirish (leaves), saantaaf (mats, made of leaves). **Pomo**: kalam kale. **Yana**: ke'watsi. **Miwok: (Northern Sierra)**: pi'pum; **(Central Sierra)**: pi'punu sa'iyi. **Pomo (Southwestern)**: qalam?. **Cahuilla**: sivily.

Citation(s): Schenck and Gifford 1952; Gifford 1957; Sapir and Spier 1943; Barret And Gifford 1933; Goodrich et al 1980; Bean and Saubel 1972.

14 *Achillea* spp.

Family: *Asperaceae*
Common Name(s): Yarrow.

Usage: Medicine: **Bear River**: compresses used to treat

cuts and sores; stomach trouble treated with an infusion of steeped plants; pains of mild rheumatism treated with steamed planes. **Karok:** soaked stalks and leaves in water and used as poultice for wounds. **Costanoan, Washo:** applied heated leaves to prevent swelling. **Costanoan:** heated leaves to hold in mouth to relieve toothache; decoction used to wash sores. **Maidu:** drank decoction to stop diarrhea, relieve bladder or kidney problems, stop aches such as stomach ache and toothache; used poultice to relieve skin irritations. **Miwok:** used leaves as poultice on cuts and wounds; tea used for stomach aches and chest congestion.

Material: **Maidu:** used leaves to produce green dye. **Shasta:** women used leaves to pad caps when carrying burden baskets.

Citation(s): Nomland 1938; Strike 1994; Rayburn 2012.

15   *Achillea millefolium*

Synonym(s): *Achillea borealis, Achillea Millefolium lanulosa.*

Common Name(s): Yarrow, Common Yarrow, Milfoil, Common Milfoil, Yerba Muela.

Usage: Material: **Shasta:** the leaves were used to pad women's caps when carrying burden baskets because of their sweet odor.

Medicine:	**Pomo (Southwestern):** juice from mashed leaves put on sores as a salve. **Costanoan:** leaves, tea, applied to cuts, bruises; plant, fresh, applied on wounds to stop bleeding. **Washo:** a poultice of mashed leaves is applied on swelling or sores. **Kawaiisu (Tehachapi):** for use on snakebite the leaves are dried, crushed to powder, and applied to wound dry. **Yuki:** tea was made from the leaves and flowers as a cure for consumption, stomach ache, and headache, and as a lotion for sore eyes. **Karok:** the stalk and the leaves were put in water and then applied to open arrow and gunshot wounds. **Miwok:** leaves and flowers steeped, the infusion drunk for bad colds; mashed leaves, either green or dried, bound to a wound were said to stop pain. **Mono Lake Paiute:** green plants smelled for colds, or by old men for headaches; small amounts taken for colds; leaves and stems boiled into a liniment for skin sores. **Costanoan:** leaves, heated, held in mouth for toothache, held over wounds to prevent swelling; decoction taken for stomachache. **Chumash:** plant, boiled, liquid held in mouth to reduce pain of toothache; plant, mashed, used as poultice for cuts, sores, stanch bleeding of a wound.
Misc:	**Tubatulabal:** considered the plants as weeds.

Citation(s): Goodrich et al 1980; Rayburn 2012; Train et al
1941; Zigmond 1981; Chesnut 1902; Curtin 1957;
Gifford 1939; Holt 1946; Schenck and Gifford 1952;
Barret and Gifford 1933; Steward 1933; Voegelin
1939; Rayburn 2012; Timbrook 2007.

Terminology: **Pomo (Southwestern):** šinam?ke·tey?. **Washo:**
<u>wem</u>-see. **Kawaiisu (Tehachapi):** togowa. **Karok:**
kuchich'apvuuy, "lizard tail;" achnatap vuyhich,
"imitation rat's tail." **Shasta:** irahatraga. **Northern
Paiute (Paviotso):** wada'a-kwasi. **Owens Valley
Paiute:** wiu'tu. **Mono Lake Paiute:** wiu'da. **Miwok
(Central Sierra):** Kamya, sepesepa; **(Southern
Sierra):** kama'iya. **Chumash (Barbareño,
Ventureño):** yepunash; **(Yenzeño):** steleq''a'emet,
masteleq 'a pistuk, matsauk"iyeena, "rattlesnake
medicine."

Citation(s): Goodrich et al 1980; Train et al 1941; Schenck and
Gifford 1952; Hold 1946; Kelly 1932; Steward 1933;
Barrett and Gifford 1933; Timbrook 2007.

Note(s): while the plant is not considered as poisonous it is
listed as containing alkaloidal and glycosidal toxic
principles; a native of Eurasia; has anti-
inflammatory effects; key actions – mild diuretic
and urinary antiseptic, reduces fever, increases
sweating, lowers blood pressure, stops internal
bleeding.

Citation(s): Kingsbury 1964; Munz and Keck 1965;
Chevallier 1996; Fuller and McClintock 1986.

16   *Achyrachaena mollis*
Family: *Asteraceae*

Usage: Food: **Yuki:** seeds gathered in May or June
for pinole, first roasted.

Citation(s): Chesnut 1902.

17 *Acmispon americanus* var. *americanus*

Synonym(s): *Lotus americana, Lotus purshianus, Lotus Purshianus* var. *glaber, Hosackia glabra*

Common Name(s): Spanish Clover.

Usage: Food: **Miwok:** green leaves were pounded with acorns that were too oily, to absorb some of the oil.

Material: **Kawaiisu (Tehachapi):** used as mat for juniper "cake." **Karok:** used for thatching houses.

Citation(s): Barrett and Gifford 1933; Zigmond 1981; Merriam 1967.

Terminology: **Miwok (Central Sierra):** pulluluku. **Kawaiisu (Tehachapi):** wa??dabigidɨnɨbɨ.

Citation(s): Barrett and Gifford 1933; Zigmond 1981.

18 *Acmispon argophyllus*

Family: *Fabaceae*

Synonym(s): *Lotus argophyllus*

Usage: Misc: **Tubatulabal:** considered weeds.

Citation(s): Voegelin 1938.

19 *Acmispon brachycarpus*

Synonym(s): *Lotus humistratus*

Common Name(s): Hill Lotus.

Usage: Medicine: **Karok:** used as medicine when a woman is in labor; the patient is washed in water in which the plant is soaked; she also drinks the water. **Maidu:** used infusion to wash women after childbirth.

Citation(s): Schenck and Gifford 1952; Strike 1994.

Terminology: **Karok:** imtanasuhanpinishik, "little bastard bush."
Citation(s): Schenck and Gifford 1952.

20   *Acmispon glaber*

Synonym(s): *Lotus scoparius*
Common Name(s): Fiddle Neck, Deerweed.

Usage:	Food:	**Tubatulabal:** used leaves.
	Material:	**Cahuilla:** used as a material in house construction. **Chumash:** made brooms, black dye (smoke) for basketry materials, thatching for sweathouses.
	Medicine:	**Costanoan:** foliage, decoction, used for cough remedy.

Citation(s): Voegelin 1938; Bean and Saubel 1972; Rayburn 2012; Timbrook 2007.
Terminology: **Cahuilla:** kiwat. **Chumash (Barbareño):** ya'I; **(Ventureño):** yai.
Citation(s): Bean and Saubel 1972.

21   *Acmispon grandiflorus*

Synonym(s): *Lotus grandiflorus*
Common Name(s): Wild Pea.

Usage:	Misc:	**Pomo (Southwestern):** named but not used.

Citation(s): Goodrich et al 1980.
Terminology: **Pomo (Southwestern):** ?ama·la má?a, "rabbit food."
Citation(s): Goodrich et al 1980.

22   *Acmispon procumbens*

Synonym(s): *Lotus procumbens*
Common Name(s): Silky California Broom.

Usage: Material: **Kawaiisu (Tehachapi):** employed as wall-filler in Winter house.
Citation(s): Zigmond 1981.
Terminology: **Kawaiisu (Tehachapi):** seena(m)bi.
Citation(s): Zigmond 1981.

23   *Acmispon strigosus*

Synonym(s): *Lotus strigosus*

Usage: Food: **Luiseño:** used for greens.
       Misc: **Miwok:** named but not used. **Tubatulabal:** considered weeds.
Citation(s): Barrett and Gifford 1933; Sparkman 1908; Voegelin 1938.
Terminology: **Miwok (Central Sierra):** cikasi. **Luiseño:** tovinal.
Citation(s): Barrett and Gifford 1933; Sparkman 1908.

24   *Aconitum columbianum*

Family: *Ranunculaceae*
Common Name(s): Aconite, Monkshood

Usage: Medicine: **Maidu:** used as pain killer or sedative for nervous disorders.
Citation(s): Strike 1994.
Note(s): all parts of the plant are considered as poisonous; eating flowers, leaves, or roots; symptoms are intense vomiting, diarrhea, muscular weakness, spasms, weak pulse, paralysis of the respiratory system, convulsions, death in a few hours.
Citation(s): Strike 1994; Hardin and Arena 1974; Kingsbury

1964.

25 *Acourtia microcephala*
Family: *Asteraceae*
Synonym(s): *Perezia microcephala*

Usage: Medicine: **Cahuilla, Chumash, Chumash (Island):** when prepared as a decoction it is used to produce a very quick passage of the bowels; to treat severe, chronic cough, colds, lung congestion, asthma, sores.
Citation(s): Barrows 1900; Strike 1994; Timbrook 2007.
Terminology: **Cahuilla:** ha-bak-a-ba. **Chumash (Barbareño):** ʾalashkhalalash.
Citation(s): Barrows; Bean and Saubel 1972; Timbrook 2007.

26 *Adenocaulon bicolor*
Family: *Asteraceae*
Common Name(s): Trail Plant

Usage: *Medicine:* **Maidu:** crushed leaves used a poultice for boils or sores.
Citation(s): Strike 1994.

27 *Adenostoma* spp.
Family: *Rosaceae*
Common Name(s): Chamise, Greasewood, Red Shank, Ribbonbush, Ribbonwood, Yerba del Pasmo.

Usage: *Material:* **Cahuilla:** used to make throwing stick, about 2 feet long, for small game, especially rabbits; made arrow foreshafts from wood; bark used for women's skirts. **Tubatulabal:** used shaft of root as a drill to make holes in the center of clam shell beads (used as a measure of wealth); gum from scale insect found on plant used as adhesive to bind arrow point to shafts, baskets to bedrock mortars.

Medicine: **Diegueño:** bathed paralyzed limbs in decoction of the plant. **Cahuilla:** boiled leaves and twigs to bathe infected, sore, or swollen parts of the body; decoction of dried leaves used as laxative; twigs were pulverized, mixed with animal fat for salve used on sores or wounds.

Citation(s): Strike 1994.

28 *Adenostoma fasciculatum*
Common Name(s): Chamise, Chamiso.

Usage: Material: **Luiseño:** foreshafts of arrows made from wood; a gum, the deposit of a scale insect is also obtained. **Cahuilla:** used limbs in house construction; branches used for arrows, building ramadas and fences; coals of wood a favorite source for roasting; branches used in conjunction with other wood for bows; branches bound together for torches; gum from a deposit of scale insect used as adhesive to bind arrows to shafts, baskets to mortars. **Chumash:** sticks, used to clean pith from Elderberry (*Sambucus*) to make hollow tubes; to make arrow shafts, pry bar "clam-gathering sticks," for abalone

Medicine: **Cahuilla:** leaves and branches boiled, solution used to bathe infected, sore, or swollen areas of body. **Esselen:** leaves, may have been used as a medicinal tea. **Chumash:** leaf, tea, women drank (perhaps for childbirth, menstrual complications).

Misc: **Pomo (Southwestern):** named but not used.

Citation(s): Barrows 1900; Sparkman 1908; Iovin 1963; Goodrich et al 1981; Bean and Saubel 1972; Breschini and Haversat 2004; Timbrook 2007.

Terminology: **Luiseño:** u'ut. **Cahuilla:** oot, u'ut. **Pomo (Southwestern):** bahg^{h}am?. **Chumash (Barbareño, Ynezeño, Purisimeño):** na'.

Citation(s): Sparkman 1908; Goodrich et al 1981; Bean and Saubel 1972; Timbrook 2007.

29   *Adenostoma sparsifolium*
        Synonym(s): *Adenostoma sparsifolia*
        Common Name(s): Chamiso Red Shanks, Ribbonwood.

Usage: Food: **Cahuilla:** used the seeds, harvested in July and August..

Material: **Cahuilla:** made rabbit stick from the wood; bark used a fibrous material for women's skirts; favorite firewood for roasting; used a head in two-piece arrow shafts. **Chumash:** powdered, taken like snuff.

Medicine: **Cahuilla:** twigs were dried and steeped for a drink to produce vomit and bowel relief; used as a common resort for pain in the stomach or intestines; ground very fine, mixed with grease and used as a slave. **Chumash:** taken for toothache; boiled, tea used to produce sweating, as a blood purifier, for ulcers, sore throats, colds, urinary disorders, wash for boils, cuts on humans ad livestock; tea drunk and applied, for toothache, gangrene, colds; leaves, or ground branches, mixed with lard or grease, salve applied to heal sores, cure sore throat; leaves sometimes mixed with sulfur and olive oil to make salve.

Citation(s): Barrows 1900; Strike 1994; Bean and Saubel 1972; Timbrook 2007.

Terminology: **Cahuilla:** henily.

Citation(s): Bean and Saubel 1972.

30 *Adiantum* spp.
Family: *Pteridaceae*

Usage: Material: **Yurok, Achomawi, Shasta, Pomo:** stem used as a black pattern in baskets. **Wintun:** used outside fiber for black overlay in baskets. **Tolowa:** used as an overlay in basketry. **Pomo:** used stems to keep holes in pierced ears open and to increased size of hole. **Karok:** used stems for decoration on ceremonial attire.

Medicine: **Maidu:** used a infusion of leaves as hair tonic.

Citation(s): Baumhoff 1958; Curtis 1924; Merriam 1967;
Merrill 1923; Strike 1994; Hudson 1893.

31 *Adiantum capillus-veneris*

Usage: Material: **Coast Yuki:** stems used for basket designs.

Medicine: **Maidu:** decoction used for bronchial problems, to ease menstrual problems, to stop hemorrhages, as a diuretic and tonic; poultice of leaves for sores and to wash insect bites. **Salinan:** used to treat liver disorders, stop hemorrhages, to stimulate poor appetites.

Citation(s): Gifford 1939; Strike 1994.
Terminology: **Yuki:** shima.
Citation(s): Gifford 1939.

32 *Adiantum jordanii*

Synonym(s): *Adiantum emarginatum*
Common Name(s): Common Maidenhair Fern.

Usage: Material: **Pomo:** used stems to keep ear-ring holes open and increasing size; stems, split lengthwise, dried, used in basketry.

Medicine: **Costanoan:** used to produce an analgesic; a decoction eased stomach trouble, purify the blood, for "pain below the shoulders;" used to help expel afterbirth and as a general post-parturition treatment; root, poultice, used to cause coagulation, for hemorrhage.

Citation(s): Chesnut 1902; Goodrich et al 1980; Strike 1994; Rayburn 2012.

33 *Adiantum aleuticum*

Synonym(s): *Adiantum pedatum, Adiantum pedatum* var. *aleuticum.*

Common Name(s): Maidenhair Fern, Five Finger Fern.

Usage: Food: **Tolowa(?):** ate roots.

Material: **Hupa:** the stems used for the glossy black decoration on baskets. **Tolowa, Whilkut, Nongatl, Lassik, Wailaki, Yurok, Achomawi, Karok, Pomo, Wintun:** stem used as black pattern in baskets. **Karok:** stem used for decoration on clothing, especially Jump Dance dress. **Pomo (Southwestern):** stem inserted into pierced ear to keep wound from closing; stem, either alone or with feathers attached, can serve as a earing. **Maidu (Northern):** used stems in basketry construction. **Bear River:** made overlay pattern design in tight baskets with stems.

Medicine: **Costanoan:** used to produce an analgesic; decoction to ease stomach trouble and to purify the blood, to expel afterbirth and for post-parturition care; used only when *Adiantum Jordanii was not available.* **Maidu:** used to poultice wounds.

Citation(s): Dixon 1905, 1907; Drucker 1937; Gifford 1967;

Goddard 1903; Merrill 1923; Nomland 1938;
O'Neale 1923; Schenck and Gifford 1952;
Goodrich et al 1980; Strike 1994; Rayburn 2012.
Terminology: **Karok:** ikritapkir, yumarekiritap, kiritapki.
Pomo (Southwestern): shamoda. **Hupa:** mukai
kinxulnewan (stems).
Citation(s): Schenck and Gifford 1952; O'Neale 1932;
Gifford 1967; Goddard 1903; Goodrich et al 1980.

34 *Aesculus californica*

Family: *Sapindaceae*
Common Name(s): Buckeye, California Buckeye.

Usage: Food: **Maidu (Nisenan):** considered good
food, sometimes buried in ground,
cooked when sprouted; roasted
underground for 36 hours or more to
extract poison. **Yana:** considered
important as food; ground up fine
with feet and leached prior to use.
**Tubatulabal; Pomo; Maidu
(Northern); Huchnom; Wintun, Hill
(Patwin); Sinkyone; Wintun, Central
(Nomlaki):** used nuts. **Pomo
(Southwestern):** nuts collected when
ripe and fell from tree, roasted in
ashes, crushed and placed in sandy
leaching basin beside stream, washed
for 5 hours, eaten without further
cooking. **Wappo:** ate nuts, prepared
as Pomo. **Yuki:** roasted, pounded,
and leached similar to acorn
preparation but with longer soaking.
Miwok: nuts eaten in time of scarcity.
Achomawi, Esselen: eaten in time of

need. **Bear River:** nuts not pounded into flour until immediately prior to use. **Atsugewi:** shelled, pounded, and soaked until juice was gone; pulpy mass was squeezed dry and eaten, often without cooking. **Salinan:** nuts eaten after leaching. **Wintun:** gathered in fall, mashed, leached. **Kawaiisu (Tehachapi):** gathered when fruit falls to ground, usually in November; seeds, broken into pieces, soaked in water for 2 days, crumbled and leached, boiled into mush.

Material: **Maidu, Southern (Nisenan):** the long straight shoots are used for arrows and basket making; nuts used for poisoning fish; used wood for hearth and drill in fire-drills. **Yuki:** wood formed the hearth for fire drilling set. **Wintun:** used for fire drills with an under block of *Juniperus* spp. **Miwok, Pomo (Southwestern):** used nuts, ground and sprinkled into water, as fish poison; used to make fire drills tick and block for fire-drill, also for bows. **Yokuts (Northern Hill):** used for fire drill hearths. **Sinkyone:** used for fire drill. **Wappo:** made bows from wood, considered second best to Manzanita (*Arctostaphylos spp.*); hand drill for fire drill. **Monache (Western Mono):** used leaves for fish poison. **Miwok, Yana:** used wood for hearth and drill in fire-drills. **Costanoan, Esselen:** fruits, used in preparation of fish poison.

Medicine: **Kawaiisu (Tehachapi):** seeds broken into pieces, mixed with water, used as suppositories for hemorrhoids. **Pomo:** used bark taken near base of tree as poultice for snakebite. **Wintun:** crushed fruit as poultice for wounds. **Costanoan:** made decoction from bark to relieve toothache, loose teeth; fruit, pulverized, salve applied to hemorrhoids. **Maidu, Pomo, Yuki:** put small piece of bark in cavity of tooth to relieve pain.

Citation(s): Beals 1933; Barrett and Gifford 1933; Chesnut 1902; Curtin 1957; Dixon 1905; Driver 1936; Du Bois 1935; Foster 1944; Garth 1953; Gayton 1948; Gifford 1967; Goldschmidt 1951; Heizer 1953; Kroeber 1932; Mason 1912; Merriam 1955; Nomland 1935, 1938; Powers 1877; Sapir and Spier 1943; Voegelin 1938; Zigmond 1981; Goodrich et al 1980; Strike 1994; Rayburn 2012; Breschini and Haversat 2004.

Terminology: **Yuki:** simpt'ol. **Kato:** laci. **Yana:** pa's I (nut). **Atsugewei:** b'ass, **Pomo (Northern, Southwestern):** bace; **(Southwestern):** bash'e; unu (tree and nuts); **(Central, Southern):** baca'; **(Southern):** be'ce. **Miwok (Central Sierra):** siwu. **Kawaiisu (Tehachapi):** paʔasvɨ.

Citation(s): Curtin 1957; Loeb 1932: Sapir and Spier 1943; Garth 1953; Barrett 1952; Gifford 1967; Barrett and Gifford 1933; Zigmond 1981; Goodrich et al 1981.

Note(s): mature growth, sprouts and mature nuts are toxic–contain the glycoside aesculin; in humans symptoms are restlessness, circulatory disturbances, involuntary urination, lack of coordination, dilated pupils, weakness, muscular

twitching, paralysis, stupor, only rare fatalities.
Citation(s): Kingsbury 1964; Hardin and Arena 1974; Fuller and McClintock 1986.

35   *Agaricus campestris*
Family: *Agaricaceae*
Common Name(s): Field Mushroom.

Usage: Food: **Pomo (Southwestern):** baked
Citation(s): Goodrich et al 1980.
Terminology: **Pomo (Southwestern):** pʰalá?cay hiċe.
Citation(s): Goodrich et al.

36   *Agaricus capestis*
Common Name(s): Common Field Mushroom

Usage: Misc.:   **Yuki:** named but somewhat superstitious about eating it as food.
Citation(s): Chesnut 1902; Strike 1994.

37   *Agaricus silvicola*
Common Name(s): Deer Mushroom, Wood Mushroom.

Usage: Food: **Pomo (Southwestern:** baked, only tops eaten.
Citation(s): Goodrich et al 1980.
Terminology: **Pomo (Southwestern):** bihšeċe.
Citation(s): Goodrich et al 1980.

38   *Agastache urticifolia*
Family: *Labiatae*
Common Name(s): Giant Hyssop.

Usage: Medicine: **Miwok:** boiled and drunk for

measles; decoction of leaves drunk to cure rheumatism and to relieve rheumatic pain.
Citation(s): Barrett and Gifford 1933; Strike 1994.
Terminology: **Miwok (Central Sierra):** lokotokoyi.
Citation(s): Barrett and Gifford 1933.

39   *Agave deserti*

Family: *Agavaceae*
Common Name(s): Agave, Mescal, Desert Agave

Usage: Food: **Cahuilla:** from April on, the cabbages and the stalk are full of sap and are then roasted; the basic food staple; the blossom were parboiled to release bitterness, then eaten or preserved by drying; leaves baked and eaten or dried and stored for future use; the larvae of the Agave Skinner Butterfly (*Megathymus stephousi*) were roasted on leaves, picked off, and eaten. **Kamia, Chemehuevi:** used plant. **Diegueño:** important food.

Material: **Cahuilla:** fibers of the leaves make the best cordage and rope, bowstrings, little brooms, and hair brushes; fibers used in nets, slings, shoes, women's skirts, mats, cactus bags, snares, baby cradles; thorns used as awls in basket making, and tattooing; dried stalks used for firewood. **Cahuilla, Diegueño:** used the leaf for foundation material in the construction of baskets.

Citation(s): Barrows 1900; Merrill 1923; Gentry 1998; Bean and Saubel 1972.
Terminology: **Cahuilla:** amul.
Citation(s): Bean and Saubel 1972.
Note(s): Agave sap can cause painful skin irritation; juice is cathartic, diuretic.
Citation(s): Strike 1994; Fuller and McClintock 1986.

40 *Agoseris aurantiaca* var. *aurantiaca*
Family: *Asteraceae*
Synonyn(s): *Agoseris gracilens*

Usage: Food: **Karok:** the juice is sucked out of the root near the crown and is chewed like chewing-gum.
Citation(s): Schenck and Gifford 1952.
Terminology: **Karok:** mikkimshakwa.
Citation(s): Schenck and Gifford 1952.

41 *Agoseris retrorsa*
Common Name(s): Mountain Dandelion.

Usage: Food: **Kawaiisu (Tehachapi):** in the Spring green leaves boiled and eaten.
Citation(s): Zigmond 1981.
Terminology: **Kawaiisu (Tehachapi):** eewaribi.
Citation(s): Zigmond 1981.

42 *Agrimonia gryposepala*
Family: *Rosaceae*
Common Name(s): Agrimony, Tall-Hairy Agrimony.

Usage: Medicine: **Maidu:** roots used to reduce fevers and to stop nosebleeds, used roots and

leaves as an astringent and mild tonic.
Citation(s): Strike 1994.

43   *Agropyron* spp.
        Family: *Poaceae*

Usage: Food: **Owens Valley Paiute:** used for food.
Citation(s): Steward 1933.
Terminology: **Owens Valley Paiute:** su'nu'u, tsu'nu'u.
Citation(s): Steward 1933.

44   *Agrostis exarata*
        Family: *Poaceae*

Usage: Food: **Modoc (Lutuami):** seeds eaten.
         Misc: **Kawaiisu (Tehachapi):** named but not
                       used.
Citation(s): Zigmond 1981; Strike 1994.
Terminology: **Kawaiisu (Tehachapi):** pa?yɨg^{w}asivɨ,
             "kangaroo-rat tail."
Citation(s): Zigmond 1981.

45   *Agrostis scabra*

Usage: Food: **Modoc (Lutuami):** seeds eaten.
Citation(s): Strike 1994.

46   *Alectoria fremontii*
        Family: *Usneaceae*
        Common Name(s): Lichen.

Usage: Food: **Wailaki:** used during times of famine.
         Material: **Maidu:** made black dye.
Citation(s): Chesnut 1902; Strike 1994.
Note(s): occur in arctic regions and upon high mountains;

occurs on branches of coniferous trees.
Citation(s): Tuckerman 1882; Schneider 1898.

47 *Alisma trivale*

Family: *Alismataceae*
Common Name(s): Water Plantain.

Usage: Food: **Maidu:** ate base after drying and boiling in several changes of water.
        Medicine: **Maidu:** used to poultice wounds, sooth skin irritations, as a counter-irritant, and stringent, to relieve bowl problems.
Citation(s): Strike 1994.

48 *Allenrolfea occidentalis*

Family: *Chenopodiaceae*
Common Name(s): Iodine Bush.

Usage: Food: **Cahuilla:** small, black seeds ground into mush or a drink; seed flour made into mush, or shaped, dried, eaten as a cookie.
Citation(s): Bean and Saubel 1972; Strike 1994.
Terminology: **Cahuilla:** hu'at.
Citation(s): Bean and Saubel 1972.

49 *Allium* spp.

Family: *Alliaceae*
Common Name(s): Wild Garlic, Wild Onion.

Usage: Food: **Maidu, Southern (Nisenan):** eaten raw; or roasted with ashes; or boiled. **Bear River:** cleaned thoroughly, peeled, wrapped in Alder leaves

(*Alnus spp.*), and cooked with Camas (*Camassia spp.*); stored for winter use. **Huchnom, Chumash:** ate bulbs. **Atsugewi:** cooked bulb. **Wintun, Hill (Patwin):** used bulbs. **Wintun:** eaten in May. **Kawaiisu (Tehachapi):** tops and roots eaten raw and fresh.

Citation(s): Beals 1933; Du Bois 1935; Foster 1944; Garth 1953; Kroeber 1932; Nomland 1938; Powers 1877; Zigmond 1981; Landberg 1965.

Note(s): confined to the northern hemisphere, extending in the New World from Mexico northwards; common in open grass lands, scrub, and desert; range from sea-level high elevations; may increase the uptake of thiamine from foods; raw onion eaten with cruciferous crops may make the vitamin B1 of the latter more readily available; some local use in preparations for pulmonary conditions, or lowering blood pressure, as anthelmintics, as intestinal antispasmodics, and externally as rubefacient or irritants; large amounts over time produce anemia, jaundice, and digestive disturbances in humans.

Citation(s): Jones and Mann 1963; Kingsbury 1964; Fuller and McClintock 1986.

50   *Allium bisceptrum*

Usage: Food: **Northern Paiute (Paviotso):** the leaves eaten as a type of relish; the seeds and root were eaten.

Citation(s): Kelly 1932.

Terminology: **Northern Paiute (Paviotso):** badis.

Citation(s): Kelly 1932.

51   *Allium bolanderi*

Usage: Food: **Karok:** used the same as *Allium Campanulatum*. **Yuki, Pomo:** gathered bulbs in late Spring and Summer.
Citation(s): Chesnut 1902; Schenck and Gifford 1952.
Terminology: **Karok:** hanach'yu.
Citation(s): Schenck and Gifford 1952.

52   *Allium campanulatum*
Synonym(s): *Allium acuminatum*
Common Name(s): Wild Onion.

Usage: Food: **Northern Paiute (Paviotso):** the leaves were eaten as a sort of relish; the seeds were eaten. **Karok:** relished by old men and women.
Citation(s): Kelly 1932; Schenck and Gifford 1952;
Terminology: **Northern Paiute (Paviotso):** gu'ka.
              Karok: hanach'yu.
Citation(s): Kelly 1932; Schenck and Gifford 1952.

53   *Allium dichlamydeum*
Common Name(s): Wild Onion.

Usage: Food: **Pomo (Southwestern):** greens and bulbs eaten raw, or cooked with other foods for flavoring.
Citation(s): Goodrich et al 1980.
Terminology: **Pomo (Southwestern):** gʰa?bá?.
Citation(s): Goodrich et al 1980.

54   *Allium howellii*
Synonym(s): *Allium hayalinum*
Common Name(s): Wild Onion.

Usage: Food: **Tubatalabal:** leaves, stalks, and heads eaten.
Citation(s): Voegelin 1938.

55   *Allium lacunosum*
Common Name(s): Wild Garlic.

Usage: Food: **Tubatalabal:** leaves, stalks, and heads eaten.
Citation(s): Voegelin 1938.

56   *Allium parvum*

Usage: Food: **Maidu (Northern):** bulbs eaten.
Citation(s): Dixon 1905.

57   *Allium peninsulare*
Common Name(s): Wild Onion.

Usage: Food: **Tubatalabal:** leaves, stalks, and heads eaten.
Citation(s): Voegelin 1938.

58   *Allium platycaule*

Usage: Food: **Northern Paiute (Paviotso):** leaves were eaten as a sort of relish; the seeds and roots were eaten.
Citation(s): Dixon 1905; Kelly 1932.
Terminology: **Northern Paiute (Paviotso):** pani'zi.
Citation(s): Kelly 1932.

59   *Allium tolmiei* var. *tomiei*
Synonym(s): *Allium pleianthum*
Usage: Food: **Northern Paiute (Paviotso):** roots

cooked overnight in earth oven and eaten; green leaves regarded as a sort of relish.

Citation(s): Kelly 1932.

Terminology: **Northern Paiute (Paviotso):** mu'a".

Citation(s): Kelly 1932.

Note(s): labeled *Allium pleianthim* by Kelly (1932) which occurs in arid transition zone – Idaho and eastern Oregon.

Citation(s): Abrams and Ferris 1960.

60 *Allium unifolium*

Common Name(s): Wild Onion.

Usage: Food: **Pomo:** root usually eaten raw; sometimes baked in underground oven. **Yuki:** bulb and base of leaves fried and eaten; bulb sometimes eaten raw; gathered in late Spring or early Summer.

Citation(s): Barrett 1952; Chesnut 1902; Curtin 1957.

Terminology: **Pomo (Central):** kaba'i. **(Eastern):** xaba'i; **(Southwestern):** se'woya. **Yuki:** shep, waihil.

Citation(s): Barrett 1952; Curtin 1957.

61 *Allium validum*

Common Name(s): Wild Onion.

Usage: Food: **Cahuilla:** bulbs eaten raw, or used as a flavoring ingredient for other foods.

Citation(s): Bean and Saubel 1972.

Terminology: **Cahuilla:** tepish.

Ciation(s): Bean and Saubel 1972.

62 *Alnus* spp.

Family: *Betulaceae*
Common Name(s): Alder.

Usage: Food: **Bear River:** used leaves during roasting of Camas (*Camassia spp.*) to make them sweet, better flavored.

Material: **Wappo:** made arrow shafts from wood.

Medicine: **Bear River:** chewed bark used to treat wounds and cuts; young Alder bark infusion used to treat fever. **Shasta:** chewed bark was placed on spot affected by poison oak.

Citation(s): Driver 1936; Holt 1946; Nomland 1938.

63 *Alnus rubra*

Synonym(s): *Alnus oregana*
Common Name(s): Alder, Red Alder.

Usage: Material: **Hupa, Nongatl, Lassik, Wailaki, Wiyot, Yurok, Pomo:** the root (fibers) used as woof in basketry. **Whilkut:** root used as woof in baskets as well as forming a brown pattern. **Hupa:** the bark used to furnish a dye to satin the stems of *Woodwardia fimbriata* a reddish brown. **Yurok, Karok:** used roots as a source of red material for basketry.

Medicine: **Pomo (Southwestern):** bark was boiled, wash used to bathe skin diseases.

Citation(s): Goddard 1903; Merrill 1923; O'Neale 1932; Schenck and Gifford 1952;

Goodrich et al 1980; Strike 1994.
Terminology: **Yurok:** were'regets (roots). **Karok:**
arkvittip, ekvit'ip (roots). **Pomo
(Southwestern):** gahe^{h}iṭi.
Citation(s): O'Neale 1932; Schenck and Gifford 1952;
Goodrich et al 1980.

64   *Alnus rhombifolia*
Common Name(s): Alder, White Alder,
Stream Alder.

Usage: Food: **Maidu:** cambium and seeds eaten.
        Material: **Hupa, Whilkut, Nongatl, Sinkyone,
                      Yurok, Karok, Shasta, Achomawi,
                      Wintun, Yuki, Pomo:** bark used a red
                      dye. **Tubatulabal:** branches used as
                      firewood. **Karok:** roots used for
                      basketry; wood used for smoking
                      salmon, eel, and deer meat. **Yuki:** fresh
                      bark used to color things; infrequently
                      deer skins. **Wailaki:** chew fresh bark
                      and color their bodies to facilitate the
                      capture of salmon, a trick first used by
                      Coyote; arrows sometimes made from
                      young shoots; soft wood available for
                      tinder. **Chumash:** wood used for
                      bowls, trays, spoons; bark, used as dye
                      for string, cordage.
        Medicine: **Maidu, Southern (Nisenan):** the
                      leaves are eaten to as a preventative
                      and as a cure from the poisonous effect
                      of plants. **Yuki:** used to produce
                      perspiration, as a blood purifier, and
                      as a check on diarrhea caused by
                      drinking bad water; allay stomach

ache; facilitate childbirth; cure consumption by checking hemorrhages; mixed tobacco (*Nicotiana spp.*) to produce emesis. **Wailaki:** dry rot from wood mixed the powdered bark of *Salix lasiolepsis* as a poultice for burns. **Pomo (Southwestern):** bark was boiled and used as a decoction to bathe a baby with skin disease. **Kawaiisu (Tehachapi):** provided medicine.

Misc.: **Miwok:** named but not used. **Kawaiisu (Tehachapi):** buckskin smoked with burning bark turned from white to yellow.

Citation(s): Chesnut 1902; Barrett and Gifford 1933; Gifford 1967; O'Neale 1932; Powers 1877; Schenck and Gifford 1952; Merrill 1923; Voegelin 1938; Zigmond 1981; Strike 1994; Timbrook 2007.

Terminology: **Pomo (Southwestern):** kachidi kale. **Karok:** kiwitip, akiwitippishchash (roots used for baskets), aiepa (dye derived from bark). **Miwok (Central Sierra):** pamalu. **Kawaiisu (Tehachapi):** pawič?uvɨ. **Chumash (Barbareño, Ynezeño, Ventureño):** mow.

Citation(s): Gifford 1967; Schenck and Gifford 1952; Barrett and Gifford 1933; Zigmond 1981; Timbrook 2007.

65 *Alnus incana* spp. *tenuifolia*
        Synonym(s): *Alnus tenuifolia*
        Common Name(s): Mountain Alder.

Usage: Food: **Maidu:** ate cambium.
Citation(s): Strike 1994.

66 *Alopecurus aequalis*
        Family: *Poaceae*

Usage: Food: **Northern Paiute (Paviotso):** seeds were eaten.
Citation(s): Kelly 1932.

67 *Amanita muscaria*
        Family: *Amanitaceae*
        Common Name(s): Amanita.

Usage: Misc:,   **Pomo (Southwestern):** considered poisonous.
Citation(s): Goodrich et al 1980.
Note(s): one of the most lethal poisons known.
Citation(s): Fuller and McClintock 1985.

68 *Amaranthus* spp.
        Family: *Amaranthaceae*
        Common Name(s): Pigweed.

Usage: Food: **Chumash:** eaten; probably of minor importance.
Citation(s): Landberg 1965; Timbrook 2007.

69 *Amaranthus albus*
Common Name(s): Tumbleweed.

Usage: Food: **Wintun, Central (Nomlaki):** used seeds.
Citation(s): Goldschmidt 1951.
Note(s): Native of tropical America.
Citation(s): Munz and Keck 1965.

70 *Amarantus blitoides*
Common Name(s): Amaranth,
Prostrate Pigweed.

Usage: Food: **Modoc (Lutuami); Maidu:** ate seeds.
Citation(s): Strike 1994.

71 *Amaranthus fimbriatus*
Common Name(s): Fringed Pigweed.

Usage: Food: **Cahuilla:** leaves boiled and eaten; seeds gathered late Summer, left on spikes until needed, seeds threshed, parched, ground into flour, made into mush..
Citation(s): Strike 1994; Bean and Saubel 1972.
Note(s): leaves are rich in vitamins, calcium, niacin, and iron.
Citation(s): Strike 1994.

72 *Amaranthus retroflexus*
Common Name(s): Careless Weed,
Rough Pigweed.

Usage: Misc:, **Kawaiisu (Tehachapi):** seed used as bait in deadfall traps to attract quail.

Citation(s): Zigmond 1981.

Note(s): native of tropical America; found in waste
places.

Citation(s): Munz and Keck 1965.

73   *Ambrosia artemisiifolia*
Family: *Asteraceae*
Synonyn(s): *Ambrosia artemisiaefolia*

Usage: Medicine: **Luiseño:** used as an emetic.
Citation(s): Sparkman 1908.
Terminology: **Luiseño:** pachavut.
Citation(s): Sparkman 1908.
Note(s): native in eastern United States.
Citation(s): Munz and Keck 1965.

74   *Ambrosia chamissonis*
Family: *Compositae*
Synonym(s): *Franseria bipinnatifida, Franseria
chamissonis* ssp. *bipinnatisecta*

Usage: Misc: **Coast Yuki:** named but not used.
Citation(s): Gifford 1939.
Terminology: **Yuki:** botkatim.
Citation(s): Gifford 1939.

75   *Ambrosia monogyra*
Family: *Asteraceae*
Synonym(s): *Hymenoclea monogyra*

Usage: Medicine: **Diegueño:** root tea drunk to relieve
pain of any bites; chewed green leaves
rubbed on any bites.
Citation(s): Strike 1994.

76 *Ambrosia psilostachya*

Common Name(s): Western Ragweed.

Usage: Medicine: **Coastanoan:** poultice of heated leaves used to ease aching joints. **Diegueño:** used decoction to bathe bad burns and sores. **Maidu:** used an infusion externally to poultice sore eyes; internally to cure colds or relieve intestinal cramps. **Costanoan:** leaves, heated, held on aching joints.

Citation(s): Strike 1994; Rayburn 2012.

77 *Amelanchier* spp.

Family: *Rosaceae*
Common Name(s): Service Berry, Sarvisberry.

Usage: Food: **Shasta:** ate berries. **Wintun:** gathered berries in Summer. **Wintun, Central (Nomlaki):** used berries.

Material: **Shasta:** arrows made from young shoots. **Modoc (Lutuami); Achomawi:** made a corset armor to wear when fighting. **Karok, Maidu:** used twigs and stems to reinforce rims of basket hoppers in which acorns (*Quercus*) were pounded. **Maidu:** made baby carriers, made black dye from berries, made digging sticks. **Karok:** bound foreshaft of wood to shaft of Arrowwood (*Philadelphus lewisii* spp. *californicus*) or Creek Dogwood (*Cornus occidentalis*) with sinew covered with sap from Western Choke

Cherry (*Prunus virginiana* var. *demissa)*; also made foreshafts of salmon harpoons.

Medicine: **Pomo:** women boiled roots, drank infusion to promote regularity of menstruation periods.

Citation(s): Du Bois 1935; Holt 1946; Goldschmidt 1951; Merriam 1967; Strike 1994.

78 *Amelanchier alnifolia*

Synonym(s): *Amelanchier pallida*
Common Name(s): Service Berry,
Western Service Berry, June Berry.

Usage: Food: **Shasta:** berries eaten either fresh or dried. **Maidu (Northern) Achmawi:** berries eaten. **Karok:** berries eaten fresh as well as dried in the sun. **Atsugewi:** berries put in Tule (*Scirpus spp.*) Basket when ripe and mashed, water added to form a paste which was eaten without cooking, or dried in berry form and stored then soaked in water when used. **Cahuilla:** berry eaten fresh; preserved by drying and could be eaten several months after gathering.

Material: **Karok:** twigs and stems used to reinforce rims of basket hoppers for pounding Acorns; a piece of twig, about 4 inches long, is inserted as a point in the end of arrow shafts (*Philadelphus Lewisii* ssp. *gordonianus*), bound with sinew smeared with gum (*Prunus virginiana* var. *demisa*). **Hupa:**

the foreshafts of arrows made from wood. **Pomo (Southwestern):** stems and foliage used a thatch for winter houses at inland villages. **Kawaiisu (Tehachapi):** straight stem fashioned into arrows.

Medicine: **Pomo (Southwestern):** roots boiled and decoction drank to check too frequent menstruation.

Misc: **Miwok:** named but not used.

Citation(s): Dixon 1905, 1907; Barrett and Gifford 1933; Garth 1953; Gifford 1967; Goddard 1903; Merriam 1967; Schenck and Gifford 1952; Zigmond 1981; Goodrich et al 1980; Bean and Saubel 1972.

Terminology: **Atsugewi:** pĭkni. **Pomo (Southwestern):** ba'kom. **Karok:** afishiip, aflshii (berry). **Miwok (Central Sierra):** yossina. **Kawaiisu (Tehachapi):** tĭva? nĭvĭ.

Citation(s): Garth 1953; Gifford 1967; Schenck and Gifford 1952; Barrett and Gifford 1933; Zigmond 1981; Goodrich et al 1980.

Note(s): fruit usually ripens in July.

Citation(s): Sargent 1922.

79   *Amelanchier utahensis*

Synonym(s): *Amelanchier venulosa*
Common Name(s): Service Berry.

Usage: Food: **Northern Paiute (Paviotso):** berry was eaten fresh, dried, boiled, or uncooked; they were crushed before being dried.

Citation(s): Kelly 1932.
Terminology: **Northern Paiute (Paviotso):** tl'gabui.
Citation(s): Kelly 1932.

80　*Ammannia coccinea*
>　Family: *Lythraceae*
>　Common Name(s): Long-leaved Ammannia.

Usage:　Food:　　　**Maidu:** seeds eaten.
Citation(s): Strike 1994.

81　*Amsinckia douglasiana*
>　Family: *Boraginaceae*
>　Common Name(s): Buckthorn Weed, Fiddleneck.

Usage:　Medicine:　**Costanoan:** used medicinally.
Citation(s): Strike 1994.
Note(s): plants contain an unknown liver toxin, possibly heat
　　　labile.
Citation(s): Kingsbury1964.

82　*Amsinckia intermedia*

Usage:　Misc:　　　**Karok:** named but not used.
Citation(s): Schenck and Gifford 1952.
Terminology: **Karok:** 'lmkaanvaaxvaah, imkanva
　　　　　　Head.
Citation(s): Schenck and Gifford 1952.
Note(s): seeds contain liver toxin, possibly heat labile.
Citation(s): Kingsbury 1964.

83　*Amsinckia menziesii*
>　Synonym(s): *Amsinckia parviflora*
>　Common Name(s): Common Fiddleneck.

Usage: Food:　　　**Atsugewi:** seeds gathered and
prepared similar to other seeds,
pounded into flour, made into cakes
and eaten without cooking. **Miwok**

(Coast): ate young leaves; used seeds for pinole. **Kawaiisu (Tehachapi), Maidu:** ate young leaves. **Chumash:** seeds, roasted, ground, make into pinole.

Citation(s): Garth 1953; Strike 1994; Timbrook 2007.

Terminology: **Atsugewi:** aiwĭci.

Citation(s): Garth 1953.

84 *Amsinckia tessellata*

Common Name(s): Fiddle Neck.

Usage: Food: **Kawaiisu (Tehachapi):** important source of greens in Spring; leaves picked, bruised between hands and eaten with salt.

Citation(s): Zigmond 1981.

Terminology: **Kawaiisu (Tehachapi):** tiva? Nɨbɨ.

Citation(s): Zigmond 1981.

85 *Amsonia tomentosa*

Family: *Apocynaceae*

Synonym(s): *Amsonia brevifolia*

Common Name(s): Short-leaved Amsonia.

Usage: Material: **Owens Valley Paiute:** made carry straps of braided three-ply twisted rope, used by men.

Citation(s): Steward 1933; Strike 1994.

Terminology: **Owens Valley Paiute:** wicivi.

Citation(s): Steward 1933.

86   *Anagallis arvensis*
            Family: *Myrsinaceae*
            Common Name(s): Scarlet Pimpernel.

Usage: Medicine: **Chumash:** plant, boiled, applied as
                        poultice to treat wounds (bound over
                        affected part); decoction, used to treat
                        infections, as a poultice, or wash, for
                        sores, to bathe eczema, ringworm.
Citation(s): Timbrook 2007.
Note(s): naturalized from Europe; if eaten, causes intense
        headache, nausea.
Citation(s): Munz and Keck 1965; Fuller and McClintock 1986.

87   *Anemopsis californica*
            Family: *Saururaceae*
            Synonym(s): *Houttuynia californica*
            Comon Name(s): Yerba Mansa, Swamp Root.

Usage: Medicine: **Yokuts:** root pounded and soaked in
                        water; the water drunk for bad
                        stomach. **Luiseño:** a decoction of the
                        root is used internally and externally.
                        Owens Valley Paiute: roots boiled
                        into tea for laxative. **Chumash:** high
                        value placed on tea of root As a
                        healing wash for cuts, ulcerated sores,
                        venereal disease; drunk for colds,
                        cough, and as a blood purifier, as a
                        relief from asthma and kidney
                        problems; hot bath for rheumatism;
                        used to ritually purify and prepare a
                        person who would be carrying 'ayip,
                        a poisonous, supernaturally powerful
                        substance made from alum and

rattlesnake. **Kawaiisu (Tehachapi):** root is broken and boiled; decoction, red in color, drunk hot for relief of colds and coughing. **Cahuilla:** the strongly aromatic, peppery roots were peeled, cut up, squeezed, boiled; decoction was used as a cure for pleurisy, stomach ulcers, chest congestion, and colds; bark gather in the Fall, boiled a deep red-wine color, drunk as a cure for ulcers; applied externally to sores as a wash. **Tubatulabal:** inhaled steam from boiling roots for colds; as a decongestant, to relieve bronchial ailments. **Costanoan:** root, decoction (tea) used to wash sores, as a general pain remedy, for menstrual cramps; plant, dried, powdered, sprinkled on wounds as a disinfectant.

Citation(s): Powers 1877; Sparkman 1908; Steward 1933; Timbrook 1987, 2007; Zigmond 1981; Bean and Saubel 1972; Strike 1994; Rayburn 2012.

Terminology: **Owens Valley Paiute:** nūpi 'tel, Sawaniva. **Luiseño:** chevnash. **Kawaiisu (Tehachapi):** cupa hnɨvɨ. **Cahuilla:** chivnish. **Chumash (Barbareño, Ynezeño):** 'onchochi; **(Obispeño):** ch'elhe' tsqono; **(Ventureño):** 'onchoshi.

Citation(s): Steward 1933; Sparkman 1908; Zigmond 1981; Bean and Saubel 1972; Timbrook 2007.

88   *Angelica* spp.
Family: *Apiaceae*

Usage: Food: **Yuki:** eaten raw like celery. **Wappo:**

roots chewed. **Maidu, Southern (Nisenan):** eaten in Spring as greens. **Shelter Cove Sinkyone:** sprouts eaten in Spring and early Summer.

Medicine: **Huchnom:** used as cough medicine; poultice put on place of ache, believed to draw out pain. **Pomo:** chewed root was tied around the head and ears in bad cases of headache and nightmare; juice mixed with saliva used as a remedy for sore eyes; chewed and swallowed for colds, colic, and especially fever; carried on the person for good luck in gambling or hunting, root after chewed is sometimes rubbed on legs to prevent snakebite. **Hupa:** used as a panacea and charm. **Sinkyone:** if a girl holds off, rub it on your hands, and if you get a chance rub her neck, and she will give in; considered strong medicine. **Bear River:** cured headaches by rubbing head with leaves, or tying leaves to head. **Owens Valley Paiute:** root decoction used to poultice sores and swelling, eased rheumatic pain and sore throats. **Costanoan:** root (plant also?) burned, smoked inhaled to cure headache; chewed root to cure stomach aches; heated sap from leaves rubbed on sores, burned twigs used to treat aching limbs; strong tea taken by people who were extremely ill. **Karok:** root, with spoken formula, used for purification after funerals; root thrown in fire before fishing ensured an abundant catch.

Citation(s): Baumhoff 1958; Chesnut 1902; Driver1936; Foster 1944; Garth 1953; Merriam 1967; Powers 1877; Steward 1933; Strike 1994; Rayburn 2012.

89   *Angelica brewerii*

Usage: Material: **Miwok:** hunters rubbed body with root to give a "pleasing odor" and to counteract the human odor before going after deer.

Medicine: **Miwok:** root chewed as headache cure and as remedy for colds; to ward off snake bite, chewed and rubbed on the body, a decoction was drunk. **Washo:** as a bronchitis remedy the root is dried and scraped, the piece then soaked in water but not boiled, the solution is given a few teaspoons at a time, twice a day, over a period of 2 weeks; as an influenza specific it is taken frequently as tea, to improve taste it may be mixed with the root of *Lomatium dissectum* var. *multifideum*; small pieces of dried root are chewed for sore throats or coughs. **Maidu:** roots used as a laxative, to relieve stomach ailments, to cure colds; poultice used to relieve pain and swellings; leaves bound to head to relieve headaches.

Citation(s): Barrett and Gifford 1933; Train et al 1941; Strike 1994.

Terminology: **Washo:** dah-had-mo-mo; dah-o-pah-phu le.

Citation(s): Train et al 1941.

90 *Angelica hendersonii*

Usage: Medicine: **Miwok:** a tea used to treat mussel poisoning. **Miwok (Coast):** tea used to treat saltwater poisoning; root chewed to ease stomach aches, eaten to cure colds; poultice used to treat cuts.

Citation(s): Strike 1994.

91 *Angelica tomentosa*

Usage: Food: **Hupa:** the fresh shoots are eaten raw. **Yana:** the stem, when fresh, was broken off, and after the outer rind had been peeled off, eaten without cooking. **Karok:** long sprouts that come in the Spring are eaten raw as greens. **Pomo (Southwestern):** young green shoots eaten raw.

Medicine: **Yana:** roots pounded and cooked, then drunk as a cure for colds, diarrhea, headaches, and other aliments; the roots, after the outer bark was removed, were pounded and laid on the as a cure for headache. **Karok:** root with spoken formula was used as a purification after a funeral. **Pomo (Southwestern):** piece of the root chewed and held in the mouth to prevent sore throat or bad breath; root boiled to make wash for bathing sores; drunk to ease menstrual cramps, to regulate menses; relieve discomfort of menopause, for stomach ache, to relieve a cold; shamans smoked shavings from wood while doctoring. **Washo:** used root infusion for bronchial ailments, chewed dried roots for sore throats and coughs.

Citation(s): Goddard 1903; Sapir and Spier 1943; Schenck and Gifford 1952; Goodrich et al 1980; Strike 1994.

Terminology: **Yana:** àtdjadji'tṗa (stems). **Hupa:** xonsílsalū, "summer salūШ." **Karok"** ishmucha (young leaves), mahimkanva "Mountain imkanva")root). **Pomo (Southwestern):** ba?ćowa.

Citation(s): Sapir and Spier 1943; Goddard 1903; Schenck and Gifford 1952; Goodrich et al 1980.

92   *Antennaria argentea*
Family: *Asteraceae*

Usage: Misc.   **Miwok:** named but not used.

Citation(s): Barrett and Gifford 1933.

Terminology: **Miwok (Central Sierra):** potokpota.

Citation(s): Barrett and Gifford 1933.

93   *Anthemis cotula*
Family: *Asteraceae*
Common Name(s): Mayweed.

Usage: Medicine: **Karok:** good medicine for pregnant women if the formula is sung. **Yuki:** fresh plants placed in bath water, which is then used for a wash for sever colds and rheumatism; juice occasionally used as a eye wash; can irritate the skin.

Citation(s): Chesnut 1902; Curtin 1957; Schenck and Gifford 1952.

Terminology: **Yuki:** pō'-muk, "man-burns." **Karok:**

nishitich, "wart plant" (flowers look like warts).

Citation(s): Curtin 1957; Schenck and Gifford 1952.

Note(s): contains an acrid substance irritating to mucous membranes and is distasteful; leaves and flowers can cause dermatitis; will blister skin; plant native of Europe.

Citation(s): Kingsbury 1964; Muenscher 1951; Pammel 1911; Munz and Keck 1965; Hardin and Arena 1974.

94 *Anthoxanthum occidentale*

Family: *Poaceae*

Synonym(s): *Torresia macrophylla, Hierochloe occidentalis*

Common Name(s): California Vanilla Grass.

Usage: Medicine: **Karok:** used as a medicine for a woman who has a miscarriage; a pregnant women drinks this to prevent the foetus from getting too large.

Citation(s): Schenck and Gifford 1952.

Terminology: **Karok:** kitikuhara.

Citation(s): Schenck and Gifford 1952.

95 *Antirrhinum nuttallianum*

Family: *Plantaginaceae*

Common Name(s): Snapdragon.

Usage: Medicine: **Diegueño:** decoction of boiled flowers drank to relieve colds.

Citation(s): Strike 1994.

96 *Apiastrum angustifolium*
Family: *Apiaceae*
Common Name(s): Wild Celery.

Usage: Food: **Luiseño:** cooked, not eaten fresh.
Cahuilla: a small seasonal food source in wet years.
Medicine: **Shasta:** burned root to cure colds.
Hupa: smoke from burning roots inhaled as a disinfectant.
Citation(s): Sparkman 1908; Bean and Saubel 1972; Strike 1994.
Terminology: **Cahuilla:** pa'kily.
Citation(s): Bean ands Saubel 1972.

97 *Apium graveolens*
Family: *Apiaceae*
Common Name(s): Common Celery.

Usage: Food: **Luiseño:** used for greens.
Medicine: **Cahuilla:** decoction employed for ailments attributed to kidney malfunction.
Citation(s): Sparkman 1908; Bean and Saubel 1972.
Terminology: **Luiseño:** pa'kil. **Cahuilla:** pa'kily.
Citation(s): Sparkman 1908; Bean and Saubel 1972.
Note(s): naturalized from Europe; can produce hypersensitivity to sunlight.
Citation(s): Munz and Keck 1965; Fuller and McClintock 1986.

98 *Apocynum* spp.
Family: *Apocynaceae*
Common Name(s): Wild Hemp, Hemp, Dogbane.

Usage: Material: **Shasta:** cord and rope made from

fibers. **Maidu (Northern):** cordage made from fibers. **Yokuts (Southern Valley):** string made from fiber. **Wappo:** pounded to separate fibers; fibers moistened in mouth, dried, wet again in mouth, then twisted on thigh to make cordage. **Wintun:** fine cordage made from fibers; used principally in making nets and seines. **Luiseño:** made apron from fibers for women. **Chumash:** used fibers to make woven headbands, net foundation for feathered ceremonial skirts and sashes. **Tubatulabal:** used fibers to make rabbit net about 200 feet long by 3-4 feet high. **Esselen:** made cordage, nets, snares.

Citation(s): Curtis 1924; Dixon 1905; Driver 1935; Gayton 1948; Strike 1994; Breschini and Haversat 2004.

99   *Apocynum cannabinum*

Common Name(s): Indian Hemp, Dogbane.

Usage: Material: **Luiseño:** a string fiber is obtained from the bark; made into a 2-ply cordage for nets. **Owens Valley Paiute:** made a gill net from fibers. **Pomo:** cordage made from the inner bark. **Tubatulabal:** string made from fibers of the shoots. **Yuki:** made fiber from tall stems; inner bark fiber made ropes and nets. **Miwok:** probably made string from the fibers. **Chumash:** stalks, cut or pulled, collected in September, dried, pounded, rolled on stones, shredded the fiber to make string; used for bindings on mats, bindings on canoes, as sewing thread, bow strings, for fishing, as 3-ply harpoon line, fishing nets, carrying nets and bags, ceremonial paraphernalia, dance regalia. **Panamint Shoshone (Koso):** made cord from fibers. **Cahuilla:** woody stems soaked in water, fibers, after washing, were a soft, silky colored yellowish-brown, were strong and durable.

Medicine: **Maidu:** used roots to treat intestinal and bronchial ailments.

Misc: **Pomo (Southwestern):** named but not used.

Citation(s): Barrett 1952; Barrett and Gifford 1933; Chesnut 1902; Coville 1892; Curtin 1957; Curtis 1924; Grant 1964; Sparkman 1908; Steward 1933; Voegelin 1938; Iovin 1963; Goodrich et al 1980; Bean and Saubel 1972; Timbrook 2007; Strike 1994.

Terminology: **Luiseño:** wicha. **Yuki:** mä. **Pomo (Southwestern):** mahša. **Chumash (Barbareño, Island, Ynezeño, Ventureñ0):** tok.

Citation(s): Sparkman 1908; Curtin 1957; Goodrich et al 1980; Timbrook 2007.

100 *Apocynum androseamifolium*

Synonym(s): *Apocynum androsaemifolium* var. *Nevadense*

Common Name(s): Mountain Hemp.

Usage: Food: **Karok:** seeds eaten raw.

Misc: **Kawaiisu (Tehachapi):** named but not

used.

Citation(s): Schenck and Gifford 1952; Zigmond 1981.

Terminology: **Karok:** apshunmunikichyam, "little
brown snake's weed." **Kawaiisu (Tehachapi):**
motoobi.

Citation(s): Schenck and Gifford 1952; Zigmund 1981.

101 *Aquilegia eximia*
Family: *Ranunculaceae*
Common Name(s): Columbine.

Usage: Misc: **Pomo (Southwestern):** flowers used in
dance wreaths for Strawberry Festival.

Citation(s): Goodrich et al 1980.

Terminology: **Pomo (Southwestern):** s'ilis'li qʰale,
"tinkle plant."

Citation(s): Goodrich et al 1980.

102 *Aquilegia formosa*

Usage: Food: **Maidu (Northern):** ate seeds. **Maidu:**
seeds, flower nectar, young leaves
eaten, boiled and ate roots. **Miwok:** in
early Spring, entire plant boiled and
eaten.

        Medicine: **Maidu:** leaves used to poultice sores;
root decoction cured diarrhea,
stomach aches, coughs, indigestion;
seeds acted as a diuretic; pulverized
seed, mixed water to form as paste,
used to rid themselves of head lice.

Citation(s): Dixon 1905; Strike 1994.

Terminology: **Northern Paiute (Paviotso):** ya'pá-
gwanábü.

Citation(s): Kelly 1932.

103 *Aquilegia formosa*

> Synonym(s): *Aquilegia truncata*.
> Common Name(s): Columbine.

Usage: Food: **Miwok:** eaten boiled as greens in early Spring.

           Misc: **Karok:** named but not used; known for its pretty blossoms.

Citation(s): Schenck and Gifford 1952; Barrett and Gifford 1933.

Terminology: **Miwok (Central Sierra):** tcuyuma.
           Karok: kishwufsansanhitihan, "leaves like kishwuf."

Citation(s): Barrett and Gifford 1933; Schenck and Gifford 1952.

Note(s): plants reputed to be diuretic and diaphoretic.

Citation(s): Stuhr 1933.

104 *Arabis* spp.

> Family: *Brassicaceae*
> Common Name(s): Rock Cress.

Usage: Food: **Maidu:** eaten.

           Medicine: **Maidu:** infusion used to cure colds.

Citation(s): Strike 1994.

105 *Aralia californica*

> Family: *Araliaceae*
> Common Name(s): Elk Clover, California Ginseng, Spikenard.

Usage: Food: **Modoc (Lutuami), Maidu:** roots eaten.

           Medicine: **Pomo:** decoction of the dried root valued for diseases of the lung and the stomach; used for consumption, for colds, and for fevers; as a skin lotion to relieve sores and itching. **Pomo (Southwestern):** root boiled and the decoction applied externally, never taken internally; sores bathed with the liquid, would cure all kinds of open sores, including itching sores. **Diegueño:** washed sores with decoction to remove scabs and expedite healing. **Maidu:** dried root decoction used for stomach and bronchial ailments.

Misc: **Karok:** named but not used.

Citation(s): Chesnut 1902; Gifford 1967; Schenck and Gifford 1952; Goodrich et al 1980; Strike 1994.

Terminology: **Pomo (Southwestern):** sitabati kale.
Karok: patarakuup.

Citation(s): Gifford 1967; Schenck and Gifford 1952; Goodrich et al 1980.

Note(s): decoction of dried roots is a stomachic, tonic, febrifuge, alterative and bronchial.

Citation(s): Stuhr 1933.

106 *Arbutus menziesii*

Family: *Ericaceae*
Common Name(s): Madroña, Madroño.

Usage: Food: **Bear River:** berries eaten raw or roasted. **Shasta:** berries eaten fresh, and sometimes dried. **Salinan:** fruit eaten. **Sinkyone:** berries roasted, stored for winter. **Hupa:** the fruit is shaken in a basket with hot rocks and then eaten. **Wailaki, Pomo:** eat berries, not kept for winter use because the berry soon decays when bruised. **Pomo (Southwestern):** berries eaten, fresh or roasted, parched and stored for winter. **Miwok:** berries used in making cider, or dried and stored for winter consumption. **Yuki:** ate berries. **Karok:** steam berries, dried, and stored; soaked in water before eating. **Esselen:** berries, occasionally eaten.

Material:
Pomo: make lodge poles. **Pomo (Southwestern):** wood was burned, but was not particularly sought for firewood. **Bear River:** made drinking cups from short, solid pieces of the limbs. **Karok:** firewood when cooking salmon during the First Salmon Ceremony; drinking cups from wood. **Esselen:** curly bark, crumbled, mixed with Black Sage and Hummingbird Sage, to make a smoking mixture.

Medicine:
Pomo: infusion of the leaves used to cure colds, stomach aches, as a lotion for skin sores, removed scabs, to facilitate healing, gargle for sore throats. **Pomo (Southwestern):** flowers used for love charm poisoning; bark boiled, tea used by women to rinse their face; tannin acts as an astringent to close pores and make skin soft and pretty. bark was boiled and the liquid used to wash skin sores, except for Poison Oak. **Yuki:** leaves and bark made into tea for sores and cuts, to promote vomiting. **Cahuilla:** leaves used to make medicine for stomach ailments. **Hupa:** boiled leaves placed on cuts to stop hemorrhages. **Maidu:** leaves rubbed on burns and aching

joints; decoction to relieve stomach aches, colds, sore throats. **Karok:** ceremonial medicine in puberty ceremonies; young girls during ceremonies threw leaves over their shoulders on the way to bathe in river for good luck. **Costanoan:** bark, peels, may have been used as tea for colds.

Citation(s): Barrett and Gifford 1933; Chesnut 1902; Curtin 1957; Gifford 1967; Goddard 1903; Holt 1946; Mason 1912; Nomland 1935, 1938; Goodrich et al 1980; Bean and Saubel 1972; Strike 1994; Rayburn 2012; Breschini and Haversat 2004.

Terminology: **Karok:** koshri'pan, koshri'pish (berry). **Hupa:** isdeau. **Yuki:** poí-ki, poí-ki-bäm. **Pomo (Southwestern):** kaba. **Miwok (Central Sierra):** mok'lkine.

Citation(s): Schenck and Gifford 1952; Goddard 1903; Curtin 1957; Gifford 1967; Barrett and Gifford 1933.

Note(s): the fruit ripens in autumn; root, bark, and leaves are astringent.

Citation(s): Sargent 1922; Stuhr 1933.

107 *Arceuthobium spp.*

Family: *Violaceae*
Common Name(s): Mistletoe.

Usage: Medicine: **Maidu, Southern (Nisenan):** grows on *Pinus edulis* and is used with the pitch of the tree as a cure for coughs, colds, and rheumatism.

Citation(s): Powers 1877.

108 *Arceuthobium campylopodum*
>
> Synonym(s): *Razoumofskya occidentalis*
> Common Name(s): Mistletoe, Pine Mistletoe.

Usage: Medicine: **Yuki:** a decoction of the plant is made as a tea for stomach ache. **Maidu:** smoke relieved colds and coughs; decoction used as a contraceptive, as a cure for stomach ache. **Bear River:** used in sweathouses to cure coughs, colds, rheumatic pain.

Citation(s): Chesnut 1902; Strike 1994.

109 *Arceuthobium campylopodum*
>
> Common Name(s): Mistletoe.

Usage: Misc: **Kawaiisu (Tehachapi):** named but not used.

Citation(s): Zigmond 1981.
Terminology: **Kawaiisu (Tehachapi):** sanapɨceeka.
Citation(s): Zigmond 1981.

110 *Arctostaphylos* spp.
>
> Family: *Ericaceae*
> Common Name(s): Manzanita.

Usage: Food: **Owens Valley Paiute:** traded from the west; mashed, soaked in water to make a non-fermented drink; used berries. **Tubatulabal, Huchnom:** used berries. **Bear River:** sun dried the berries; packed in deep baskets between layers of leaves. **Yuki:** eaten directly from the bush, or allowed to dry; sometimes made into cider. **Shasta:** dried berries and made into meal, also crushed berries to make "cider." **Yokuts (Southern Valley):** made cider from fruit. **Monache (Western Mono):** berries gathered, mashed, and made into cider. **Wintun:** berries pounded into coarse flour, dried and parched; fine flour made into cakes, coarse made into cider. **Wintun, Hill (Patwin):** dried, ground, and eaten as "bread;" made drink by pouring water on pounded berries. **Wintun, Central (Nomlaki):** used berries. **Maidu, Southern (Nisenan):** were gathered half-ripe, then dried, broken up, placed in porous baskets, water poured through; after standing a few hours, the cider was drunk. **Sinkyone:** berries roasted, stored for winter. **Cahuilla:** berries gathered from June until September; pulp mashed, mixed with water, strained into drink, or, soaked in water without crushing; sundried and stored in ollas for future use; seeds ground into meal from which a mush or cakes were prepared; a primary food due to the large quantities that were available. **Karok:** cooked dried, pulverized berries with salmon eggs. **Kawaiisu (Tehachapi):** blossoms stepped for tea, chia (*Salvia Columbariae*) seeds sometimes added to tea; ripe berries made into a gelatinous substance and eaten. **Pomo:** tea used to relieve diarrhea; blossom tea cured dizziness. **Maidu:** held green fruit in mouth, occasionally nibbling to quench thirst. **Costanoan:** fruit, dried, stored for Winter use. **Esselen:** berries occasionally eaten. **Chumash:** berries, raw, ground, eaten as pinole, mixed with water (in historic times, with milk).

Material: **Wappo:** made bows from wood, considered best, most expensive. **Miwok:** forked branches formed part of deer-headdress used in stalking prey; made acorn mush stirrer. **Cahuilla:** wood preferred for hot-fire and long-lasting coals; branches often employed in house construction; stems used to make pipes and other small tools. **Karok, Maidu:** made utensils from wood. **Chumash:** hairpins of wood used to secure men's topknot; fish dried in smoke of wood fire. **Pomo:** burned wood during ceremonials due to its bright light. **Hupa:** small branches, approximately 4 inches long, made into smoking pipes for everyday use, 12 inches long for ceremonial use.

Medicine: **Shasta:** pounded berries used to stop diarrhea. **Cahuilla:** leaves occasionally mixed with tobacco or steeped in water to make tea used in curing diarrhea or poison oak rash. **Maidu:** Leaves or sap poultice used on sores; leaves used as a diuretic, or to relieve stomach aches; crushed berries used to stop diarrhea; decoction of berries or leaves used to treat bronchial problems. **Miwok (Southern Sierra):** berry tea sucked off hawk feathers on feathered wand to cure stomach disorders, the feather held the healing power not the tea. **Wintun:** new-born girl's umbilical cord placed in bush to insure that the child would be mild and pleasant. **Costanoan:** plant parts, decoction, used for bladder ailments. **Chumash:** berries, boiled, decoction use to treat Poison Oak rash.

Citation(s): Barrett and Gifford 1933; Beals 1933; Clark 1904; Driver 1936; Du Bois 1935; Foster 1944; Gayton 1948; Gifford 1932; Holt 1946; Kroeber 1932; Goldschmidt 1951; Nomland 1935; Steward 1933; Voegelin 1938; Bean and Saubel 1972; Strike 1994; Rayburn 2012; Breschini and Haversat 2004; Timbrook 2007.

Terminology: **Owens Valley Paiute:** apōsō'gwa. **Yokuts:** a'ptu (berry). **Cahuilla:** kelel. **Chumash (Barbareño):** sq'oyon; **(Ynezeño):** sq'o'yon; **(Ventureño):** tsqoqo'n.

Citation(s): Gayton 1948; Bean and Saubel 1972.; Steward 1933; Timbrook 2007.

Note(s): the leaves, and probably the bark, of Arctostaphylos uva-ursi, and probably the other species of *Arctostaphylos* as well, contain chemicals which are diuretic and have an antiseptic action on the urinary tract.

Citation(s): Trease and Evans 1966; Stuhr 1933.

111 *Arctostaphylos canescens*

Usage: Food: **Karok:** the berries were eaten when dried.

Material: **Karok:** tobacco pipes are made from twigs.

Citation(s): Schenck and Gifford 1952.

Terminology: **Karok:** ohusukamfas, "The-fas-that-
looks-down-toward-the-ocean."
Citation(s): Schenck and Gifford 1952.

112 *Arctostaphylos columbiana*
Common Name(s): Hairy Manzanita.

Usage: Material: **Pomo (Southwestern):** wood favored
for tools such as awl handles; burned at
dances and ceremonies because it
makes a bright light.

Medicine: **Pomo (Southwestern):** bark was boiled
and the decoction was drunk to check
diarrhea.

Misc: **Pomo (Southwestern):** berries eaten
because too sour.

Citation(s): Gifford 1967; Goodrich et al 1980.
Terminology: **Pomo (Southwestern):** balaikale, bakai (berry).
Citation(s): Gifford 1967.

113 *Arctostaphylos glandulosa*
Common Name(s): Eastwood Manzanita.

Usage: Food: **Pomo (Southwestern):** berries boiled to
kill worms; dried, pounded, stored for
later use in pinole; mixed with water
and eaten.

Medicine: **Pomo (Southwestern):** bark boiled,
drunk for diarrhea.

Material: **Pomo (Southwestern):** favored wood
for making awl handles.

Misc: **Pomo (Southwestern):** named but not
used.

Citation(s): Gifford 1967; Goodrich et al 1980.
Terminology: **Pomo (Southwestern):** kaya kale.

Citation(s): Gifford 1967; Goodrich et al 1980.

Note(s): note the different statements by the several authors between Misc: and other categories – this may be a temporal difference in their informants knowledge [author].

114 *Arctostaphylos glauca*

Common Name(s): Manzanita.

Usage: Food: **Maidu, Southern (Nisenan):** the berries are the favorite article of food; eaten raw, or pounded into flour in a basket, the seeds separated out, the flour made into mush, or sacked and laid away for the Winter; cider is made by soaking the flour in water several hours, then draining the liquid off; the dried leaves are mixed with *Nicotiana Bigelovii* and smoked. **Huchnom:** berries used to make bread. **Miwok:** principle berries used by the tribes lower down in the foothills. **Cahuilla:** berries eaten raw, or dried, pounded into flour, and mixed with water. **Kawaiisu (Tehachapi):** picked ripe in August-September, may be eaten fresh, but if waxy and sticky, they are discarded; soak in water and drunk; may be mixed with Chia (*Salvia clumbariae*).

Material: **Cahuilla:** used branches and trunks as house building materials.

Citation(s): Barrows 1900; Clark 1904; Powers 1877; Zigmond 1981.

Terminology: **Kawaiisu (Tehachapi):** kiinarabï.

Citation(s): Zigmond 1981.

Note(s): seeds contain:
        Ash   3.2%
        Protein 24.9%
        Oil   55.7%
Citation(s): Earle and Jones 1962.

115 *Arctostaphylos manzanita*
Common Name(s): Manzanita, Parry Manzanita.

Usage: Food: **Hupa:** fresh fruit eaten when ripe in midsummer, gather in large quantities and dried on the sane by the river; dried fruit pounded and seeds separated out; the flour was eaten dry without cooking; the seeds soaked in water and the liquid was drunk without fermentation. **Pomo:** used berries. **Karok:** dried and eaten; soaked in water for drink; mixed with salmon eggs and cooked. **Shasta:** berries used to make cider. **Miwok:** berries made into cider, or dried and stored for winter. **Yuki:** gathered berries in July and August when ripe; eaten raw or cooked or parched and used in pinole; cider made from ripe fruit, berries crushed and strained through sieve basket-an equal amount of water added to juice; dried berries stored for winter.

Material: **Hupa:** used to make tobacco pipes. **Karok:** make canes, spoons, tobacco pipes, scraping stick for acorn soup, and reels for string (netting needles).

Medicine: **Pomo:** boiled leaves used to wash body and head; cure headaches by washing

head; leaves used to check diarrhea.
Maidu: chew leaves into thick curd and
place on sores for healing effect.

Citation(s): Barrett 1952; Barrett and Gifford 1933;
Chesnut 1902; Curtin 1957; Dixon 1907;
Goddard 1903; Schenck and Gifford 1933.

Terminology: **Yuki:** kö-öch'-öl. **Hupa:** dinŪШ. **Karok:** fas'ip,
"fas tree;" fas (berry). **Miwok (Central Sierra):**
mo'kosŪ. **Pomo (Southern):** gaiya'; **Central,
Southern):** kaiye'; **(Southwestern):** xĪye', kĪye' (bush
and wood - all cases).

Citation(s): Curtin 1957; Goddard 1903; Schenck and
Gifford 1952; Barret and Gifford 1933; Barrett 1952.

116   *Arctostaphylos viscida* spp. *mariposa*

Usage: Food: **Miwok:** berries eaten and made into
cider.

Citation(s): Merriam 1955.

117   *Arctostaphylos nevadensis*
Common Name(s): Pine-Mat Manzanita.

Usage: Food: **Karok:** the berries are eaten and dried.
Material: **Karok:** tobacco pipes are made from
twigs.

Citation(s): Schenck and Gifford 1952.

Terminology: **Northern Paiute (Paviotso):** koda'bü. **Karok:**
apuunfas, "Ground fas."

Citation(s): Schenck and Gifford 1952; Kelly 1932.

118 *Arctostaphylos parryana*

Synonym(s): *Arctostaphylos parryi*
Common Name(s): Manzanita.

Usage: Food: **Luiseño:** the bulb of the berries is ground and used.
Citation(s): Sparkman 1908.
Terminology: **Luiseño:** kolul.
Citation(s): Sparkman 1908.

119 *Arctostaphylos patula*

Common Name(s): Green Manzanita.

Usage: Food: **Karok:** the black berries are good when dried; they are not good when picked fresh; picked when ripe in September or later. **Atsugewi:** gathered in July and August, stored in pits, pounded up when needed, sift and made into fine flour which was molded into biscuit-like cakes and put away until needed; cider was made by adding water to pounded berries. **Kawaiisu (Tehachapi):** berries are inedible.

Material: **Karok:** tobacco pipes made from twigs.
Medicine: **Atsugewi:** leaves are pounded and boiled, solution good for cuts, burns, skin rashes, bruises.
Misc: **Miwok:** named but not used.
Citation(s): Barrett and Gifford 1933; Garth 1953; Schenck and Gifford 1952; Zigmond 1981; Strike 1994.
Terminology: **Miwok (Central Sierra):** palapala.
Karok: fao'uruhsa, "Round manzanita;" pahaav. **Atsugewi:** we^hyar. **Kawaiisu (Tehachapi):** hɨrɨrɨbiivɨ.

Citation(s): Barrett and Gifford 1933; Schenck and
Gifford 1952; Garth 1953; Zigmond 1981.

120 *Arctostaphylos pringlei* var. *drupacea*

Usage: Food: **Kawaiisu (Tehachapi):** picked ripe in
August-September; may be eaten fresh,
but if waxy and sticky, they are
discarded; soaked in water and drunk;
may be mixed with Chia (*Salvia
columbariae*).

Citation(s): Zigmond 1981.

Terminology: **Kawaiisu (Tehachapi):** kɨɨnarabɨ.

Citation(s): Zigmond 1981.

121 *Arctostaphylos pungens*
Common Name(s): Manzanita.

Usage: Food: **Maidu (Northern):** berries used in
making "manzanita cider."

Citation(s): Dixon 1905.

Note(s): seeds contain:

Protein 2.5%

Oil 5.6%

Citation(s): Earle and Jones 1962.

122 *Arctostaphylos tomentosa*

Usage: Food: **Pomo:** ground berries into a very fine
meal and made biscuit about 4-5 inches
in diameter, eaten raw or dried; made
cider. **Yuki:** eat berries, raw or dried;
made cider. **Miwok:** made berries into
cider, or dried and stored for winter.

Citation(s): Barrett 1952; Barrett and Gifford 1933; Chesnut

1902.

> Terminology: **Miwok (Central Sierra):** e'ye. **Pomo (Southeastern):** nōbakai.
> Citation(s): Barrett and Gifford 1933; Barrett 1952.
> Note(s): food values – berries:

	(wet)	(dry)
Moisture	12.8%	
Sugar		
Reducing	1.6%	21.3%
Non-reducing	0.0%	0.0%
Starch	0.0%	0.0%
Hemicellulose	13.1%	15.0%

> Citation(s): Yanovsky 1938.

123 *Arctostaphylos uva-ursi*

Common Name(s): Sand-Berry; Bearberry.

> Usage: Food: **Wiyot:** gathered and eaten after being placed in a basket with hot coals and shaken until nearly ready to pop. **Coast Yuki:** berries eaten raw.
> Citation(s): Gifford 1939; Loud 1918.
> Terminology: **Wiyot:** shogowi. **Yuki:** hyulkut.
> Citation(s): Loud 1918; Gifford 1939.
> Note(s): food values – berries:

	(wet)	(dry)
Moisture	58.0%	
Sugar		
Reducing	7.9%	18.8%
Non-reducing	1.3%	3.1%
Starch	0.0%	0.0%
Hemicellulose	8.4%	20.0%

> Citation(s): Yanovsky 1938.

124 *Arctostaphylos viscida*

Usage: Food: **Miwok:** berries eaten and made into cider; dried and stored for winter.
Citation(s): Barrett and Gifford 1933; Merriam 1955.
Terminology: **Miwok (Central Sierra):** e'ye.
Citation(s): Barrett and Gifford 1933.

125 *Argemone munita*

Family: *Papaveraceae*
Synonym(s): *Argemone platyceras*
Common Name(s): Prickly Poppy.

Usage: Medicine: **Washo:** grind the seed into an oily paste to make a slave for burns, sores, or cuts. **Kawaiisu (Tehachapi):** ripe seeds roasted "like coffee," mashed and applied salve on burns.
Misc: **Mono Lake Paiute:** named but not used.
Citation(s): Steward 1933; Train et al 1941; Zigmond 1981; Strike 1994.
Terminology: **Mono Lake Paiute:** tsa'gida'ᵃ. **Kawaiisu (Tehachapi):** Caaruwagadɨbɨ.; caaru "rattle."
Citation(s): Steward 1933; Zigmond 1981.
Note(s): seeds of various *Argemone spp.* are considered to be narcotic; if ingested the alkaloid in the seeds is capable of producing dilation of the capillaries, leading to leakage of fluid.
Citation(s): Stuhr 1933; Kingsbury 1964.

126 *Aristida spp.*

Family: *Poaceae*
Common Name(s): Triple-awned Grass.

Usage: Food: **Luiseño:** seeds eaten.
Citation(s): Strike 1994.

127   *Aristolochia californica*
           Family: *Aristolochiaceae*
           Common Name(s): Dutchman's Pipe.

Usage: Medicine: **Miwok:** plant was steeped and the
                       decoction drunk to cure colds; flowers
                       used to treat snake bite.
Citation(s): Barrett and Gifford 1933; Strike 1994.
Terminology: **Miwok (Central Sierra):** okise.
Citation(s): Barrett and Gifford 1933.
Note(s): in small doses, it is a stimulant, tonic, diaphoretic,
          promotes the appetite.
Citation(s): Grieve 1931.

128   *Arnica* spp.
           Family: *Asteraceae*

Usage: Misc: **Mono Lake Paiute:** named but not
                       used.
Citation(s): Steward 1933.
Terminology: **Mono Lake Paiute:** si'daġwana'da.
Citation(s): Steward 1933.

129   *Arnica discoidea*
           Common Name(s): Rayless Arnica.

Usage: Medicine: **Maidu:** flowers and roots used to heal
                       wounds, to induce vomiting, and as a
                       narcotic.
Citation(s): Strike 1994.

130 *Artemisia* spp.

Family: *Asteraceae*
Common Name(s): Sagebrush.

Usage: Food: **Luiseño:** fresh, tender shoots peeled and eaten raw; seeds eaten.

Material: **Cahuilla:** used as roofing in houses; used in outdoor granary construction. **Mono Lake Paiute:** used as covering over temporary shelters.

Medicine: **Atsugewi:** bark was chewed and mixed with deer manure to make a small pack which was placed over jaw where toothache was and set afire; leaves and stems pounded up and boiled to make a tea which was drunk to prevent blood poisoning. **Yokuts (Northern Hill):** plant rubbed between hands to extract juice which was then rubbed on for rheumatism pains; sometimes the plant was soaked in water and the solution was used instead. **Wappo:** leaves chewed or eaten to cure diarrhea.

Citation(s): Curtis 1924; Garth 1953; Gayton 1948; Driver 1936; Sparkman 1908; Heizer 1960; Strike 1994; Bean and Saubel 1972.

131 *Artemisia californica*

Common Name(s): Coastal Sagebrush.

Usage: Material: **Chumash:** wood, horizontal piece of fire-drill fire starting apparatus, foreshafts of arrows, fuel for cooking; branches, windbreaks, brush barricades.

Medicine: **Cahuilla:** used in girl's puberty rites;

decoction drunk just before commencement of each menstrual period, accompanied by dietary restrictions, no salt, grease, or meat, given to newborn babies one day after birth to purge their bodies. **Costanoan:** leaves, held against tooth to reduce pain, tied over wounds (to reduce pain); decoction, used to bathe patients with colds, coughs, rheumatism, taken internally to treat asthma (accompanied by poultices on patient's back and chest). **Chumash:** leaves, tied to forehead to treat headache, or by rubbing leaves in water, and wetting hair and head, leaves carried inside hat to feel good, cool; leaves, boiled, steam inhaled to treat paralysis, decoction used for Poison Oak treatment; plant used to treat coughs; plant burned to fumigate house after a funeral; bundles of the plant erected during Winter solstice ceremony.

Citation(s): Strike 1994; Rayburn 2012; Timbrook 2007.

Terminology: **Chumash (Barbareño):** we'wey; **(Inland, Ynezeño):** wewey; **(Obispeño):** tilho; **(Ventureño):** wewe'y.

Citation(s): Timbrook 2007.

132 *Artemisia cana* ssp. *bolanderi*

Usage: Material: **Tubatulabal:** used for roasting piñon cones (*Pinus monophylla*); brush bed made of limbs or of *Artemisia tridentata*.

Citation(s): Voegelin 1938.

133 *Artemisia douglasiana*

Synonym(s): *Artemisia heterophylla,*
Artemisia vulgaris var. *heterophylla.*
Common Name(s): Wormwood, Mugwort,
California Mugwort.

Usage: Food: **Pomo (Southwestern):** leaves dried, rubbed between hands to make tobacco.

Material: **Miwok:** leaves worn in nostrils of mourners when crying, the pungent odor clearing the head. **Luiseño:** small boy's arrows were sometimes made from the plant.

Medicine: **Miwok:** decoction of leaves drunk to cure rheumatism; leaves inserted into nostrils to cure headache; rubbed on body to keep ghosts away; small balls containing the plant and other "medicine" plants attached at intervals to prevent dreaming of the dead; with such a necklace one might venture forth at night without fear of ghosts; necklace worn for a month after a death of close relatives of the deceased; malevolent shamans or "poisoners" reputed to carry poison in the leaves of plant to avoid personal injury; corpse handlers rubbed themselves with leaves to keep from being haunted by the ghost of the deceased; women tied bundles of leaves around bodies to promote blood circulation after childbirth. **Luiseño:** the plant was used medicinally. **Pomo:** used as a cure for colic and colds and bronchitis; a decoction used internally for stomach ache, headache, diarrhea, and some kinds of fever; externally used as a head wash to relieve headache and as a wash for sore eyes; leaves tied in bundles around body to cure rheumatism, and after childbirth to promote circulation of the blood. **Miwok (Coast):** used to cure stomach aches, to reduce fevers, to treat eye ailments. **Pomo (Southwestern):** used in childbirth; warmed leaves placed on a baby's umbilicus after the navel cord has been severed with an obsidian knife; repeated for four days until cord came off; a boiled decoction of the leaves was drunk to stop excessive menstruation. **Yuki:** dry or green leaves brewed into a strong hot drink for pains or "any other trouble inside," rheumatic or back pains; cuts, bruises, wounds, sores and sore eyes; bathed with a tea and poultices are applied until a cure is affected; in difficulties attending childbirth, the patient is steamed over a shallow pit containing plant material; used for back pains by steaming until sweating occurs; crushed plant material rubbed in during tattooing to give a green color; hot coals wrapped in leaves to poultice an earache; ashes of plant applied as poultice to neck to cure coughs, sore throats; whole plant used as poultice for cuts, bruises, skin sores, back pains. **Karok:** branches are mixed with branches of *Umbellularia californica* and

Pseudotsuga menziesii laid in a pit of hot rocks, the patients in placed over this on a blanket and allowed to steam, good for colds and about any kind of sickness; a drink is administered to a woman to relieve the pains of childbirth, with a special charm. **Washo:** boiled leaves used as a wash for headaches; for rheumatism the boiled leaves are applied as a liniment. **Chumash:** cauterized wounds of patients with small cones of leaves placed on skin and slowly burned down; tea of leaves is drunk for headache and asthma; also taken for severe fright; used as remedy for wounds, skin lesions, rheumatism Poison Oak rash; leaves, applied as a plaster to sore muscles, as a poultice for headache, toothache, soothing bath for measles. **Kawaiisu (Tehachapi):** soak plant (except root) in water; infusion used as a bath after menstruation flow is stopped - said to prevent aging prematurely; as hair wash before a girl took Jimsonweed, it would prevent hair from falling out; a bath for both parents after childbirth; a boy did not eat meat of first kill but together with parents chewed meat mixed with plant and spit it into a fire; if not done he would never kill deer again and become a transvestite; decoction of leaves used to treat headaches, colic, bronchitis, rheumatic pain. **Wappo:** chewed leaves to cure diarrhea. **Yokuts:** rubbed

crushed leaves on body to relieve
rheumatic pain; used infusion to bathe
sore eyes. **Costanoan:** leaves, heated,
held over ear for earache; decoction,
used for compresses on wounds, for
pain from rheumatism, to treat asthma,
to treat urinary problems. **Esselen:**
leaves, boiled, tea consumed to prevent
Poison Oak; leaves, poultice, used as
poultice of Poison Oak rash.

Citation(s): Barrett and Gifford 1933; Chesnut 1902; Curtin
1957; Gifford 1967; Schenck and Gifford 1952;
Sparkman 1908; Train et al 1941; Iovin 1963;
Timbrook 1987; Zigmond 1981; Goodrich et al 1980;
Strike 1994; Rayburn 2012; Breschini and Haversat
2004; Timbrook 2007

Terminology: **Pomo (Southwestern):** ka'pula' "Wormwood."
Karok: kaat. **Luiseño:** pakoshish. **Yuki:** pun'-kuni.
Washo: paal-<u>lume</u>-it. **Kawasiisu:** tugusoovi.
Chumash (Barbareño, Ynezeño): molosh; **(Isalnd):**
qlogol; **(Obispeño):** tpinusmu'; **(Ventureño):** molish.

Citation(s): Gifford 1967; Schenck and Gifford 1952;
Sparkman 1908; Curtin 1957; Train et al 1941;
Zigmond 1981; Goodrich et al 1980; Timbrook 2007.

134 *Artemisia drancunculus*

Synonym(s): *Artemisia drancunuloides*

Usage: Food: **Luiseño:** the seeds are eaten.

Medicine: **Luiseño:** the plants is used for
medicinal purposes. **Kawaiisu
(Tehachapi):** leaves boiled, infusion
used to wash limbs for relief of
rheumatism. **Costanoan:** root,
decoction, used for dysentery, for infant

colic, for urinary problems.
Citation(s): Sparkman 1908; Zigmond 1981; Strike 1994;
Rayburn 2012.
Terminology: **Luiseño:** wachish. **Kawaiisu (Tehachapi):** pḭsivḭ.
Citation(s): Sparkman 1908; Zigmond 1981.

135 *Artemisia ludoviciana*
Synonym(s): *Artemisia gnaphalodes*
Common Name(s): Sagebrush, Wormwood.

Usage: Food: **Luiseño:** fresh, tender shoots peeled and eaten raw; seeds eaten.

    Material: **Shasta:** handful of leaves used to wash body in preparation for burial; stems used as warp and woof in baskets; used as roofing and wattled in closely to form walls; shoots were peeled, notched, and pointed, and straightened with teeth to make arrows; made into basket granaries to store dried seed pod of *Prosopis juliflora* var. *Torreyana.* **Cahuilla:** stems used as warp and woof in basketry.

    Medicine: **Miwok:** made necklaces of small bundles to keep disease away. **Yurok:** make tea for sore eyes and itching skin; used it as lotion for eyes. **Washo:** a boil leaf decoction is an internal treatment for heavy colds, head colds, coughs; the solution is also used as a cooling, aromatic wash for headaches. **Pomo:** leaves, pounded or chewed, applied to cut end of newborn's umbilical cord, renewed each time baby was washed; poultice of leaves applied for bruises, rheumatic pain, placed in nose to relieve cold symptoms; leaf decoction used to sooth headaches, bathe burns, relieve sore eyes, used internally to cure colds, colic, stomach aches, headaches, diarrhea, fevers, to stop excessive menstruation; sap used to relieve sores caused by poison oak. **Serrano:** leaf decoction applied for headaches, drunk to cure colds, coughs. **Hupa, Gabrielino:** used to relieve aches, colds. **Costanoan:** leaves, decoction, used for fever, chills; poultice used for wounds.

Citation(s): Sparkman 1908; Holt 1946; Barrows 1900; Merriam 1955, 1967; Merrill 1923; Train et al 1941; Bean and Saubel 1963, 1972; Strike 1994; Rayburn 2012.

Terminology: **Washo:** auga-<u>lem</u>-lu. **Cahuilla:** hang-al.

Citation(s): Train et al 1941; Barrows 1900.

Note(s): in medicine it does act as a stimulant tonic.

Citation(s): Stille and Maisch 1880.

136 *Artemisia ludoviciana* var. *incompta*

Synonym(s): *Artemisia vulgaris, Artemisia vulgaris* var. *discolor*

Usage: Medicine: **Owens Valley Paiute:** tops boiled, applied to gonorrheal sores; for female backache, knee ache, and other troubles; a fire was built on the ground, removed, and the warm ground covered with the plant on which the patient lay.

Citation(s): Steward 1933.

Terminology: **Miwok (Northern Sierra):** kitci'ñu; **(Central Sierra):** kitciño. **Owens Valley Paiute:** kō'sidava, kósidapa. **Shasta:** wasa'.

Citation(s): Barrett and Gifford 1933; Steward 1933; Holt 1946.

Note(s): the plant is considered emmenagogue and
antispasmodic.

Citation(s): Stuhr 1933; Grieve 1931.

137 *Artemisia tridentata*

Common Name(s): Common Sage,
Wormwood, Sage, Basin Sagebrush, Big
Sage.

Usage: Food: **Owens Valley Paiute:** unimportant
because seeds are bitter; used generally
mixed with other seeds; in times of
food shortage, seeds roasted, ground
into flour, and eaten with water.
Cahuilla: seeds gathered in late Fall,
August through October, and pounded
into pinole. **Luiseño:** ate peeled fresh
shoots.

Material: **Owens Valley Paiute:** used as the base
of a fire-drill. **Tubatulabal:** used for
roasting piñon (*Pinus monophylla*) cones;
brush bed made of this material or
Artemisia cena. **Kawaiisu (Tehachapi):**
preferred material for hearth and
foreshaft of composite firedrill; for
roasting of pinyons; menstruating
women used wood for head-scratcher
only, else hair would fall out and face
wrinkle; pounded bark serves as lining
or wrapper inside winter shoes.
Cahuilla: shoots laid across rafters for
roofing material, or used in wall
construction, at times.

Medicine: **Cahuilla:** tea made from leaves used for stomach complaints; burned branches to fumigate sickroom, disinfect implements used in childbirth. **Mono Lake Paiute:** tea of leaves drunk to produce sweating during fever. **Washo:** boiled green leaves made into tea for the treatment of colds and as a general tonic; burned branches to fumigate sickroom, disinfect implements used in childbirth. **Kawaiisu (Tehachapi):** boiled with the fumes inhaled for relief of headache and colds. **Owens Valley Paiute:** chewed leaves to relieve stomach aches, cure colds. **Salinan:** leaves used to poultice burns, wounds, facial blemishes. **Modoc (Lutuami):** chewed leaves, or boiled and ate, to relieve stomach aches, drank infusion to reduce fever, relieve pain; leaves applied as poultice for aches, pains. **Diegueño:** leaves applied as poultice for aches, pains.

Citation(s): Barrows 1900; Steward 1933; Train et al 1941; Voegelin 1938; Zigmond 1981; Strike 1994; Bean and Saubel 1972.

Terminology: **Owens Valley Paiute:** sāwā'va, sāwā'vi, sāwāvu. **Mono Lake Paiute:** sawapi. **Washo:** da-bel, tah-bul. **Cahuilla:** wik-wut, wikwat. **Kawaiisu (Tehachapi):** sohovɨ.

Citation(s): Steward 1933; Train et al 1941; Barrows 1900; Zigmond 1981.

Note(s): the plant is considered to be diaphoretic, fabrifuge, mild cathartic, diuretic.

Citation(s): Stuhr 1933.

138 *Arthrocnemum subterminale*

Family: *Chenopodiaceae*
Synonym(s): *Salicornia subterminalia, Salicornia subterminalis*

Usage: Food: **Cahuilla:** seeds, available June to October, crushed into fine meal on a metate.
Citation(s): Barrows 1900; Bean and Saubel 1972.
Terminology: **Cahuilla:** hoat.
Citation(s): Bean and Saubel 1972.
Note(s): common on salt marshes and elsewhere in coastal regions; indicators of aridity, and invaders of waste lands and of root crop fields.
Citation(s): Tsukada 1967.

139 *Arundo donax*

Family: *Poaceae*
Common Name(s): Giant Reed.

Usage: Material: **Chumash:** side shoots, made arrow mainshafts; thickest stems split, scraped, to make sharp edge, pointed end knife; stone throwers used in sham fights between men and boys; made flutes, split-stick clapper, musical bow (introduced by Yaqui Indians from Sonora, Mexico).
Citation(s): Timbrook 2007.
Terminology: **Chumash (Ventureño):** shukepesh 'ishaq.
Citation(s): Timbrook 2007.

140 *Asarum spp.*

Family: *Aristolochiaceae*

Usage: Food: **Maidu:** used root to make refreshing drink.

Medicine: **Sinkyone:** leaves ground and soaked in water for pain in the stomach; sick person drinks plenty and vomits, after awhile they get hungry and eat. **Achomawi:** believed to be best remedy for cuts and boils; leaves put on fresh, not cooked or heated. **Maidu:** decoction of root used for stomach aches, colds; mashed root used as poultice on open sores, painful joints; fresh leaves put on boils, cuts. **Pomo:** warmed leaves as poultice to draw boils to head; made skin lotion for skin rashes. **Pomo (Southwestern):** leaf decoction used to wash sores; leaf poultice used on toothaches. **Yuki:** shamans used warm leaf poultice as part of curing ceremony; poultice used for general aches and pains.

Citation(s): Baumhoff 1958; Merriam 1967; Strike 1994.

Note(s): *Asarum* spp. plants are considered stimulant, carminative, diuretic, diaphoretic.

Citation(s): Grieve 1931.

141 *Asarum caudatum*

Common Name(s): Wild Ginger.

Usage: Medicine: **Pomo (Southwestern):** the leaves were warmed and used as a poultice at night to draw a boil to a head; applications were repeated. **Coast Yuki:** leaves used by shaman, being warmed and put on affected part as a poultice.

Citation(s): Gifford 1939, 1967; Goodrich et al 1980.

Terminology: **Pomo (Southwestern):** mô'bo kale' "Swelling plant." **Yuki:** balshata.

Citation(s): Gifford 1939, 1967; Goodrich et al 1980.

Note(s): *Asarum spp.* plant considered stimulant, carminative, diuretic, diaphoretic.

Citation(s): Grieve 1931.

142 *Asclepias* spp.

Family: *Apocynaceae*
Synonym(s): *Gomphocarpus* spp.
Common Name(s): Lowland Milkweed, Rock Milkweed, Milkweed, Milk Plant.

Usage: Food: **Karok, Chumash:** juices used for chewing gum. **Yokuts (Northern Hill:** collected sap from plants and chewed gum for pleasure. **Wappo:** leaves and roots eaten. **Tubatulabal:** collected sap in buckwheat (*Eriogonum elongatum*) stem, then laid on a fire to congeal the sap before chewing.

Material: **Monache (Western Mono):** made carrying net of woven fibers. **Maidu, Southern (Nisenan):** when dried the bark is stripped off and twisted into strands, then cords; used for string, cords, and nets. **Achomawi:** stems used for making string and cord. **Atsugewi:** twine made from fibers. **Yokuts (Southern Valley), (Northern Hill):** made string from fibers. **Wintun:** fiber used for string, principally in making dipnets and seines. **Wappo:** made string from fibers, used especially for nets, stronger than hemp. **Miwok:** used fibers to make cord for dipnets. **Luiseño:** made women's aprons from fibers. **Washo:** fibers used principally in fish nets. **Chumash:** fiber, made carrying nets, tumplines, cradle bands, men's belts; men often wore hairnets made from fiber cordage, decorated with shells. **Diegueño:** made brushes from fibers, used to paint patterns on clay pottery; made bowstrings from fibers, in cold weather legging of hide were bound around legs with fiber cordage; cordage used to twine rabbit skin into large blankets used as capes and/or blankets. **Esselen:** fibers, used to make string.

Medicine: **Wintun:** white juices of stalks used to treat poison oak. **Maidu, Southern (Nisenan):** for toothaches the plant is heated as hot as can be borne, placed in mouth against the offending member, and tightly gripped between the teeth. **Esselen:** leaves, stems, decoction, may have been used medicinally.

Citation(s): Curtis 1924; Driver 1936; Du Bois 1935; Garth 1953; Gayton 1948; Merriam 1967; Powers 1877; Steward 1933; Strike 1994; Breschini and Haversat 2004.

Note(s): it is likely that many of the species have some degree of toxicity; a resinoid, several glycosides, and a small amount of alkaloid.

Citation(s): Kingsbury 1964.

143 *Asclepias californica*

Common Name(s): California Milkweed.

Usage: Medicine: **Kawaiisu (Tehachapi):** used to treat spider bite; plant dried, ground to powder, wound moistened and powder applied.
Citation(s): Zigmond 1981; Strike 1994.
Terminology: **Kawaiisu (Tehachapi):** huk^{w}abita matasuk^{w}iyeena, "spider medicine."
Citation(s): Zigmond 1981; Strike 1994.

144 *Asclepias cordifolia*

Common Name(s): Milkweed, Purple Milkweed.

Usage: Food: **Shasta:** gum was chewed..
Material: **Miwok:** made string from fibers.
Medicine: **Miwok:** root used as medicine. **Maidu:** dried roots used as an emetic, diuretic; decoction relieved symptoms of colds; sap used to cure worts.
Citation(s): Barrett and Gifford 1933; Dixon 1907; Strike 1994.

145 *Asclepias eriocarpa*

Common Name(s): Milkweed, True Milkweed.

Usage: Food: **Luiseño:** A chewing gum is made from the sap which exudes from the stems when cut. **Karok:** juice gathered and congealed, used as chewing gum; putting fat or deer grease on it makes it hold together better in chewing.
Material: **Pomo:** cordage made from bark fiber. **Luiseño:** a string fiber is obtained from

the stems; made 2-ply cordage for nets.
Maidu: used juice to hold soot in place during tattooing; made fiber into rope and string.

Medicine: **Maidu:** sticky juice used as healing lotion for cuts and sores. **Costanoan:** decoction drunk to relieve colds, respiratory ailments; roots dried, pulverized, powder inhaled to cause sneezing; plant burned, smoke inhaled for asthma treatment; milky juice used to reduce corns; slave used to treat colds. **Wintun:** sap used to cure sores, including those from poison oak.

Misc: **Yuki:** named, but plant considered as undesirable and of diabolical importance of the Whites.

Citation(s): Barrett 1952; Chesnut 1902; Curtis 1924; Schenck and Gifford 1952; Sparkman 1908; Strike 1994; Rayburn 2012.

Terminology: **Karok:** mitimshaxiri, "popping gum."
Luiseño: tokmut.

Citation(s): Schenck and Gifford 1952; Sparkman 1908.

146 *Asclepias erosa*

Usage: Food: **Tubatulabal:** chewing gum made from plant juice collected in 4-inch tubular sections of jointed stalk of *Eriogonum elongatum;* sap used to treat sores, rashes caused by poison oak. **Cahuilla:** gum is collected, heated over fire, then chewed. **Kawaiisu (Tehachapi):** milky juice, drained from split stalk, is boiled until it thickens and chewed like "gum."

Citation(s): Barrows 1900; Voegelin 1938; Zigmond 1981; Strike 1994.

Terminology: **Cahuilla:** keat. **Kawaiisu (Tehachapi):** avibo?ogadɨbɨ, "Having chalklike water."

Citation(s): Sparkman 1908; Barrows 1900; Zigmond 1908.

147 *Asclepias fascicularis*

Synonym(s): *Asclepias mexicana*

Common Name(s): Milkweed, Narrow-Leaf Milkweed.

Usage: Food: **Pomo:** eat blossoms. **Owens Valley Paiute:** occasionally eaten. **Miwok:** eaten boiled as greens; when boiled it goes to pieces; sometimes added to Manzanita (*Arctostaphylos spp.*) Cider to thicken it.

Material: **Cahuilla:** gum for adhesive purposes is obtained from white sap. **Kawaiisu (Tehachapi):** stems principle source of cordage; stems dry about August and cut off at base; fibers separated by pounding or chewing; used for Tule mats, belts, handles, and straps.

Citation(s): Barrows 1900; Chesnut 1902; Barrett and Gifford 1933; Steward 1933; Zigmond 1981.

Terminology: **Miwok (Central Sierra):** isatawü.

Owens Valley Paiute: wiütava. **Chumash (Barbareño, Ynezeño):** 'okhponush; **(Ventureño):** 'usha'ak.

Citation(s): Barrett and Gifford 1933; Steward 1933; Timbrook 2007.

148　*Asclepias speciosa*

> Synonym(s): *Asclepias Giffordi*
> Common Name(s): Milkweed.

Usage:　Food:　　**Owens Valley Paiute:** only used occasionally.

　　　　Material:　**Monache (Western Mono):** made carrying nets of woven fibers. **Miwok:** made strong twine from the bark, which was sometimes made into fish nets. **Maidu (Northern):** cordage of various sizes made from the fibers. **Pomo (Southwestern):** 2-ply string made from stem fibers; fibers shredded to make women's skirts.

　　　　Medicine:　**Miwok:** decoction of the root taken in small doses to cure venereal disease; juice of root applied to warts.

Citation(s): Barrett and Gifford 1933; Clark 1904; Dixon 1905; Steward 1933; Goodrich et al 1980; Strike 1994.

Terminology: **Owens Valley Paiute:** avanava. **Miwok: (Central Sierra):** sū'kennû, tumuka, tī'gunu'; **(Plains):** tī'gūn. **Pomo (Southwestern):** ši?dp qʰále, "milk plant."

Citation(s): Steward 1933; Barrett and Gifford 1933; Goodrich et al 1980; Strike 1994.

149　*Astragalus* spp.

> Family: *Fabaceae*
> Common Name(s): Rattlesnake Weed; Loco Weed.

Usage:　Food:　　**Cahuilla:** in the Summer the pods are pounded up and used as a spice.

Citation(s): Barrows 1900.

Terminology: **Cahuilla:** Kash-lem.
Citation(s): Barrows 1900.

150 *Astragalus didymocarpus*
Common Name(s): Locoweed, Rattleweed,
Common Dwarf Locoweed.

Usage: Misc: **Kawaiisu (Tehachapi):** named but not used.
Citation(s): Zigmond 1981.
Terminology: **Kawaiisu (Tehachapi):** muhu(m(bi.
Citation(s): Zigmond 1981.

151 *Astragalus douglasii*

Usage: Misc: **Kawaiisu (Tehachapi):** named but not used.
Citation(s): Zigmond 1981.
Terminology: **Kawaiisu (Tehachapi):** paalaagavizi (paalaa "insane"); paalaagadɨbɨ.
Citation(s): Zigmond 1981.

152 *Astragalus pachypus*
Common Name(s): Locoweed, Rattleweed.

Usage: Medicine: **Kawaiisu (Tehachapi):** root boiled, decoction drunk hot to relieve menstrual pains, for a month thereafter "you don't eat meat, lard, or salt. If you do, you'll die."
Citation(s): Zigmond 1981; Strike 1994.
Terminology: **Kawaiisu (Tehachapi):** nookamatɨbɨ.
Citation(s): Zigmond 1981.

153 *Astragalus purshii*
   Common Name(s): Wooly-Pod.

   Usage: Medicine: **Kawaiisu (Tehachapi):** same as 152 but
                        not considered as strong.
   Citation(s): Zigmond 1981.

154 *Athyrium felix-femina*
   Family: *Woodsiaceae*
   Common Name(s): Fern, Lady Fern.

   Usage: Food: **Maidu:** peeled rhizomes, roasted
                        overnight, eaten; new shoots eaten.
            Medicine: **Maidu:** rhizome decoction used to
                        relieve pain, urinary difficulties;
                        mashed stem decoction used to
                        alleviate labor pains.
   Citation(s): Strike 1994.

155 *Atriplex* spp.
   Family: *Chenopodiaceae*
   Common Name(s): Salt Bush.

   Usage: Food: **Owens Valley Paiute:** used as food.
            Material: **Panamint Shoshone (Koso):** made
                        arrow foreshafts 5 inches long.
   Citation(s): Coville 1892; Steward 1933.
   Terminology: **Owens Valley Paiute:** sunuva, tüsi, kã'ãp,
                kã'ava. **Chumash (Barbareño):** 'i'laq; **(Ventureño):**
                mo'.
   Citation(s): Steward 1933.
   Note(s): common on salt marshes and elsewhere in
            coastal regions; indicators of aridity.
   Citation(s): Tsukada 1967.

156 *Atriplex canescens*

Common Name(s): Hoary Saltbush, Four-Wing, Wingscale, Shadscale.

Usage: Food: **Cahuilla:** ate fresh leaves as greens, or boiled with other foods to add flavor. **Serrano:** leaf tea used as cathartic. **Diegueño:** used to relieve pain.

Material: **Kawaiisu (Tehachapi):** used to make arrowpoints.

Citation(s): Zigmond 1981; Strike 1994.

Terminology: **Kawaiisu (Tehachapi):** murunavɨ.

Citation(s): Zigmond 1981.

157 *Atriplex confertifolia*

Common Name(s): Spiny Saltbush, Shade Scale.

Usage: Material: **Kawaiisu (Tehachapi):** used to make arrowpoints.

Citation(s): Zigmond 1981.

Terminology: **Kawaiisu (Tehachapi):** murunavɨ.

Citation(s): Zigmond 1981.

158 *Atriplex lentiformis*

Common Name(s): Salty Sage.

Usage: Food: **Cahuilla:** seeds gathered in July to September, ground and cooked with salt and water; parched, ground into flour, mixed with water to make mush or small cakes; stored in great abundance.

Material: **Cahuilla:** leaves and roots crushed and rubbed into articles requiring cleaning; branch used to make arrow foreshafts approximately 5 inches long.

Medicine: **Cahuilla:** flowers, stems, leaves were crushed, steamed, and inhaled for relief of nasal congestion; fresh leaves chewed to relieve head colds, stepped in water used as eyewash, to soothe wounds; dried leaves were smoked for same purpose

Citation(s): Barrows 1900; Bean and Saubel 1972; Strike 1994.

Terminology: **Cahuilla:** ká-sil, qashil.

Citation(s): Barrows 1900; Bean and Saubel 1972.

159 *Atriplex rosea*
Common Name(s): Red Scale.

Usage: Misc: **Kawaiisu (Tehachapi):** named but not used.

Citation(s): Zigmond 1981.

Terminology: **Kawaiisu (Tehachapi):** sïïkatïbï.

Citation(s): Zigmond 1981.

Note(s): naturalized from Eurasia.

Citation(s): Munz and Keck 1965.

160 *Atriplex semibaccata*
Common Name(s): Australian Saltbush.

Usage: Food: **Cahuilla:** small sweet and salty berry eaten fresh.

Citation(s): Bean and Saubel 1972.

Note(s): introduced from Australia.

Citation(s): Munz and Keck 1965.

161 *Atriplex serenana*
Common Name(s): Bracted Saltbush, Bract-Scale.

Usage: Food: **Kawaiisu (Tehachapi):** leaves boiled and fried in grease.
Citation(s): Zigmond 1981.
Terminology: **Kawaiisu (Tehachapi):** sïïkatïbï.
Citation(s): Zigmond 1981.

162 *Atriplex torreyi*

Usage: Food: **Kamia:** collected seed capsules, pounded in wooden mortar, winnowed, soaked in water prior to cooking.
Citation(s): Gifford 1931.

163 *Avena* spp.
Family: *Poaceae*
Common Name(s): Wild Oats.

Usage: Food: **Bear River:** gathered seeds by means of an oval elongated fan. **Yokuts (Northern Hill):** seeds gathered and parched, made into flour, shaped into cakes and cooked in hot ashes; eaten hot or cold; favorite accompaniment with Manzanita (*Artostaphylos* spp.) cider. **Luiseño:** parched with husks and pounded into meal. **Salinin; Maidu, Southern (Nisenan):** gathered. **Wintun, Central (Nomlaki):** used. **Yuki:** used seeds in pinole. **Pomo (Southwestern):** grain used in pinole; gathered in June-July. **Chumash:** seeds, collected by

hand or a seedbeater, ground on a flat
rock to removed chaff, pounded in
mortar, mixed with ground Chia (*Salvia
columbariae*) and water for a hot weather
drink; plant, used as livestock fodder.

Citation(s): Beals 1933; Curtin 1957; Gayton 1948; Goldschmidt
1951; Mason 1912; Nomland 1938; Sparkman 1908;
Goodrich et al 1980; Timbrook 2007.

Terminology: **Wiyot:** rakwiyidāgʿeral. **Yuki:** wa-shet-
ki-lich. **Chumash (Barbareño):** ʾlucheʾesh; **(Ynezeño):**
shushtewesh, "sharp, a spine that digs into your
skin."

Citation(s): Loud 1918; Curtin 1957; Timbrook 2007.

Note(s): all *Avena* species introduced from Europe as early as
1835.

Citation(s): Munz and Keck 1965; Bean and Saubel 1972.

164   *Avena barbata*

Usage: Food: **Miwok:** seeds parched, pulverized, and
made into soup.

Citation(s): Barrett and Gifford 1933.

Terminology: **Miwok (Central Sierra):** aweni; loan word from
the Spanish "avena."

Citation(s): Barrett and Gifford 1933.

Note(s): see 163.

seeds contain:

Ash	2.0%
Protein	16.2%
Oil	11.5%

Citation(s): Earle and Jones 1962.

165 *Avena fatua*

Common Name(s): Wild Oat.

Usage: Food: **Maidu Southern (Nisenan):** the seeds were eaten, but very sparingly. **Luiseño:** the seeds were ground into flour and eaten. **Pomo:** used seeds for pinole. **Kawaiisu (Tehachapi):** seeds gathered, pounded, boiled, and eaten. **Cahuilla:** seeds gathered July through September, parched, ground into flour, mixed with other weed seeds in mush.

Citation(s): Barrett 1952; Chesnut 1902; Powers 1877; Sparkman 1908; Zigmond 1981; Bean and Saubel 1972.

Terminology: **Luiseño:** arus, urus. **Pomo (Central):** bita'baa; **(Southwestern):** mucha'. **Kawaiisu (Tehachapi):** sik^{k}e?evɨ.

Citation(s): Sparkman 1908; Barrett 1952; Gifford 1967; Zigmond 1981.

Note(s): see 163.

166 *Avena sativa*

Common Name(s): Cultivated Oat.

Usage: Food: **Karok:** used seeds.

Citation(s): Schenck and Gifford 1952.

Terminology: **Karok:** ikravapu, "pounded." **Pomo (Southwestern):** mucha'.

Citation(s): Schenck and Gifford 1952; Gifford 1967.

Note(s): see 163.

167 *Azolla filiculoides*
>   Family: *Azollaceae*

> Usage: Misc: **Tubatulabal:** considered as weeds.
> Citation(s): Voegelin 1938.

168 *Baccharis glutinosa*
>   Familiy: *Asteraceae*
>   Common Name(s): Salt Marsh Baccharis, Seep
>   Willow, Water Wally.

> Usage: Material: **Luiseño:** the wood mostly used for drilling fire.
> Medicine: **Luiseño, Maidu:** a decoction of the leaves was used to bathe sores and wounds. **Costanoan:** leaves, decoction used to bathe sores and wounds, as a disinfectant, kidney aid, general cure-all; crushed leaves mixed with animal fast, heated, applied to boils; stems, pulverized, applied to (sprinkled on) wound as disinfectant.
> Citation(s): Sparkman 1908; Strike 1994; Rayburn 2012.
> Terminology: **Luiseño:** morwaxpish.
> Citation(s): Sparkman 1908.

169 *Baccharis pilularis*

> Usage: Material: **Chumash:** branchlets, used to brush away the tiny spines when collecting Prickly-Pear Cactus.
> Medicine: **Costanoan:** used as general remedy for various illnesses. **Miwok (Coast):** heated leaves, used to poultice swellings. **Chumash:** leaves, boiled,

applied to part affected by Poison Oak
rash, considered the best remedy.

Misc: **Coast Yuki:** named but not used.

Citation(s): Gifford 1939; Strike 1994; Rayburn 2012; Timbrook
2007.

Terminology: **Yuki:** wesuk.

Citation(s): Gifford 1939.

170 *Baccharis pilularis* var. *consanguinea*

Synonym(s): *Baccharis consanguinea*

Usage: Material: **Pomo:** valuable for arrow foreshafts.

Citation(s): Kniffen 1939.

171 *Baccharis plummerae*

Common Name(s): Plummer's Baccharis.

Usage: Medicine: **Chumash:** ground, rubbed on any
aching part; tea, drunk for kidney
problem.

Citation(s): Timbrook 2007.

172 *Baccharis salicifolia* ssp. *salicifolia*

Synonym(s): *Baccharis viminea, Baccharis glutinosa*

Common Name(s): Mule Fat, Water Wally, Seep
Willow.

Usage: Material: **Chumash:** wood, favorite stick for
making firesticks; stems, used in
making fish traps. **Cahuilla:** limbs and
branches favorite material used in
house construction.

Medicine: **Cahuilla:** leaves were steeped, used as
eyewash; at onset of baldness, hair was
washed in a solution made from the

leaves to prevent further loss; leaves and stems boiled, decoction used as a female hygienic agent. **Maidu:** leaves boiled, decoction used as female hygienic; leaves and twigs infusion used to wash hair, promote hair growth, prevent hair loss. **Costanoan:** leaves and twigs infusion used to wash hair, scalp, promote hair growth, prevent hair loss. **Kawaiisu (Tehachapi):** straight stems made into one-piece arrow for hunting small game.

Citation(s): Bean and Saubel 1972; Strike 1994: Timbrook 2007; Barrows 1900; Zigmond 1981; Rayburn 2012.

Terminology: **Cahuilla:** paq'ily, pá-ki. **Kawaiisu (Tehachapi):** pogosɨvɨ. **Chumash (Barbareño, Ynezeño);** shu'; **(Obispeño):** twalilí'; **(Ventureño):** wita'y.

Citation(s): Bean and Saubel 1972; Barrows 1900; Zigmond 1981; Timbrook 2007.

173 *Balsamorhiza spp.*

Family: *Asteraceae*

Usage: Food: **Atsugewi:** seeds parched, winnowed, ground in flour, made in cakes, eaten without cooking.

Citation(s): Strike 1994.

174 *Balsamorhiza deltoidea*

Common Name(s): Balsam Root.

Usage: Food: **Atsugewi:** seeds gathered in July; parched in flat tray, skin removed by abrasion with rock, winnowed, and ground with mono/metate; flour molded into cakes and eaten without cooking. **Maidu:** ate seeds, peeled root, young shoots, leaves, and sap.

Medicine: **Kawaiisu (Tehachapi):** root split and boiled; decoction drunk warm "like tea" to relieve coughing and colds.

Citation(s): Garth 1953; Zigmond 1981; Strike 1994.

Terminology: **Atsugewi:** uitsĭnyami. **Kawaiisu (Tehachapi):** Wiita(m)bĭ.

Citation(s): Garth 1953; Zigmond 1981.

175 *Balsamorhiza hookeri*

Synonym(s): *Balsamorhiza hirsuta*
Common Name(s): Sunflower.

Usage: Food: **Atsugewi:** seeds gathered in July; parched in flat tray, skin removed by abrasion with rock, winnowed, and ground with mono/metate; flour molded into cakes and eaten without cooking.

Medicine: **Washo:** root decoction used for female complaints.

Citation(s): Garth 1953; Train et al 1941; Strike 1994.

Terminology: **Atsugewi:** tamtiye. **Washo:** augn-<u>lem</u>-lu.

Citation(s): Garth 1953; Train et al 1941.

176 *Balsamorhiza sagittata*

Common Name(s): Balsam Root, Sunflower.

Usage: Food: **Miwok:** seeds eaten. **Atsugewi:** seeds gathered in July; parched in flat tray, skin removed by abrasion with rock, winnowed, and ground with

	mono/metate; flour molded into cakes and eaten without cooking.
Material:	**Miwok:** if a hunter did not get a deer, he steamed himself and bathed with a decoction of the pulverized root.
Medicine:	**Miwok:** root ground, boiled, cooled, and drunk for rheumatism, headache, or other pain; cure for rheumatism due to profuse perspiration which followed drinking a small cupful of liquid. **Washo:** burning the root in a room after an illness is thought to be a good fumigant.

Citation(s): Barrett and Gifford 1933; Garth 1953;
Train et al 1941.

Terminology: **Miwok (Central Sierra):** ho'tcŌtca. **Atsugewi:** iwitsĬnyami. **Washo:** <u>shugil</u>-artus, sugilatse.

Citation(s): Barrett and Gifford 1933; Garth 1953;
Train et al 1941.

Note(s): food values – root:

	(wet)	(dry)
Moisture	7.0%	
Sugar		
Reducing	10.6%	11.4%
Non-reducing	9.3%	10.1%
Hemicellulose	10.8%	11.6%
Ether Extract	5.8%	
Protein		3.5%
Crude Fiber		49.2%
Ash		6.0%

Citation(s): Yanovsky 1938.

177   *Balsamorhiza X terebinthacea*

Usage: Food: **Northern Paiute (Paviotso):** the roots were eaten raw, either fresh or dried; sometimes they were roasted in the ground and pounded first.

Citation(s): Kelly 1932.

Terminology: **Northern Paiute (Paviotso):** bikwa'ida.

Citation(s): Kelly 1932.

Note(s): the roots are diuretic; this hybrid was discussed by Weber (1953) but is not listed in Baldwin (2012).

Citation(s): Stuhr 1933.

178   *Barbarea orthoceras*
Family: *Brassicaceae*
Common Name(s): Winter Cress.

Usage: Food: **Maidu:** stems, leaves eaten raw, or cooked.

Citation(s): Strike 1994.

179   *Barbarea vulgaris*

Usage: Misc: **Miwok:** named but not used.

Citation(s): Barrett and Gifford 1933.

Terminology: **Miwok (Central Sierra):** soswina.

Citation(s): Barrett and Gifford 1933.

Note(s): introduced from Eurasia.

Citation(s): Munz and Keck 1965.

180   *Batis maritima*
Family: *Berberidaceae*
Common Name(s): Saltwort

Usage: Food: **Diegueño:** boiled and eaten; chewed

fresh stems, leaves for their moisture
content.
Citation(s): Strike 1994.

181 *Beckmannia syzigachne*
Family: *Poaceae*
Common Name(s): Slough Grass.
Usage: Food: **Modoc (Lutuami):** seeds eaten.
Citation(s): Strike 1994.

182 *Berberis* spp.
Family: *Berberidaceae*
Common Name(s): Oregon Grape.

Usage: Food: **Yana:** berries pounded with a pestle
                    into fine flour, made into mush with
                    water and eaten.
         Material: **Hupa:** the root is used sometimes to
                    dye *Xerophyllum tenax* a bright shade of
                    yellow.
         Medicine: **Achomawi:** made into a tea which is
                    highly regarded as a blood purifier; it
                    should be taken for a month.
Citation(s): Goddard 1903; Merriam 1967; Sapir and
         Spier 1943.

183 *Berberis aquifolium*
Common Name(s): Barberry, Mountain
                Grape.

Usage: Material: **Hupa:** the root used as a yellow dye.
         Medicine: **Karok:** root is boiled and liquid drunk
                    for all kinds of sickness; leaves and root
                    are used to steam anyone with "yellow
                    fever," colds. **Cahuilla:** tea made from

root used to stimulate poor appetites, used as general tonic.

Misc: **Karok:** berries are said to be poisonous, not eaten.

Citation(s): Merrill 1923; Schenck and Gifford 1952; Strike 1994.

Terminology: **Karok:** ŒiŒuná'aay, "Oregon Grape."

Citation(s): Schenck and Gifford 1952.

Note(s): small does appear to increase the appetite and confine the bowels; rhizome and roots are bitter tonic and alternative; improves digestion and absorption. [author's note - note the difference into the two items for Karok - perhaps difference informants at different times?]

Citation(s): Stille and Maish 1880; Stuhr 1933; Grieve 1931.

184 *Berberis aquifolium* var. *repens*

Synonym(s): *Berberis repens*

Common Name(s): Creeping Barberry.

Usage: Medicine: **Northern Paiute (Paviotso):** drink made from boiled roots used. **Maidu:** decoction of stems used to heal skin afflictions, drunk as a laxative; general tonic made from bark. **Modoc (Lutuami):** infusion of leave used to bathe eyes; infusion of crushed roots used for mouth sores, throat infections. **Yuki:** root tea used as a general tonic and to treat stomach problems.

Citation(s): Kelly 1932; Strike 1994.

Terminology: **Northern Paiute (Paviotso):** oha'nazagodidi.

Citation(s): Kelly 1932.

185   *Berberis dictyota*

Common Name(s): Barberry.

Usage: Medicine: **Kawaiisu (Tehachapi):** root boiled; decoction drunk for relief of gonorrhea. **Maidu:** cambium used to prepare a general health tonic.

Citation(s): Zigmond 1981; Strike 1994.

186   *Berberis nervosa*

Common Name(s): Oregon Grape.

Usage: Material: **Hupa:** the bark was used for yellow dye. **Yurok, Karok:** used roots for yellow dye. **Hupa, Karok:** dyed porcupine quills yellow, used for pattern on basketry.

Medicine: **Maidu:** roots and leaves decoction used as tonic, to stimulate poor appetites.

Misc: **Karok:** named but not used (?).

Citation(s): Merrill 1923; O'Neale 1932; Schenck and Gifford 1952; Strike 1994.

Terminology: **Karok:** Œunkinpirish, "bile plant."

Citation(s): Schenck and Gifford 1952.

Note(s): note the discrepancy in the references to the **Karok**; it would appear the Schenck and Gifford either missed something, or that different informants told the several ethnographers different things [author's note].

187   *Berberis pinnata*

Common Name(s): California Barberry.

Usage: Medicine: **Miwok:** leaf chewed, or decoction of the root drunk for heartburn, ague, consumption, chills, fevers, breathing problems, and rheumatism; the leaves chewed to prevent, ague, etc.; root chewed and liquid placed on cuts, wounds, and abrasions; cuts and bruises washed with a decoction of the root; root chewed, when traveling, to ward off diseases; chewed bark, or root, inserted into wound to prevent swelling; bark used as laxative. **Yana:** root decoction used as general treatment for illness.

Citation(s): Barrett and Gifford 1933.

Terminology: **Miwok (Central Sierra):** holo'metu̱.

Citation(s): Barrett and Gifford 1933.

188 *Betula occidentalis*

Family: *Bignoniaceae*
Synonym(s): *Betula fontanalis, Betula fontinalis*
Common Name(s): Water Birch.

Usage: Material: **Owens Valley Paiute:** used in making bows.

Citation(s): Steward 1933.

Terminology: **Owens Valley Paiute:** kugujava.

Citation(s): Steward 1933.

189 *Bidens laevis*

Family: *Asteraceae*
Synonym(s): *Bidens levis*
Common Name(s): Bur Marigold.

Usage: Food: **Owens Valley Paiute:** seeds used.

Citation(s): Steward 1933.

190   *Bistorta bistoroides*

        Family: *Polygonaceae*
        Synonym(s): *Polygonum bistoroides*
        Common Name(s): Knotweed, Alpine Smartweed.

Usage: Food: **Maidu:** young leaves eaten.
          Medicine: **Miwok, Maidu:** root was mashed and used as a poultice on sores and boils.
Citation(s): Barrett and Gifford 1933; Dixon 1905; Strike 1994.
Terminology: **Miwok (Central Sierra):** kaima.
Citation(s): Barrett and Gifford 1933.

191   *Blennosperma nanum*

        Family: *Asteraceae*
        Synonym(s): *Blennosperma californicum*

Usage: Food: **Maidu, Southern (Nisenan):** the seed is gathered by sweeping through the plant with a long-handled basket or gourd; the seeds is parched a little, then beaten into flour, and eaten without further cooking, or, made into bread, or, mush.
Citation(s): Powers 1877.

192   *Bloomeria crocea*

        Family: *Themidaceae*
        Common Name(s): Golden Stars.

Usage: Food: **Cahuilla, Luiseño:** corm eaten raw at almost any time of year.
          Material: **Kawaiisu (Tehachapi):** coated inside of seed-gathering baskets with mashed corms to prevent small seeds from falling through.
Citation(s): Bean and Saubel 1972; Strike 1994.

193 *Bloomeria crocea* var. *aurea*

Synonym(s): *Bloomeria aurea*

Usage: Food: **Cahuilla, Luiseño:** corm eaten raw at almost any time of year.

Material: **Kawaiisu (Tehachapi):** starchy "glue" yielded when fibrous skin of corm is removed, used on seed baskets to close interstices.

Citation(s): Sparkman 1908; Zigmond 1981; Strike 1994.

Terminology: **Luiseño:** kawichhal. **Kawaiisu (Tehachapi):** kucidabɨ.

Citation(s): Sparkman 1908; Zigmond 1981.

194 *Bolboschoplectus fluviatilis*

Family: *Cyperaceae*
Synonym(s): *Scirpus maritimus, Scirpus fluviatilis*
Common Name(s): Bulrush.

Usage: Material: **Pomo:** used root (dyed) and root bark as black pattern in baskets; used central fibers as overlay in baskets; **(Southwestern):** root used as black in basketry. **Panamint Shoshone (Koso):** used root (dyed) as black pattern in baskets.

Citation(s): Curtis 1924; Merrill 1923; Goodrich et al 1980.

Terminology: **Pomo (Northern, Central, Eastern):** tsīwi'c; **(Southwestern):** s'iwiš.

Citation(s): Barrett 1933; Goodrich et al 1980.

195 *Bolboschoenus robustus*

Synonym(s): *Scirpus robustus*
Common Name(s): Triangular Stemmed Tule.

Usage: Food: **Pomo:** the roots and the tender young shoots were eaten as greens.

Material: **Modoc (Lutuami):** mats made from this plant formed the middle of three layers that covered the Summer house. **Pomo:** used stem as warp and woof in baskets; used shredded material to make tule skirts and armless, knee-length capes for rainy weather; also in weaving mats and for house thatching (top course) and for padding under sleeping mats. **Yokuts:** used stem as foundation in baskets. **Wintun:** balls of clay wrapped with plant used as fish seine sinkers. **Kawaiisu (Tehachapi):** used in coiled basketry; provides a brown color.

Citation(s): Barrett 1910, 1952; Curtis 1924; Merrill 1923; Zigmond 1981.

Terminology: **Pomo (Eastern):** gūca′l; **(Southeastern):** kaa′l. **Kawaiisu (Tehachapi):** cokovišivɨ.

Citation(s): Barrett 1952; Zigmond 1981.

196 *Boletus* spp.

Family: *Boletaceae*

Usage: Food: **Pomo:** a yellow and green variety eaten raw.

Citation(s): Chesnut 1902.

197 *Boletus edulis*

Usage: Food: **Pomo (Southwestern):** appears after first rains; cooked on hot stones and eaten.
Citation(s): Gifford 1967; Goodrich et al 1980.
Terminology: **Pomo (Southwestern):** chepokol.
Citation(s): Gifford 1967; Goodrich et al 1980.

198 *Boykinia occidentalis*
Family: *Saxifragaceae*
Synonym(s): *Therofon elatum, Boykinia elata*
Common Name(s): Brook Foam.

Usage: Material: **Karok:** the leaves are dried and sometimes worn inside basket caps for fragrance.
Medicine: **Yuki:** roots used medicinally.
Citation(s): Chesnut 1902; Schenck and Gifford 1952; Strike 1994.
Terminology: **Karok:** mafukafich, "make-believe kafich."
Citation(s): Schenck and Gifford 1952.

199 *Brandega bigelovi*
Family: *Cucurbitaceae*

Usage: Medicine: **Tubatulabal:** mashed, burned ripe seeds rubbed on the umbilicus of newborn babies to speed healing.
Citation(s): Strike 1994.

200 *Brasenia schrèberi*
        Family: *Cabombaceae*
        Common Name(s): Water Shield.

        Usage: Food: **Maidu:** young stems and leaves eaten raw; rhizomes peeled, boiled, dried, and crushed into flour.
                Medicine: **Maidu:** rhizomes used to cure dysentery, stomach ache.
        Citation(s): Strike 1994.

201 *Brassica* spp.
        Family: *Brassicaceae*
        Common Name(s): Mustard.

        Usage: Food: **Yokuts:** greens were cooked. **Pomo (Southwestern):** flowers and young leaves eaten raw, or cooked after leaves washed, boiled, or fired. **Chumash:** young plants, used as greens.
                Misc: **Miwok:** named but not used.
        Citation(s): Barrett and Gifford 1933; Gayton 1948; Goodrich et al 1980; Timbrook 2007.
        Terminology: **Miwok (Central Sierra):** wimeya. **Pomo (Southwestern):** kulucícca.
        Citation(s): Barrett and Gifford 1933; Goodrich et al 1980.

202 *Brassica nigra*
        Common Name(s): Black Mustard.

        Usage: Food: **Luiseño:** much used for greens.
        Citation(s): Sparkman 1908.
        Note(s): seeds contain allyl isothiocyanate, considered potentially toxic; may also be potentially goitrogenic

(natural substances which prevent the thyroid from accumulating inorganic iodide normally); naturalized from Europe.

Citation(s): Kingsbury 1964; Munz and Keck1965.

203 *Brickellia californica*

Family: *Asteraceae*
Common Name(s): California Boneset.

Usage: Food: **Maidu:** leaves used to make tea.
       Medicine: **Maidu:** made a lotion to sooth infant's infected sores; used to cure spasms, bronchial problems.
       Misc: **Tubatulabal:** considered as weeds.

Citation(s): Voegelin 1938; Strike 1994.

204 *Briza minor*

Family: *Poaceae*
Common Name(s): Quaking Grass.

Usage: Food: **Karok:** used seeds (?).
       Material: **Miwok:** young girls and boys wore flowers in their ears, with the flower forward, the stem passing through the hole in the ear lobe.

Citation(s): Barrett and Gifford 1933; Schenck and Gifford 1952.

Terminology: **Karok:** `apsun'axraan, "snake tracks."
             Miwok (Central Sierra): seppute.

Citation(s): Schenck and Gifford 1952; Barrett and Gifford 1933.

Note(s): introduced from Europe.

Citation(s): Munz and Keck 1965.

205　*Brodiaea* spp.

Family: *Themidaceae*
Synonym(s): *Calliprora spp.*
Common Name(s): Nigger-Toe, Grass Nuts, Camas.

Usage:　Food:　　　　**Miwok:** roots cooked and eaten. **Wappo; Wintun, Central (Nomlaki):** used roots. **Shelter Cove Sinkyone:** bulbs cooked and eaten. **Maidu:** ate young seed pods; **Southern (Nisenan):** eaten raw, or roasted in ashes, or boiled. **Salinan:** eaten. **Wintun:** gathered in May and June; some stored for winter. **Yokuts:** valued for their nutty flavor; roasted and eaten whole or mashed into flour. **Bear River:** eaten raw, considered a great delicacy; stored raw in tree trunks or hung inside house for immediate use; Winter supply assured by roasting before packing in baskets. **Cahuilla:** dug; large corms kept for food, small corms replanted; eaten raw or cooked, boiled for half an hour. **Chumash:** eaten.

　　　　Material:　　**Wintun:** used as paste for sinew on bows; bind paint pigments when used on bows.

Citation(s): Barrett and Gifford 1933; Baumhoff 1958; Clark 1904; Driver 1936; Du Bois 1935; Gayton 1948; Goldschmidt 1951; Mason 1912; Nomland 1938; Powers 1877; Bean and Saubel 1972; Strike 1994; Landberg 1965.

Terminology: **Miwok (Central Sierra):** tene, wata´. **Yokuts:** tsa´lu. **Tubatulabal:** hocal (bulb). **Cahuilla:** mehawot (?).

Citation(s): Barrett and Gifford 1933; Gayton 1948;
Voegelin 1938; Bean and Saubel 1972; Strike 1994.

206 *Brodiaea coronaria*

Synonym(s): *Brodiaea grandiflora*
Common Name(s): Indian Potato, Harvest
Brodiaea.

Usage: Food: **Yana:** roasted in hot ashes and eaten.
Wiyot: the bulb was eaten. **Miwok:**
bulb dug about the first of May when
shoots just appear above the ground;
cooked in earth oven. **Atsugewi:** roots
boiled in water and sometimes cooked
in earth oven. **Pomo (Southwestern):**
bulbs cooked in hot ashes or boiled for
food.

Citation(s): Barrett and Gifford 1933; Garth 1953;
Loud 1918; Sapir and Spier 1943;
Goodrich et al 1980.

Terminology: **Yana:** ts·`l, ´lumals·unna. **Wiyot:**
topŌdërŌs, bòderūc. **Miwok (Central Sierra):** walla.
Atsugewi: wra'ya. **Pomo (Southwestern):** hi?bú?la.

Citation(s): Sapir and Spier 1943;Loud 1918; Barrett
and Gifford 1933; Garth 1953; Goodrich et al 1980.

207 *Brodiaea elegans*

Synonym(s): *Hookera coronaria*
Common Name(s): Harvest Brodiaea.

Usage: Food: **Pomo:** corms are very sweet after
roasting for a day. **Yuki:** corms are
roasted for a day. **Miwok:** only largest
corms eaten, smaller were replanted.

Citation(s): Barrett 1952; Chesnut 1902; Strike 1994.

Terminology: **Pomo (Central):** koko'bū.
Citation(s): Barrett 1952.
Note(s): food values – corms:

	(wet)	(dry)
Moisture	43.8%	
Sugar		
Reducing	1.5%	2.7%
Non-reducing	14.3%	25.4%
Starch	31.9%	0.0%
Hemiicellulose	11.9%	4.9%

Citation(s): Yanovsky 1938.

208 *Brodiaea minor*

Usage: Food: **Yana:** bulbs were steamed in earth oven about an hour and eaten.
Citation(s): Sapir and Spier 1943.
Terminology: **Yana:** p'uli'ls˙unna.
Citation(s): Sapir and Spier 1943.

209 *Bromus carinatus*

Family: *Gramineae*
Synonym(s): *Bromus virens*
Common Name(s): Wild Grass, Ripgut Grass.

Usage: Food: **Maidu, Southern (Nisenan):** the seeds are eaten; parched a little, then beaten into flour; eaten without further cooking, or made into bread and mush. **Miwok:** seeds pulverized, then eaten as pinole.
Citation(s): Barrett and Gifford 1933; Powers 1877; Strike 1994.
Terminology: **Miwok (Central Sierra):** tcuppayu.
Citation(s): Barrett and Gifford 1933.

210 *Bromus carinatus* var. marginatus
　　　　　Synonym(s): *Bromis marginatus*
　　Usage:　Food:　　　**Yuki, Pomo:** seeds used in pinole.
　　Citation(s): Chesnut 1902.

211 *Bromus diandrus*
　　　　　Synonym(s): *Bromus rigidus, Bromus maximus*

　　Usage: Food:　　　**Karok:** gathered and prepared the same
　　　　　　　　　　　as *Bromus hordeaceus*. **Luiseño:** the seeds
　　　　　　　　　　　are eaten.
　　Citation(s): Schenck and Gifford 1952; Sparkman 1908.
　　Terminology: **Karok:** aktipannaria. **Luiseño:** woshhat.
　　Citation(s): Schenck and Gifford 1952; Sparkman 1908.
　　Note(s): common weed in waste places, field, etc., at low
　　　　　　elevations.
　　Citation(s): Munz and Keck 1965.

212 *Bromus diandrus*
　　　　　Synonym(s): *Bromus rigidus* var. *Gussonei*

　　Usage:　Food:　　　**Miwok (Central Sierra):** seeds are
　　　　　　　　　　　pulverized and eaten.
　　Citation(s): Barrett and Gifford 1933;
　　Terminology: **Miwok (Central Sierra):** sū'llū.
　　Citation(s): Barrett and Gifford 1933.
　　Note(s): especially common in northern California;
　　　　　　introduced from Europe.
　　Citation(s): Munz and Keck 1965.

213 *Bromus hordeaceus*
　　　　　Synonym(s): *Bromus mollis*

　　Usage:　Food:　　　**Karok:** food grain gathered about the
　　　　　　　　　　　first of July; parched and ground; meal

mixed with water into gruel and eaten without further cooking.

Citation(s): Schenck and Gifford 1952.

Terminology: **Karok:** ikravapu, "pounded."

Citation(s): Schenck and Gifford 1952.

Note(s): common weed in waste places; native of Europe.

Citation(s): Munz and Keck 1965.

214 *Bromus madritensis* ssp. *rubens*

Synonym(s): *Bromus rubens*

Usage: Misc: **Tubatulabal:** considered weeds. **Miwok:** named but not used; stickers considered nuisance to people wearing moccasins.

Citation(s): Barrett and Gifford 1933; Voegelin 1938.

Terminology: **Miwok (Central Sierra):** tcuppayu.

Citation(s): Barrett and Gifford 1933.

Note(s): common troublesome weed in waste and cultivated ground at low elevations, especially in central and southern California; introduced from southern Europe.

seeds contain:

Ash   2.0%

Protein 17.9%

Oil   2.0%

Citation(s): Munz and Keck 1965; Earle and Jones 1962.

215 *Bromus orcuttianus*

Common Name(s): Bromegrass.

Usage: Misc: **Kawaiisu (Tehachapi):** named but not used.

Citation(s): Zigmond 1981.

Terminology: **Kawaiisu (Tehachapi):** pawahavɨ.

Citation(s): Zigmond 1981.

216 *Bromus tectorum*
>   Common Name(s): Cheat Grass, Downy Cheat.

Usage: Food: **Cahuilla:** considered a famine food; seeds gathered in quantity during food shortage, cooked into gruel.
Citation(s): Bean and Saubel 1972.
Note(s): introduced from Europe.
Citation(s): Munz and Keck 1965.

217 *Bursera microphylla*
>   Family: *Burseraceae*
>   Common Name(s): Elephant Tree, Elephant Trunk.

Usage: Medicine: **Cahuilla:** rubbed sap for curing on body of patient; cure for skin disease; plant considered to have great "power," sap too dangerous to keep openly in household, always hidden.
Citation(s): Bean and Saubel 1972.
Terminology: **Cahuilla:** eneneka.
Citation(s): Bean and Saubel 1972.

218 *Calamagrostis nutkaensis*
>   Family: *Poaceae*

Usage: Food: **Coast Yuki:** seeds eaten.
Citation(s): Gifford 1939.
Terminology: **Yuki:** wehil.
Citation(s): Gifford 1939.

219   *Calandrinia* spp.
Family: *Montiaceae*

Usage: Food: **Yokuts:** reddish-purple flowers rolled into a ball with seeds and eaten like candy. **Maidu:** ate leaves and stems raw, or cooked. **Miwok:** considered important part of diet.

Material: **Miwok (Coast):** flowers crushed, rubbed on faces for color.

Medicine: **Chumash:** seeds important as ceremonial offerings.

Citation(s): Gayton 1948; Strike 1994.
Terminology: **Yokuts:** Ka'sy⌊n (seeds and plant).
Citation(s): Gayton 1948.

220   *Calandrinia breweri*

Usage: Food: **Chumash:** seeds, gathered in Spring, toast, grind, made balls, eaten; a highly esteemed food; high value trade item.

Citation(s): Timbrook 2007.
Terminology: **Chumash (Barbareño, Ynezeño, Ventureño):** khutash.
Citation(s): Timbrook 2007.

221   *Calandrinia ciliata*
Synonym(s): *Calandrinia caulescens*
Common Name(s): Red Maids.

Usage: Food: **Luiseño:** used when tender for greens; the seeds are also eaten. **Miwok:** about the end of May the entire plant was pulled up and allowed to dry to separate the seeds; the seeds were parched and pulverized. **Chumash:** seeds, gathered in Spring, toast, grind, made balls, eaten; a highly esteemed food; high value trade item.

Citation(s): Barrett and Gifford 1933; Sparkman 1908; Timbrook 2007.

Terminology: **Miwok (Northern Sierra, Southern Sierra):** ko'tca. **Luiseño:** puchakla.

Citation(s): Barrett and Gifford 1933; Sparkman 1908.

222 *Callitriche* spp.

Family: *Plantaginaceae*
Common Name(s): Water Chickweed.

Usage: Medicine: **Maidu:** used to relieve urinary problems.

Citation(s): Strike 1994.

223 *Calocedrus decurrens*

Family: *Cupressaceae*
Synonym(s): *Libocedrus decurrens*
Common Name(s): Incense Cedar, Post Cedar

Usage: Material: **Karok:** boards made from wood; the bounds and twigs are sometimes used for brooms. **Yuki:** leaves used as basin in acorn leaching; smaller limbs sometimes used for bows. **Miwok:** covered conical house with bark (winter house); made bows from branches; treated wood with deer marrow for several days to make it flexible when dry for bows 3-4 feet long, back with sinew, attached with tree pitch and glue made from boiled deer bones; strips of bark used to cover conical dwellings; shredded bark used to line moccasins during long journeys, and in cold weather. **Wintun:** covered conical houses with bark. **Cahuilla:** constructed conical-shaped bark houses in the mountain for temporary use to camp and gathered and process acorns; slabs were used in more permanent construction. **Maidu:** used root as decorative overly in coiled baskets.

Medicine: **Yuki:** a decoction of leaves occasionally used to relieve stomach troubles. **Modoc (Lutuami):** resin chewed to relieve colds, coughs, to soothe mouth, throat pain, to treat infections.

Citation(s): Barrett and Gifford 1933; Chesnut 1902; Bean and Saubel 1957; Clark 1904; Merriam 1955; Schenck and Gifford 1952; Strike 1994

Terminology: **Miwok (Central Sierra, Northern Sierra, Southern Sierra):** mō'nōku. **Karok:** ichiwanaeiach. **Cahuilla:** yulil.

Citation(s): Barrett and Gifford 1933; Schenck and Gifford 1952; Bean and Saubel 1972.

224   *Calochortus* spp.

Family: *Liliaceae*
Synonym(s): *Cyclobothra spp.*
Common Name(s): Beaver-Tail Grass-Nut, Mariposa Lily.

Usage: Food: **Pomo:** corms eaten raw or after roasting in ashes. **Shasta:** eaten to a considerable extent; considered one of the principal edible roots. **Maidu, Southern**

(Nisenan): eaten raw, or roasted in ashes, or boiled. Cahuilla: bulbs eaten raw or cooked in fire pit. Chumash: bulbs, after flower stalks had died, collected, roasted, eaten.

Citation(s): Barrett 1952; Curtis 1924; Dixon 1907; Powers 1877; Bean and Saubel 1972; Timbrook 2007.

Terminology: Chumash (Barbareño): nakhaykha'y, kh'a'w, matakh; (Ynezeño): pilash; (Ventureño): 'utapits, 'utapikets.

Citation(s): Timbrook 2007.

225 *Calochortus amabilis*

Common Name(s): Cat's Ear, Fairy Lantern.

Usage: Food: Pomo (Southwestern): bulb baked and eaten.

Citation(s): Goodrich et al 1981.

Terminology: Pomo (Southwestern): wit^{h}i?.

Citation(s): Goodrich et al 1980.

226 *Calochortus luteus*

Common Name(s): Yellow Mariposa Lily.

Usage: Food: Miwok: bulbs were usually dug when the bud appeared on the plant in April; roasted about 20 minutes in ashes and eaten. Pomo (Southwestern): bulb baked and eaten.

Citation(s): Barrett and Gifford 1933; Goodrich et al 1980.

Terminology: Miwok (Central Sierra): tcikimtci susa.

Pomo (Southwestern): sik^{h}oló?lo.

Citation(s): Barrett and Gifford 1933; Goodrich et al 1980.

Note(s): food values – bulbs:

	(wet)	(dry)
Moisture	77.2%	
Sugar		
Reducing	2.4%	10.5%
Non-reducing	3.6%	15.8%
Hemicellulose	5.9%	25.9%
Starch	4.0%	17.5%
Ether Extract	0.6%	
Protein		9.1%
Ash		6.2%

Citation(s): Yanovsky 1938.

227 *Calochortus macrocarpus*

Usage: Food: **Northern Paiute (Paviotso):** the bulbs were skinned and eaten in the Spring, but they were never plentiful enough to be dried.

Citation(s) Kelly 1932.

Terminology: **Northern Paiute (Paviotso):** ko·gi´.

Citation(s): Kelly 1932.

228 *Calochortus palmeri*

Usage: Food: **Tubatulabal:** used bulbs.

Citation(s): Voegelin 1938.

229 *Calochortus pulchellus*

Common Name(s): Golden Lantern.

Usage: Food: **Karok:** bulbs baked in earth oven and eaten. **Pomo:** bulbs are eaten raw or roasted in ashes for about an hour. **Yuki:** ate bulbs.

Citation(s): Chesnut 1902; Curtin 1957; Schenck and Gifford

1952.

Terminology: **Karok:** xavin. **Pomo (Central):** bicē´bū,
"deer potato." **(Southeastern):** neē´bun.

Citation(s): Schenck and Gifford 1952; Barrett 1952.

230 *Calochortus tolmiei*

Synonym(s): *Calchortus caeruleus* var.
Maweanus, Calochortus maweanus

Common Name(s): Cat's Ears.

Usage: Food: **Hupa:** bulbs are roasted in ashes or
boiled in baskets. **Pomo, Pomo
(Southwestern):** bulbs are eaten in
small quantities, mostly by children.
Yuki: the bulbs are eaten raw; were
enjoyed for their sweetness.

Citation(s): Chesnut 1902; Curtin 1957; Goddard 1903;
Goodrich et al 1980.

Terminology: **Yuki:** pōsh-huntel. **Pomo
(Southwestern):** kúška šima, "cat ear."

Citation(s): Curtin 1957; Goodrich et al 1980.

231 *Calochortus venustus*

Common Name(s): White Mariposa Lily.

Usage: Food: **Tubatulabal:** used bulbs. **Miwok:** bulbs
dug in August, roasted for about 20
minutes in ashes of fire and eaten.
Pomo: bulbs eaten.

Citation(s): Barrett and Gifford 1933; Chesnut 1902; Voegelin
1938.

Terminology: **Miwok (Central Sierra):** tcikimtci.

Citation(s): Barrett and Gifford 1933.

232 *Calochortus vestae*

Synonym(s): *Calochortus luteus oculatus*

Usage: Food: **Pomo:** bulbs eaten. **Pomo (Southwestern):** bulbs baked, eaten. **Miwok:** bulbs dug in April, roasted for about 20 minutes in ashes of fire and eaten.

Citation(s): Barrett 1952; Barrett and Gifford 1933; Goodrich et al 1980.

Terminology: **Pomo (Central):** batū´m, **(Southwestern):** sikʰdó?lo.

Citation(s): Barrett 1952; Goodrich et al 1980.

233 *Caltha leptosepala*

Family: *Ranunculàceae*
Synonym(s): *Caltha howellii*
Common Name(s): Marsh Marigold.

Usage: Food: **Maidu:** young leaves cook, eaten.

Citation(s): Strike 1994.

Note(s): *Caltha palustris,* and probably others of the species, is noted to have the alkaloids delphinine, delphineidine, ajacine, and others. These alkaloids cause stomach upset, nervous symptoms. Depression, may be fatal if eaten in large quantities. The young plant is apparently less poisonous or not poisonous at all.

Citation(s): Kingsbury 1964; Hardin and Arena 1974.

234 *Calycadenia multiglandulosa*

Family: *Asteraceae*
Synonym(s): *Hemizoniu multiglandulosa*
Common Name(s): Sticky Calycadenia, Rosin Weed.

Usage: Food: **Maidu:** seeds eaten.

Medicine: **Yana:** the black seeds were cooked, dried, and pounded up and eaten as a cure for the chills.

Citation(s): Sapir and Spier 1943; Strike 1994.

Terminology: **Yana:** mā'l'ila (seeds).

Citation(s): Sapir and Spier 1943.

235 *Calycanthus* spp.

Family: *Calycanthaceae*

Usage: Material: **Pomo:** most highly prized for arrow shafts.

Citation(s): Kniffen 1939.

236 *Calycanthus occidentalis*

Synonym(s): *Butneria occidentalis*

Common Name(s): Spice Bush, Sweet Shrub, Western Sweet-Scented Shrub.

Usage: Material: **Pomo:** used stem and bark from fresh shoots in basketry; pithy shoots values for making arrows. **Miwok:** made shafts of arrows from young shoots; highly prized for making main shafts of arrows.

Medicine: **Pomo (Southwestern):** bark peeled or scraped off with mussel shell, then boiled, drunk for severe cold, as an expectorant, sore throat, stomach ache.

Misc: **Karok:** named but not used.

Citation(s): Barrett and Gifford 1933; Chesnut 1902; Gifford 1967; Merrill 1923; Schenck and Gifford 1952; Goodrich et al 1980; Strike 1994..

Terminology: **Miwok (Central Sierra):** so'ksokoTu̲.

Pomo (Southwestern): shoné. **Karok:** oshoxurip.
Citation(s): Barrett and Gifford 1933; Gifford 1967;
Schenck and Gifford 1952; Goodrich et al
1980.

237 *Calyptridium monandrum*

Family: *Montiaceae*
Common Name(s): Sand Grass, Pussypaws.

Usage: Food: **Kawaiisu (Tehachapi):** seed gathered.
Citation(s): Zigmond 1981.
Terminology: **Kawaiisu (Tehachapi):** maastabɨ (?).
Citation(s): Zigmond 1981.

238 *Calyptridium umbellatum*

Usage: Misc: **Miwok:** named but not used.
Citation(s): Barrett and Gifford 1933.
Terminology: **Miwok (Central Sierra):** paipaiyu.
Citation(s): Barrett and Gifford 1933.

239 *Calystegia occidentalis* ssp. *fulcrata*

Family: *Convolvulaceae*
Synonym(s): *Convolvulus fulcratus, Calystegia fulcrata*

Usage: Medicine: **Karok:** used as a love medicine and
requires a sung formula.
Citation(s): Schenck and Gifford 1952.
Terminology: **Karok:** axapakataturahitihan, "Climb-
up-on-a-straw."
Citation(s): Schenck and Gifford 1952.

240 *Calystegia longipes*

Common Name(s): Paiute Morning Glory.

Usage: Medicine: **Kawaiisu (Tehachapi):** root boiled and brew drunk for relief of gonorrhea, taken every morning, noon, and evening until cure has been affected.
Citation(s): Zigmond 1981.
Terminology: **Kawaiisu (Tehachapi):** tanavɨnɨ(m)bɨ.
Citation(s): Zigmond 1981.

241 *Calystegia macrostegia* ssp. *cyclostegia*

Common Name(s): Coastal Morning Glory.

Terminology: **Chumash (Barbareño):** s'epsu' 'i'ashk'à', "Coyote's basket-hat;" **(Ynezeño):** s'epsu' 'aquqa'w; **(Ventureño):** 'almakhmal 'I suninakhshep.
Citation(s): Timbrook 2007.

242 *Camassia spp.*

Family: *Agavaceae*
Common Name(s): Camas.

Usage: Food: **Sinkyone:** ate bulbs. **Tolowa:** used bulbs.
Citation(s): Drucker 1937; Nomland 1935.

243 *Camassia quamash*

Synonym(s): *Camassia esculenta*
Common Name(s): Camas, Blue Camas.

Usage: Food: **Shasta:** bulbs eaten to a considerable extent. **Miwok, Maidu (Northern Sierra):** bulbs eaten. **Northern Paiute (Paviotso):** the bulbs were cooked overnight in an earthen oven and eaten or dried. **Yuki:** bulbs baked overnight in earth oven and eaten.

Citation(s): Clark 1904; Dixon 1905, 1907, Curtin 1957; Kelly 1932.

Terminology: **Northern Paiute (Paviotso):** pa·si´go´·, tapa´ko´. **Yuki:** a´-lich.

Citation(s): Kelly 1932; Curtin 1957.

Note(s): native of rich meadows, very wet in Winter and Spring, but dry in the Summer; water often stands on the surface at flowering time.

Citation(s): Bailey 1950.

244 *Camassia leichtlinii* var. *suksdorf*
Synonym(s): *Quamasia Leichtlinii*

Usage: Food: **Yuki:** bulbs dug in June or July with a digging stick; sometimes boiled, generally roasted in an earth oven. **Maidu:** eaten after long cooking.

Citation(s): Chesnut 1902; Strike 1994.

Note(s): food values – bulbs:

	(wet)	(dry)
Moisture	69.5%	
Sugar		
Reducing	0.5%	1.6%
Non-reducing	11.6%	36.6%
Hemicellulose	4.9%	15.8%
Starch	0.0%	0.0%
Protein		11.7%
Ash		3.2%

Citation(s): Yanovsky 1938.

245 *Camerion angustifolium* ssp. *circumvagum*
  Synonym(s): *Epilobium angustifolium*
  Common Name(s): Fireweed.

Usage: Food: **Maidu:** young shoots, leaves, flower
                     buds eaten.
        Medicine: **Maidu:** infusion, or poultice, used to
                     treat pain, muscle cramps, bruises,
                     sores; leaf decoction used for diarrhea,
                     sore throats.
Citation(s): Strike 1994.

246 *Camissonia contorta*
  Synonym(s): *Oenothera contorta*

Usage: Misc: **Tubatulabal:** considered weeds.
Citation(s): Voegelin 1938.

247 *Camissoniopsis bistorta*
  Family: *Onagraceae*
  Synonym(s): *Oenothera bistorta*

Usage: Misc: **Tubatulabal:** considered weeds.
Citation(s): Voegelin 1938.

248 *Cantharellus cibarius*
  Family: *Cantharellaeae*
  Common Name(s): Chanatarelle.

Usage: Food: **Pomo (Southwestern):** baked.
Citation(s): Goodrich et al 1980.
Terminology: **Pomo (Southwestern):** qʰale másinče·.
Citation(s) Goodrich et al 1980.

249 *Capsella bursa-pastoris*

Family: *Brassicaceae*
Common Name(s): Shepherd's Purse.

Usage: Food: **Cahuilla:** gathered January to June as greens; edible seeds collected June to August.

Medicine: **Costanoan:** tea used for dysentery. **Cahuilla:** tea made from plant to cure dysentery, warned against taking more than two cups.

Citation(s): Bean and Saubel 1972; Rayburn 2012.
Note(s): naturalized from Europe.
Citation(s): Munz and Keck 1965.

250 *Cardamine californica*

Family: *Brassicaceae*
Synonym(s): *Dentaria californica*
Common Name(s): Toothwort.

Usage: Food: **Maidu:** ate roots.
Citation(s): Strike 1994.

251 *Cardamine nuttallii*

Synonym(s): *Dentaria tenella* var. *palmata*
Common Name(s): Toothwort.

Usage: Food: **Maidu:** all parts eaten.
Citation(s): Strike 1994.

252 *Cardamine oligosperma*

Family: *Brassicaceae*
Common Name(s): Bitter Cress.

Usage: Food: **Maidu:** eaten raw, or cooked.

Medicine: **Maidu:** used to remove intestinal obstructions.
Citation(s): Strike 1994.

253   *Carduus* spp.
Family: *Asteraceae*
Common Name(s): Thistle.

Usage: Food: **Luiseño:** used as greens; the buds were eaten raw.
Misc: **Miwok:** named but not used.
Citation(s): Barrett and Gifford 1933; Sparkman 1908.
Terminology: **Miwok (Central Sierra):** yotcta.
            Luiseño: chochawish.
Citation(s): Barrett and Gifford 1933; Sparkman 1908.

254   *Carex* spp.
Family: *Cyperaceae*
Common Name(s): Sedge.

Usage: Food: **Maidu, Modoc (Lutuami):** root, stems, shoots eaten.
Material: **Pomo:** roots used as wrapping and woof in basketry. **Pomo (Southwestern):** roots used as a sewing element in coiled baskets; grass cut, dried, and bound for torches. **Washo, Wailaki, Yokuts, Tubatulabal:** used root as wrapping in basketry. **Wailaki:** used leaves in weaving hats and flexible baskets. **Miwok:** floor where spectators sat in dance house covered with leaves. **Yuki:** used in basketry. **Wintun:** used roots as weft in baskets. **Esselen:** roots, used in basketry.

Citation(s): Barrett and Gifford 1933; Chesnut 1933; Curtin
1957; Curtis 1924; Gifford 1967; Merrill 1923; Strike
1994; Breschini and Haversat 2004.
Terminology: **Miwok (Central Sierra):** kissi. **Yuki:** ti.
Citation(s): Barrett and Gifford 1933; Curtin 1957.

255 *Carex aquatilis*

Usage: Misc: **Mono Lake Paiute:** named but not
used.
Citation(s): Steward 1933.
Terminology" **Mono Lake Paiute:** vaka'novü.
Citation(s): Steward 1933.

256 *Carex barbarae*
Common Name(s): Slough Grass.

Usage: Material: **Pomo:** used root as wrapping, woof,
and foundation in basketry; used root
bark as black pattern in baskets. **Yuki,
Wappo:** used root as wrapping in
basketry; used root as black pattern in
baskets. **Monache (Western Mono):**
used root as wrapping, woof in
basketry; root bark as black pattern in
baskets. **Yokuts:** used root as wrapping
and woof in basketry. **Miwok,
Panamint Shoshone (Koso),
Kitanemuk (Tejon):** used root as
wrapping in baskets. **Pomo
(Southwestern):** used as sewing
material in basketry.
Citation(s): Chesnut 1902; Merrill 1923; Goodrich et al
1980; Barrett 1908.
Terminology: **Miwok (Southern Sierra):** pa'iwa, **(Plains,**

Northern Sierra, Central Sierra): sū´lī. Pomo
(Northern, Central, Eastern): kūh´m.
(Southwestern): qa?díhqʰo'.
Citation(s): Barrett and Gifford 1933; Barrett 1933;
Goodrich et al 1980.

257 *Carex densa*
Common Name(s): Basket Sedge.

Usage: Material: **Miwok (Coast):** used in basketry.
Citation(s): Strike 1994.

258 *Carex douglasii*
Common Name(s): Sedge.

Usage: Food: **Kawaiisu (Tehachapi):** enlarged base of
the stem eaten raw.
Citation(s): Zigmond 1981.
Terminology: **Kawaiisu (Tehachapi):** tu(m)busi.
Citation(s): Zigmond 1981.

259 *Carex leptopoda*

Usage: Misc: **Karok:** named but not used.
Citation(s): Schenck and Gifford 1952.
Terminology: **Karok:** katikuxara.
Citation(s): Schenck and Gifford 1952.

260 *Carex mendocinoensis*
Common Name(s): Sedge.

Usage: Material: **Pomo:** used root as wrapping in
baskets.
Citation(s): Merrill 1923; Hudson 1893.

261 Carex obnupta

Usage: Material: **Coast Yuki:** inner white portion of root used for basketry decoration.
Citation(s): Gifford 1939.
Terminology: **Yuki:** sii.
Citation(s): Gifford 1939.

262 Carpobrotus chilensis

Family: *Aizoaceae*
Synonym(s): *Mesembryanthemum aequilaterale,*
Mesembryanthemum chilense
Common Name(s): Sea Fig, Fig Marigold.

Usage: Food: **Chumash:** fruit, eaten raw, sweet tasting, a little salty. **Luiseño:** the fruit is eaten. **Pomo (Southwestern):** fruit eaten raw.
Citation(s): Timbrook 2007; Gifford 1967; Sparkman 1908; Goodrich et al 1980.
Terminology: **Chumash (Barbareño, Ynezeño):** sto'yots'; **(Ventureño):** shtamhɨl, shtoyho'os. **Luiseño:** panavut. **Pomo (Southwestern):** kataicha.
Citation(s): Timbrook 2007; Sparkman 1908; Gifford 1967; Goodrich et al 1980.
Note(s): introduced, perhaps from South Africa; probably native to South Africa.
Citation(s): Timbrook 2007; Baldwin et al 2012.

263 Castilleja spp.

Family: *Orobanchaceae*
Synonym(s): *Castilleia spp.*
Common Name(s): Painted Cup, Indian Paintbrush.

Usage: Food: **Miwok:** seeds gathered in June, dried

and stored for Winter; parched, pounded, and eaten dry. **Maidu:** ate flowers.

Medicine: **Tubatulabal:** entire plant boiled, decoction used to treat rashes caused by Poison Oak (*Rhus diversiloba*). **Kawaiisu (Tehachapi):** used to treat skin problems. **Maidu:** used to cure bronchial problems, as a diuretic; decoction used as laxative; root decoction used as blood tonic; piece of root dropped in ear, cured earache.

Citation(s): Barrett and Gifford 1933; Strike 1993.

Terminology: **Miwok (Central Sierra):** ponko. **Owens Valley Paiute:** pi'tcinüpuva. **Chumash (Barbareño):** swo's 'i'ask'á', "Coyote's headdress-pin," ste'led' 'Ipistuk', "ground squirrel's tail;" **(Ynezeño):** mashqupshlét 'akhukha'w, "Coyote's rectum," stelek 'ipistúk.

Citation(s): Barrett and Gifford 1933; Steward 1933; Timbrook 2007.

Note(s): various species of this genus are secondary or facultative selenium absorbers.

Citation(s): Kingsbury 1964.

264 *Castilleja affinis*

Usage: Medicine: **Costanoan:** decoction used as wash on sores; powdered plant plied to infected sores.

Citation(s): Strike 1994; Rayburn 2012.

265 *Castilleja applegatei*

Synonym(s): *Castilleia pinetorum*

Usage: Food: **Miwok:** occasionally sipped nectar from flowers.
Citation(s): Barrett and Gifford 1933.
Terminology: **Miwok (Central Sierra):** litcitci, "Hummingbird."
Citation(s): Barrett and Gifford 1933.

266 *Castilleja attenuatus*

Synonym(s): *Orthocarpus attenuatus*
Common Name(s): Valley Tassels.

Usage: Food: **Miwok:** seeds dried, parched, and pulverized; eaten dry; root eaten.
Medicine: **Costanoan:** decoction used as cough medicine.
Citation(s): Barrett and Gifford 1933; Strike 1994.
Terminology: **Miwok (Central Sierra):** tummu.
Citation(s): Barrett and Gifford 1933.

267 *Castilleja densiflorus*

Synonym(s): *Orthocarpus densiflorus*

Usage: Misc: **Pomo (Southwestern):** flowers used in dance wreaths at Strawberry Festival.
Citation(s): Goodrich et al 1980.
Terminology: **Pomo (Southwestern):** s'amo·hu?uy, "fly eyes."
Citation(s): Goodrich et al 1980.

268 *Castilleja exserta*

Synonym(s): *Orthocarpus purpurascens*

Usage: Misc: **Pomo (Southwestern):** flowers used in dance wreaths at Strawberry Festival.
Citation(s): Goodrich et al 1980.
Terminology: **Pomo (Southwestern):** s'amo·hu?uy, "fly eyes."

Citation(s): Goodrich et al 1980.

269 *Castilleja foliolosa*

Common Name(s): Indian Paint Brush.

Usage: Food: **Cahuilla:** children sucked nectar from the flower.

Citation(s): Bean and Saubel 1972.

270 *Castilleja miniata*

Synonym(s): *Castilleia miniata*

Usage: Misc: **Tubatilabal:** considered as weeds.

Citation(s): Voegelin 1938.

271 *Castilleja peirsonii*

Synonym(s): *Castilleia parviflora* var. *douglasii*
Common Name(s): Indian Paint Brush.

Usage: Food: **Miwok:** occasionally sipped nectar from flowers.

Misc: **Karok:** children sometimes play with flowers treating them as woodpecker scalps. **Coast Yuki:** named but not used.

Citation(s): Barrett and Gifford 1933; Gifford 1939; Schenck and Gifford 1952.

Terminology: **Miwok (Central Sierra):** litcitci, "Hummingbird." **Karok:** funahich, "Little Woodpecker-Head." **Yuki:** kauwu', balchi kauwu', "Butterfly Kauwu'," (flower).

Citation(s): Barrett and Gifford 1933; Schenck and Gifford 1952; Gifford 1939.

272 *Castilleja rubicundula*

Synonym(s): *Orthocarpus lithospermoides*
Common Name(s): Paint Brush, Coyote's Tail.

Usage: Misc: **Wailaki:** named but not used.
Citation(s): Chesnut 1902.

273 *Caulanthus crassicaulis*

Family: *Brassicaceae*

Usage: Food: **Panamint Shoshone (Koso):** leaves and stems eaten.
Citation(s): Coville 1892.

274 *Caulanthus coulteri*

Common Name(s): Wild Cabbage.

Usage: Food: **Kawaiisu (Tehachapi):** leaves gathered in early Spring before flowers appear, boiled, salted, fried in grease, and eaten.
Citation(s): Zigmond 1981.
Terminology: **Kawaiisu (Tehachapi):** pišɨɨta(m)bɨ.
Citation(s): Zigmond 1981.

275 *Caulanthus inflatus*

Common Name(s): Squaw Cabbage, Desert Candle.

Usage: Food: **Kawaiisu (Tehachapi):** the stem is roasted in a pit-oven, removed, and eaten.
Citation(s): Zigmond 1981.
Terminology: **Kawaiisu (Tehachapi):** pišɨɨla(m)bɨ.
Citation(s): Zigmond 1981.

276   *Ceanothus* spp.

Family: *Rhamnaceae*
Common Name(s): Buckbrush, Chaparral, California
Lilac, Wild Lilac.

Usage: Material: **Atsugewi:** used withes to tie bark on
sides of acorn granaries. **Yokuts
(Northern Hill):** mush stirrers made
from green wood, heated and bent into
shape. **Cahuilla:** used for firewood.
        Medicine: **Costanoan:** decoction used as hair
wash.
Citation(s): Garth 1953; Gayton 1948; Bean and Saubel 1972;
       Rayburn 2012.
Note(s): the leaves, bark, and root are astringent and tonic.
Citation(s): Sargent 1922.

278   *Ceanothus cordulatus*

Usage: Food: **Maidu (Northern):** berries used to
some extent (?).
        Misc: **Miwok:** named but not used.
Citation(s): Barrett and Gifford 1933; Dixon 1905.
Terminology: **Miwok (Central Sierra):** cenebe.
Citation(s): Barrett and Gifford 1933.

279   *Ceanothus cuneatus*

Common Name(s): Mountain Mahogany,
Buckbrush.

Usage: Material: **Owens Valley Paiute:** made into
digging sticks. **Yuki:** rigid branches
used to make fish dams. **Miwok:** used
in openwork baskets; sometimes made
digging sticks 3-4 feet long. **Achomawi:**

used for making the slender needle used in piercing the lobe of young girls ears; after this had been worn for about a month replaced with a larger one made of *Rhus trilobata*. **Kawaiisu (Tehachapi):** twigs used for foreshafts of two-piece arrows.

Misc: **Karok:** named but not used.

Citation(s): Barrett and Gifford 1933; Chesnut 1902; Merriam 1955, 1967; Schenck and Gifford 1952; Steward 1933; Zigmond 1981,

Terminology: **Miwok (Central Sierra):** paiwa. **Karok:** poh'rip. **Kawaiisu (Tehachapi):** tuu(m)bɨ.

Citation(s): Barrett and Gifford 1933; Schenck and Gifford 1952; Zigmond 1981.

280   *Ceanothus integgerrimus*

Common Name(s): Deer Brush, Snow-Brush, Sweet-Brush, Buckbrush.

Usage: Food: **Maidu:** used in pinole. **Maidu (Northern):** berries used to some extent.

Material: **Maidu:** stem used in basketry. **Miwok:** stem used as warp in basketry. **Karok:** young shoots are used to make baskets.

Medicine: **Hupa:** sticks are placed from each of the fire-places toward the center during the burial ceremony. **Karok:** when anybody dies, they make a medicine of the leaves and twigs; a different medicine is made for a woman who has suffered an injury in childbirth.

Citation(s): Chesnut 1902; Dixon 1905; Goddard 1903; Merriam 1955; Merrill 1923; Schenck and Gifford 1952.

Terminology: **Hupa:** tsēLitsŌ. **Miwok (Central Sierra),**
Southern Sierra: ūsū'nni. Karok: kisiriip.
Citation(s): Goddard 1903; Barrett and Gifford 1933;
Schenck and Gifford 1952.

281 *Ceanothus leucodermis*
Common Name(s): Wild Lilac, Chaparral
White Thorn.

Usage: Material: **Kawaiisu (Tehachapi):** children made
"soap" by dipping the viscid fruit in
water and rubbing them between their
hands.
Citation(s): Zigmond 1981.
Terminology: **Kawaiisu (Tehachapi):** manarnī(m)bɨ.
Citation(s): Zigmond 1981.

282 *Ceanothus megacarpus*
Common Name(s): Bigpod Ceanothus.

Usage: Material: **Chumash:** wood, made wedges for
splitting canoe planks.
Citation(s): Timbrook 2007.
Terminology: **Chumash (Barbareño, Ynezeño, Ventureño):**
sekh.
Citation(s): Timbrook 2007.

283 *Ceanothus obliganthus*
Synonym(s): *Ceanothus divaricatus*
Common Name(s): Hairy Ceanothus.

Usage: Material: **Chumash:** wood, used for building
fences, corrals, offertory poles at
shrines, digging sticks with fire-
hardened ends, awls for basketry, pry

bars for collecting abalone, wedges for splitting canoe planks.

Misc: **Tabatulabal:** named but not used.

Citation(s): Voegelin 1938; Timbrook 2007.

284 *Ceanothus sanguineus*

Common Name(s): Oregon Tea Tree, Redstem Ceanothus.

Usage: Medicine: **Modoc (Lutuami):** tea used for coughs, colds, chewed leaves used as poultice for rheumatic pain; seeds used as an emetic.

Citation(s): Strike 1994.

285 *Ceanothus spinosus*

Common Name(s): Greenbark Ceanothus.

Usage: Material: **Chumash:** wood, used for building fences, corrals, offertory poles at shrines, digging sticks with fire-hardened ends, awls for basketry, pry bars for collecting abalone, wedges for splitting canoe planks.

Citation(s): Timbrook 2007.

Terminology: **Chumash (Barbareño, Ynezeño, Ventureño):** washiko.

Citation(s): Timbrook 2007.

286 *Ceanothus thyrsiflorus*

Synonym(s): *Ceanothus griseus*

Common Name(s): Blue Brush, California Lilac, Carmel Ceanothus, Carmel Creeper.

Usage: Material: **Esselen:** flowers, used to produce a soap.

Medicine: **Yurok:** leaves and twigs boiled to wash newborn babies.

Misc: **Pomo (Southwestern):** flower, fresh or dried, mixed with water to make a soapy lather for washing hands, face, and body; flower used in dance wreaths at Strawberry Festival; cocoon found on this bush used for ceremonial rattle. **Coast Yurok:** named but not used.

Citation(s): Gifford 1939; Merriam 1967; Goodrich et al 1980; Breschini and Haversat 2004.

Terminology: **Pomo (Southwestern):** se?e kíli, "bush black."
Yuki: sik. **Pomo (Southwestern):** se?e kīli, "bush black."

Citation(s): Goodrich et al 1980; Gifford 1939.

287 *Ceanothus velutinus*

Common Name(s): Tobacco Brush.

Usage: Food: **Maidu (Northern):** berries used to some extent (?).

Material: **Karok:** place a few leaves in their basket caps to smell good.

Medicine: **Achomawi:** leaves made into tea for fever and coughs.

Citation(s): Dixon 1905; Merriam 1967; Schenck and Gifford 1952.

Terminology: **Karok:** oyuhorrarip.

Citation(s): Schenck and Gifford 1952.

288 *Celtis reticulata*

Family: *Cannabaceae*
Synonym(s): *Celtis douglasii*

Common Name(s): Arizona Hackberry.

Usage: Misc: **Kawaiisu (Tehachapi):** named but not
used.
Citation(s): Zigmond 1981.
Terminology: **Kawaiisu (Tehachapi):** nawakkimahavɨ.
Citation(s): Zigmond 1981.

289 *Centromadia fitchii*

Synonym(s): *Hemizonia fitchii*
Common Name(s): Fitch's Spikeweed.

Usage: Food: **Miwok:** seeds eaten in the form of
mush.
Citation(s): Barrett and Gifford 1933.

290 *Centromadia pungens*

Synonym(s): *Hemizonia pungens*

Usage: Misc: **Tubatulabal:** considered weeds.
Kawaiisu (Tehachapi): named but not
used.
Citation(s): Voegelin 1938; Zigmond 1981.
Terminology: **Kawaiisu (Tehachapi):** managadɨbɨ.
Citation(s): Zigmond 1981.

291 *Centrostegia thurberi*

Family: Polygonaceae
Synonym(s): *Chorizanthe thurberi*
Common Name(s): Chorizanthe, Spine
Flower.

Usage: Misc: **Kawaiisu (Tehachapi):** named but not
used.
Citation(s): Zigmond 1981.

Terminology: **Kawaiisu (Tehachapi):** tïvimanavi, "ground
thorn."
Citation(s): Zigmond 1981.

292 *Cephalanthus occidentalis*
Family: *Rubiaceae*
Synonym(s): *Cephalanthus occidentalis* var. *californica*
Common Name(s): California Buttonbush.

Usage: Medicine: **Maidu:** root and bark decoction used as
tonic, laxative, emetic, diuretic, cough
relief, sooth sore eyes; chewed bark ro
cure toothache.
Citation(s): Strike 1994.

293 *Cerastium arvense*
Family: *Caryophyllaceae*

Usage: Misc: **Tubatulabal:** considered as weeds.
Citation(s): Voegelin 1938.

294 *Ceratophyylum demersum*
Family: *Ceratophyllaceae*
Common Name(s): Hornwort.

Usage: Medicine: **Maidu:** soothing lotion used on sore or
inflamed skin.
Citation(s): Strike 1994.

295 *Cercis occidentalis*
Family: *Fabaceae*
Common Name(s): Redbud, Western Red Bud.

Usage: Food: **Maidu, Wintun:** may have used seed
pods and seeds.

Material: **Maidu, Southern (Nisenan):** used as a woof in basketry. **Wailaki, Yuki, Wappo, Achomawi, Atsugewi, Pomo, Washo, Wintun, Maidu, Miwok, Yokuts, Monache (Western Mono), Kawaiisu (Tehachapi), Panamint Shoshone (Koso):** used bark as rim binding in basketry. **Lassik, Kato, Yuki, Wappo, Achomawi, Atsugewi, Yana, Pomo, Washo, Wintun, Maidu, Miwok, Monache (Western Mono), Kawaiisu (Tehachapi), Panamint Shoshone (Koso):** used bark as red pattern in basketry. **Yuki, Wappo, Pomo, Maidu:** used bark (dyed) as black pattern in basketry. **Kato, Yuki, Pomo, Yana, Washo, Wintun, Maidu, Miwok, Yokuts, Monache (Western Mono), Chemehuevi, Panamint Shoshone (Koso), Tubatulabal:** used sapwood as wrapping in basketry. **Washo, Maidu, Miwok, Yokuts, Monache (Western Mono), Panamint Shoshone (Koso):** used sapwood as woof in basketry. **Pomo, Washo, Wintun, Maidu, Yokuts, Monache (Western Mono):** used stem as foundation in basketry. **Washo, Miwok, Yokuts, Monache (Western Mono), Panamint Shoshone (Koso):** used stem as warp in basketry. **Kawaiisu (Tehachapi):** used stem as rim hoop in basketry. **Pomo (Southwestern):** switches used in basketry; strips of bark used for brown, or brown peeled to show white. **Wintun:** in Spring, girls tied bunches of the blossoms to their shoulders and waists in ceremonial dance of womanhood.

Medicine: **Maidu:** used buds to relieve dysentery, diarrhea.

Misc: **Karok:** named but not used.

Citation(s): Barrett 1908, 1917; Barrett and Gifford 1933; Chesnut 1902; Curtin 1957; Curtis 1924; Merriam 1955, 1967; Merrill 1923; Dixon 1905; Powers 1877; Schenck and Gifford 1952; Goodrich et al 1980; Strike 1994.

Terminology: **Karok:** saxayamsurip, "Down-by-the-river-honeysuckle." **Yuki:** cha'-hā. **Miwok (Northern Sierra):** lüli, **(Central Sierra):** TapáTap̲u̲, **(Southern Sierra):** tapátapa. **Pomo (Southwestern):** háyhia qʰale.

Citation(s): Schenck and Gifford 1952; Curtin 1957; Barrett and Gifford 1933; Goodrich et al 1980.

Note(s): there are seven species of redbud native to North America, southern Europe, and Asia, but only this species is native to California. The plant is drought tolerant and found in 22 counties.

seeds contain:

Ash   2.6%
Protein   18.6%
Oil   8.8%

Citation(s): Anderson 1991; Earle and Jones 1962.

296   *Cercocarpus betulodies*

Family: *Rosaceae*

Synonym(s): *Cercocarpus betulfolius, Cercocarpus parvifolius*

Common Name(s): Hard Tack, Mountain Mahogany, Ironwood Tree, Mahogany.

Usage: Material: **Monache (Western Mono:** used in house construction; made pack straps of the braided inner bark, **Miwok:** used long poles to make two-pronged harpoon used in taking salmon. **Achomawi:** used for digging sticks and for spear points. **Karok:** used for digging sticks. **Wailaki:** used for arrow tips and digging sticks, spears, and war and fighting clubs. **Yuki:** canes and arrows made from straight branches. **Maidu:** used inner bark to produce a purple dye. **Chumash:** sometimes made digging sticks, used especially for digging brodiaea bulbs.

Medicine: **Kawaiisu (Tehachapi):** root boiled to make a wine-colored brew; drunk to relieve coughing. **Diegueño:** bark used as tea for respiratory problems; pulverized bark mixed with used as a laxative.

Citation(s): Barrett and Gifford 1933; Chesnut 1902; Curtin 1957; Gifford 1932; Merriam 1967; Schenck and Gifford 1952; Zigmond 1981; Strike 1994; Timbrook 2007.

Terminology: **Karok:** weiip. **Monache (Western Mono):** sätip. **Yuki:** um-se. **Owens Valley Paiute:** tunap (sp?). **Kawaiisu (Tehachapi):** sɨna?ruubĭ. **Chumash (Barbareño):** pɨch.

Citation(s): Schenck and Gifford 1952; Gifford 1932; Curtin 1957; Steward 1933; Zigmond 1981; Timbrook 2007.

297 *Cercocarpus ledifolius*

Common Name(s): Mountain Mahogany.

Usage: Material: **Maidu (Northern):** made stick armor of straight round sticks; braided straps made from inner bark.

Medicine: **Kawaiisu (Tehachapi):** bark is boiled, decoction drunk for relief of gonorrhea; shrubs exudation dried, ground to a powder, and applied as an earache remedy; bark and leave decoction used for women's gynecological problems.

Citation(s): Dixon 1905; Zigmond 1981; Strike 1994.
Terminology: **Kawaiisu (Tehachapi):** keezima?ahnitɨbɨ.
Citation(s): Zigmond 1981.

298 *Chaenactis douglasii*

Family: *Asteraceae*
Common Name(s): Pincushion Flower.

Usage: Medicine: **Northern Paiute (Paviotso):** mashed leaves applied as a poultice for sprains and to reduce swellings. **Maidu:** decoction used for indigestion, coughs, colds; mashed leaves used on rattlesnake bite.

Misc: **Kawaiisu (Tehachapi):** named but not used.

Citation(s): Kelly 1932; Zigmond 1981; Strike 1994.
Terminology: **Northern Paiute (Paviotso):**
bawa'natizua. **Kawaiisu (Tehachapi):** tu?urɨbɨ.
Citation(s): Kelly 1932; Zigmond 1981.

299 *Chaenactis glabriuscula*

Common Name(s): Pin Cushion.

Usage: Food: **Cahuilla:** seeds gather June until August; parched, ground into flour, mixed with other seeds, to form mush.
Citation(s): Bean and Saubel 1972.

300 *Chaenactis santolinoides*
Common Name(s): Perennial Pincushion Flower.

Usage: Medicine: **Kawaiisu (Tehachapi):** root boiled, decoction drunk warm for relief of sore chest and shoulders, and for internal soreness.
Citation(s): Zigmond 1981; Strike 1994.

301 *Chaenactis stevioides*

Usage: Misc: **Tubatulabal:** considered as weeds.
Citation(s): Voegelin 1938.

302 *Chamaebatia foliolosa*
Family: *Rosaceae*
Common Name(s): Mountain Misery.

Usage: Medicine: **Miwok:** leaves made into hot tea and drunk for rheumatism and for diseases manifested by skin eruptions; leaves also used as an ingredient in medicines for the treatment of venereal diseases; for coughs and colds a decoction, sometimes with other herbs mixed in, was drunk.
Citation(s): Barrett and Gifford 1933.
Terminology: **Miwok (Northern Sierra, Central Sierra):** kiTkitī'su.

Citation(s): Barrett and Gifford 1933.

303 *Chamaecyparis lawsoniana*
Family: *Cupressaceae*
Common Name(s): Port Orford Cedar,
Lawson Cypress.

Usage: Material: **Karok:** planks are used in building sweathouses; blocks made into pillows or headrests; branches are sometimes used as brooms.
Citation(s): Schenck and Gifford 1952; Strike 1994.
Terminology: **Karok:** kupurrip.
Citation(s): Schenck and Gifford 1952.
Note(s): the wood is light, hard, strong, very close grained, abounding in fragrant resin, durable, easily worked, light yellow, or almost white, with hardly distinguishable sapwood.
Citation(s): Sargent 1922.

304 *Cheilanthes covillei*
Family: *Pteridaceae*
Common Name(s): Lip Fern.

Usage: Food: **Kawaiisu (Tehachapi):** stems and leaves are brewed to produce non-medicinal tea.
         Medicine: **Hupa:** used by women during childbirth.
         Misc: **Tubatulabal:** considered as weeds.
Citation(s): Voegelin 1938; Zigmond 1981; Strike 1994.
Terminology: **Kawaiisu (Tehachapi):** tɨ(m)binawivɨ.
Citation(s): Zigmond 1981.

305 *Chenopodium* spp.

Family: *Chenopodiaceae*
Common Name(s): Pigweed.

Usage: Food: **Luiseño:** used as a pot-herb. **Bear River:** gathered seeds by means of an oval, elongated basketry fan.
Material: **Owens Valley Paiute:** woven into mats for the covering of the outside of winter valley houses.
Citation(s): Curtis 1924; Nomland 1938; Steward 1933.

306 *Chenopodium album*

Synonym(s): *Chenopodium alba*
Common Name(s): Lamb's Quarters, Pigweed, White Pigweed, White Goosefoot,

Usage: Food: **Luiseño:** leaves are boiled and eaten as greens. **Northern Paiute (Paviotso):** seeds are eaten. **Owens Valley Paiute:** harvested late in August. **Miwok:** eaten boiled as greens; sometimes stored and dried for later use. **Kawaiisu (Tehachapi):** upper leaves boiled, "rinsed" in cold water, fried in grease and salt.
Medicine: **Owens Valley Paiute:** single leaf chewed as an emetic.
Citation(s): Barrett and Gifford 1933; Kelly 1932; Sparkman 1908; Steward 1933; Zigmond 1981.
Terminology: **Northern Paiute (Paviotso):** wa·'ta' (seed). **Mono Lake Paiute, Owens Valley Paiute:** wā'tā; ā'tā. **Miwok (Central Sierra):** somala. **Luiseño:** ket.

Citation(s): Kelly 1932; Steward 1933; Barrett and
Gifford 1933; Sparkman 1908.
Note(s): common weed in waste and fallow places below 6,000
feet; naturalized from Europe; frequently found to
contain high, potentially dangerous concentrates of
nitrate.
Citation(s): Munz and Keck 1965; Kingsbury 1964.

307 *Chenopodium berlandieri* var. *sinuatum*
Common Name(s): Pitseed Goosefoot.

Usage: Food: **Chumash:** seeds, eaten; young plants,
leaves, eaten as greens.
Citation(s): Timbrook 2007.
Terminology: **Chumash (Barbareño):** we'lel; **(Yenezeño,
Ventureño):** welel.
Citation(s): Timbrook 2007.

308 *Chenopodium californicum*
Common Name(s): Soap Plant.

Usage: Food: **Luiseño:** the seeds are used for food.
         Material: **Luiseño:** the root is used for soap.
                     Tubatulabal: root pulverized by
                     rubbing on rock and used for shampoo.
                     Cahuilla: carrot-like, hard root was
                     stored and grated on a rock and used
                     for soap; seeds were parched and
                     ground into flour. **Kawaiisu
                     (Tehachapi):** stout fleshy root mashed
                     and put in water produces a soapy
                     lather. **Chumash:** root, scraped, or
                     grated, for used in washing hair, skin,
                     clothing.
         Medicine: **Kawaiisu (Tehachapi):** leaves and stem

decoction used an emetic. **Costanoan:** root, decoction, used as poultice for numb or paralyzed limbs. **Chumash:** root bark, boiled, tea used for treatment of consumption.

Misc: **Tubatulabal:** considered as a weed (?).

Citation(s): Barrows 1900; Sparkman 1908; Voegelin 1938; Zigmond 1981; Bean and Saubel 1972; Strike 1994; Rayburn 2012; Timbrook 2007.

Terminology: **Luiseño:** kahawut. **Kawaiisu (Tehachapi):** ookovɨbɨ. **Cahuilla:** ki'awet. **Chumash (Barbareño):** su'núk'; **(Ynezeño):** 'akhwayɨsh; **(Ventureño):** choch.

Citation(s): Sparkman 1908; Zigmond 1981; Bean and Saubel 1972; Rayburn 2012; Timbrook 2007.

Note(s) note the difference of opinion of/by the **Tubatulabal** – may be a shift through time [author].

309 *Chenopodium fremontii*
Common Name(s): Goosefoot, Pigweed.

Usage: Food: **Owens Valley Paiute:** used for food. **Cahuilla:** used seeds, ground into flour and baked into cakes.

Citation(s): Barrows 1900; Steward 1933; Bean and Saubel 1972.

Terminology: **Owens Valley Paiute:** kö'yo. **Cahuilla:** kit, ka-at.

Citation(s): Steward 1933; Barrows 1900.

Note(s): seeds contain:

protein   12-17%
ash   1-4.6%
oil   7.3-27.7%

Citation(s): Bean and Saubel 1972.

310 Chenopodium nevadense

Usage: Food: **Northern Paiute (Paviotso):** the seeds were eaten.
Citation(s): Kelly 1932.
Terminology: **Northern Paiute (Paviotso):** üyü'p (seed).
Citation(s): Kelly 1932.

311 Chilopsis linearis
Family: *Bignoniaceae*
Common Name(s): Desert Willow.

Usage: Food: **Cahuilla:** seed pods eaten, not a major food source.
Material: **Cahuilla:** primarily a source of wood for house frames and granaries; its strength and pliability made it attractive for the making of bows; long limbs used to reach fruits and nuts to high to grasp; bark fiber used in making nets, shirts, breechclouts.
Citation(s): Bean and Saubel 1972; Strike 1994.
Terminology: **Cahuilla:** qaankish.
Citation(s): Bean and Saubel 1972.

312 Chimaphila menziesii
Family: *Ericaceae*
Common Name(s): Western Pipsissiwa.

Usage: Food: **Maidu:** leaves smoked.
Medicine: **Maidu:** used tonic made with leaves, used to cure many illnesses and as a general tonic.
Citation(s): Strike 1994.

313 *Chimaphilia umbellata*

Synonym(s): *Chimaphilia umbellata* var. *Occidentalis*
Common Name(s): Prince's Pine.

Usage: Medicine: **Karok:** placed in the bed as a remedy for backache; also boiled, the infusion is drunk, or, the patient is steamed with the infusion.
Citation(s): Schenck and Gifford 1952.
Terminology: **Karok:** hunyeip rukwtixa, "that which grows in the oaks."
Citation(s): Schenck and Gifford 1952.
Note(s): fresh leaves, when bruised and applied to the skin, act as vesicants and rubefacient; as a decoction, diuretic, astringent, tonic, alterative.
Citation(s): Grieve 1931; Stille and Maisch 1880; Stuhr 1933.

314 *Chlorogalum spp.*

Family: *Agavaceae*
Common Name(s): Soaproot.

Usage: Food: **Bear River:** root cleaned throughly and peeled, then wrapped in Alder (*Alnus* spp.) leaves and cooked in ashes with Camas (*Camassia* spp.); not stored for Winter. **Yokuts (Southern Valley):** eaten roasted. **Yuki:** shoots roasted in ashes and eaten like celery. **Sinkyone:** eaten (roasted?). **Wappo:** ate root.

Material: **Bear River:** used root for washing hair, basketry, and body. **Yuki:** root was soaked, pounded and shredded and fastened to a three-foot stick to form a brush. **Huchnom:** used root as soap; washed in the morning. **Yokuts**

		(Southern Valley): used for washing; **(Northern Hill):** made brushes from fiber. **Monache (Western Mono), Wintin:** pounded into a mash used as fish poison.
	Medicine:	**Bear River:** fever patients washed with root tied in small bundles. **Wintun:** pounded into a mash and applied as a poultice on poison oak. **Costanoan:** bulb, roasted, used as poultice for sores; bulb, fresh, crushed, rubbed on body to cure pain, cramps.

Citation(s): Driver 1936; Du Bois 1935; Foster 1944; Gayton 1948; Nomland 1938; Rayburn 2012.

315 *Chlorogalum angustifolium*

Usage:	Food:	**Karok:** cooked overnight in a earth oven and then eaten by separating the layers of the bulb much as artichokes are handled.
	Material:	**Karok:** the fibers that remain after the bulbs are eaten are used to make a small brush; the bulb, pounded and mixed with water is used as a detergent for washing clothes.

Citation(s): Schenck and Gifford 1952.
Terminology: **Karok:** xanchusa.
Citation(s): Schenck and Gifford 1952.

316 *Chlorogalum parviflorum*

Usage: Food: **Luiseño:** the bulb was eaten.
Citation(s): Sparkman 1908.

Terminology: **Luiseño:** kenut.
Citation(s): Sparkman 1908.

317 *Chlorogalum pomeridianum*

Synonym(s): *Kaothoe pomeridiana* (misspelled
Lathoe in Voegelin 1938).
Common Name(s): Soap-Plant, Soap Root,
Amole.

Usage: Food:

Miwok: soaked and baked in earth oven. **Maidu, Southern (Nisenan):** eaten in times of great scarcity; roasted underground 36 hours or more to extract poison. **Hupa:** cooked for about two days in an underground pit. **Karok:** cooked overnight in a earth oven and then eaten by separating the layers of the bulb much as artichokes are handled. **Yuki:** shoots gathered in March. **Chumash:** eaten.

Material:

Maidu, Southern (Nisenan): crushed bulbs mixed with hot water, used to remove tar from widows coming out of mourning; used for poisoning fish; baked bulbs provided glue. **Maidu (Northern):** both men and women frequently washed their heads with the root. **Yokuts:** used fiber for household brushes and for washing; used root as fish poison. **Luiseño:** the fibers covering the bulb are used to make a brush. **Hupa:** made a brush from the fibers taken from the sheath of the bulb bound with buckskin to sweep up scattered meal during the grinding of acorns;

bulbs used as soap. **Karok:** the fibers that remain after the bulbs are eaten are used to make a small brush; the bulb, pounded and mixed with water is used as a detergent for washing clothes. **Shasta:** made brushes to brush off the fine ground acorn meal from the catch tray. **Yuki:** made brush from fibers on the outside of bulb; sometimes used as bedding; leaves used to cover dough when baking acorn bread; green leaves used to be pricked into skin to form green tattoo marks; juice used as glue to attach arrow feathers, and to color bows black when mixed with soot; bulbs pounded on stone and rubbed between hands to make soap for washing hair, skin, clothing, and baskets; used as a fish poison. **Pomo:** used as fish poison. **Pomo (Southwestern):** bulb used as soap for washing body, hair, utensils, and baskets; pounded and mashed bulb used as a fish drug when stream was low; fibers used to make a scrub brushes and hair brushes; used as a detergent it prevented lice, especially if the suds were left on the head awhile. **Miwok:** used juice of plant to catch fish; juice used to seed-proof baskets; made brush from fibers of the dry, outer layers of the root. **Chumash:** bulbs, dug in springtime, crushed, mixed with water for washing hair, clothing, for preparing hides other than deer skin; bulb, juice, used by women to make bangs lie flat against foreheads, to make hair glossy; bulb, crushed, used as fish poison; made brushes from the fibers. **Cahuilla:** made a sifter for separating the finer parts of ground seeds of Mesquite (*Prosopis* spp.) from the coarser, and for removing the iron pyrite grains after milling on stone grinding slabs; coarse husk of fiber from bulb tied to together to make a cleaning brush to sweep out acorn mortars, baskets, and other containers; used bulb, rich in saponin, crushed, rubbed in water, for soap. **Kato:** used to make center-post slippery for public performances of an activity called "Chushinpuhlpeategh" by the society of magicians. **Tubatulabal:** bulb pounded and used as shampoo. **Kawaiisu (Tehachapi):** bulb, without fibrous coat, boiled and rubbed on a rock, liquid starch applied to seed-gathering baskets to close off the interstices; fibers used to make brushes. **Miwok, Owens Valley Paiute:** baked root in hot ashes, dipped it in water, rubbed on conical carrying baskets to keep seeds from slipping out. **Costanoan:** bulb used to produce foam used for washing.

Medicine: **Maidu, Southern (Nisenan), Bear River, Costanoan, Pomo, Yana, Yuki:** bulbs used to heal and cleanse old sores, being heated and laid on hot. **Pomo, Wintun:** bulbs, pulverized used to poultice rashes, sores caused by Poison Oak. **Maidu:** decoction used as a laxative, diuretic, remedy for stomach aches. **Yokuts:** considered as a healing and cleansing medicine. **Wailaki:** rub fresh bulb on the body for cramps and rheumatism; a decoction of the bulb as a diuretic and laxative and for stomach aches caused by gas. **Chumash:** brewed a medicinal tea; buried bulb used while giving it thanks for its healing properties. **Costanoan:** stem, pounded, made wash used to reduce dandruff.

Citation(s): Barrett and Gifford 1933; Barrows 1900; Chesnut 1902; Curtin 1957; Clark 1904; Curtis 1924; Dixon 1905, 1907; Gifford 1967; Goddard 1903; Grant 1964; Heizer 1953; Merriam 1923; Powers 1877; Schenck and Gifford 1952; Sparkman 1908; Voegelin 1938; Zigmond 1981; Goodrich et al 1980; Bean and Saubel 1972; Strike 1994; Landberg 1965; Rayburn 2012; Timbrook 2007.

Terminology: **Pomo (Southwestern):** ha·ûm, ha?am?.
Miwok (Northern Sierra): sôpa, **(Central Sierra):** palawi., **(Plains):** so'pa. **Yuki:** nösh, ku-lum. **Wiyot:** kātserā. **Karok:** imyuha. **Kawaiisu (Tehachapi):** winižibɨ. **Cahuilla:** mocee (?). **Chumash (Barbareño):** qi'w; **(Ynezeño):** kot', chunuy (fiber); **(Obispeño):** tqupa'; **(Ventureño):** pash.

Citation(s): Gifford 1967; Barrett and Gifford 1933; Curtin 1957; Loud 1918; Schenck and Gifford 1952; Zigmond 1981; Goodrich et al 1980; Bean and Saubel 1972; Timbrook 2007.

Note(s):

Analysis of bulb:

Moisture	73.13%
Ash	0.70%
Saponin*	6.95%

* the soapy principle of the plant.
Citation(s): Trimble 1890.

318   *Chlorogalum pomeridianum* var. *divaricatum*
Common Name(s): Little Soap Root.

Usage: Food: **Maidu, Southern (Nisenan):** eaten raw,
or roasted in ashes, or boiled.
Citation(s): Powers 1877.

319   *Chorizanthe staticoides*
Family: *Polygonaceae*
Synonym(s): *Chorisanthe staticoides*
Common Name(s): Turkish Rugging.

Usage: Medicine: **Tubatulabal:** infusion used as lotion to
relieve skin abscesses.
Misc: **Tubatulabal:** considered weeds.
Citation(s): Voegelin 1938; Strike 1994.
Note(s): note the difference between the two references -
probably different time periods, or different
informants [author].

320   *Chorizanthe uniaristata*

Usage: Misc: **Kawaiisu (Tehachapi):** named but not
used.
Citation(s): Zigmond 1981.
Terminology: **Kawaiisu (Tehachapi):** manavi, "thorns."
Citation(s): Zigmond 1981.

321   *Chorizanthe watsonii*
Synonym(s): *Chorisanthe watsonii*

Usage: Misc: **Tubatulabal:** considered weeds.

Citation(s): Voegelin 1938.

322 *Chrysolepsis chrysophylla*
Family: *Fagaceae*
Synonym(s): *Castanea chrysophylla, Castanopsis chrysophylla*
Common Name(s): Giant Chinquapin.

Usage: Food: **Hupa:** the nuts, when found, are eaten.
Karok: the nuts are stored if there are enough, otherwise they are eaten when found. **Pomo (Southwestern):** nuts collected in the Fall, cracked open, and eaten raw; stored in shell for winter.
Citation(s): Gifford 1967; Goddard 1903; Schenck and Gifford 1952; Goodrich et al 1980.
Terminology: **Karok:** sonyisip (plant); sonyisi (nut).
Pomo (Southwestern): kamidish, "water" (ka), "nut" (midish), gʰamí?diš gʰale, "Water-nut tree."
Citation(s): Schenck and Gifford 1952; Gifford 1067; Goodrich et al 1980; Strike 1994.
Note(s): nut are dark purple-red, sweet and edible.
Citation(s): Sargent 1922.

323 *Chrysolepsis sempervirens*
Synonym(s): *Castanopsis sempervirens*
Common Name(s): Bush Chinquapin, Sierra Chinquapin.

Usage: Food: **Kawaiisu (Tehachapi):** hunters sometimes eat raw seeds when in the field. **Maidu:** nuts roasted, eaten.
Citation(s): Zigmond 1981; Strike 1994.
Terminology: **Kawaiisu (Tehachapi):** pikahipaka?agadɨbɨ.
Citation(s): Zigmond 1981.

324 *Chrysothamnus* spp.

Family: *Asteraceae*

Usage: Material: **Tubatulabal:** armfuls piled over house frame, then covered with mud.

Misc: **Mono Lake Paiute:** named but not used.

Citation(s): Steward 1933; Voegelin 1938.

325 *Chrysothamnus nauseosa* var. *oreophila*

Usage: Food: **Northern Paiute (Pavitoso):** "chewing gum" made from chewing the root that occurs at the surface level.

Citation(s): Kelly 1932.

Terminology: **Northern Paiute (Paviotso):** sigu'p'.

Citation(s): Kelly 1932.

326 *Chrysothamnus nauseosa* var. *speciosa*

Synonym(s): *Chrysothamnus nauseosus* var. *occidentalis, Chrysothamnus nauseosus* var. *albicaulis*

Usage: Material: **Karok:** girls tie the stems and flowers on the end of their hair rolls as imitation mink skins, like the skins used in the Deerskin Dance.

Citation(s): Schenck and Gifford 1952.

Terminology: **Karok:** oxuichpachi.

Citation(s): Schenck and Gifford 1952.

327 Chylismia brevipes

Family: *Onagraceae*
Synonym(s): *Oenothera brevipes*

Usage: Food: **Panamint Shoshone (Koso):** used as food.
Citation(s): Coville 1892.

328 Chylisma claviformis

Synonym(s): *Oenothera claviformis*
Common Name(s): Evening Primrose, Desert Primrose.

Usage: Food: **Cahuilla:** used a greens; caterpillar of The White Line Sphinx Moth (*Celerio lineata*) that frequented plant were eaten, head chopped off, insides cleaned out, boiled, or parboiled, or dried in the sun.
Citation(s): Bean and Saubel 1972
Terminology: **Cahuilla:** tesaval.
Citation(s): Bean and Saubel 1972.

329 Chylisma walkerii ssp. tortilis

Synonym(s): *Oenothera scapoidea, Oenothera scapoidea* var. *seorsa*

Usage: Misc: **Tubatulabal:** considered weeds.
Citation(s): Voegelin 1938.

330 Cicuta dougalsii

Family: *Aplaceae*
Common Name(s): Water Hemlock.

Usage: Medicine: **Kawaiisu (Tehachapi):** root is mashed, put on hot stone, and sore limbs laid directly over it; used an analgesic. **Maidu:** root poultice used on sores caused by thorns, rattlesnake bites, any sore or swollen part.

Misc: **Kawaiisu (Tehachapi):** considered poisonous.

Citation(s): Zigmond 1981; Strike 1994.

Terminology: **Kawaiisu (Tehachapi):** pavɨbogovɨ.

Citation(s): Zigmond 1981.

Note(s): the roots are toxic in all stages of growth except the very youngest; symptoms appears within 15 minutes to more than an hour as a violent convulsant, delirium is encountered in humans, death may occur as quickly as 15 minutes after an ingestion of a lethal amount or not until after 8 hours; the root is extremely poisonous, one mouthful is sufficient to kill a grown man.

Citation(s): Kingsbury 1964; Hardin and Arena 1974.

331 *Cirsium* spp.

Family: *Asteraceae*
Common Name(s): Thistle.

Usage: Food: **Chumash:** may have used several kinds of thistles for food.

Medicine: **Costanoan:** stems, pounded, pulp used for face sores, to dry infected sores; stems(?), raw, chewed to treat stomach pain; root, decoction, given for asthma.

Citation(s): Rayburn 2012; Timbrook 2007.

332   *Cirsium occidentale*

Common Name(s): Coyote's Tail, Thistle.

Usage: Food: **Northern Paiute (Pavitoso):** the peeled stems of young plants were eaten. **Tubatulabal:** used plant. **Kawaiisu (Tehachapi):** stems are skinned and eaten raw in Spring. **Maidu:** peeled taproots eaten raw, or cooked.

Misc: **Cahuilla:** named but not used.

Citation(s): Kelly 1932; Voegelin 1938; Zigmond 1981; Bean and Saubel 1972; Strike 1994.

Terminology: **Kawaiisu (Tehachapi):** Čiiyavɨ. **Cahuilla:** ya'I he'ash, "wet-pet."

Citation(s): Zigmond 1981; Bean and Saubel 1972.

333   *Cirsium occidentale* var. *californicum*

Synonym(s): *Cardus californicus, Cirsium californicum*

Usage: Food: **Kawaiisu (Tehachapi):** in the Spring stems are skinned and eaten raw.

Material: **Miwok:** stems laid about a sleeping place to keep away snakes and lizards which bite the sleeper.

Citation(s): Barrett and Gifford 1933; Zigmond 1981.

Terminology: **Miwok (Central Sierra):** swala. **Kawaiisu (Tehachapi):** ciɨyavɨ.

Citation(s): Barrett and Gifford 1933; Zigmond 1933.

334   *Cirsium occidentale* var. *candidissimum*

Synonym(s): *Cirsium pastoris*

Usage: Food: **Northern Paiute (Paviotso):** peeled stems eaten raw.

Citation(s): Kelly 1932.

Terminology: **Northern Paiute (Paviotso):** iza$^{\prime\text{la}}$kwasi$^\prime$, "coyote's tail."
Citation(s): Kelly 1932.

335 *Cirsium scariosum*

Synonym(s): *Cirsium acaulescens, Cirsium drummondii*
Common Name(s): Thistle, Sunflower.

Usage:　Food:　　**Northern Paiute (Paviotso):** the roots were eaten raw, or, if plentiful, were roasted in the earth oven. **Atsugewi:** stalks were eaten raw when young and tender. **Cahuilla:** ate the bud at the base of the thistle.
Citation(s): Garth 1953; Kelly 1932; Bean and Saubel 1972.
Terminology: **Atsugewi:** ca$^\prime$a$^\bullet$ko. **Northern Paiute (Pavitoso):** ko$^\prime$Tciü.
Citation(s): Garth 1953; Kelly 1932.
Note(s): food values – root:

	(wet)	(dry)
Moisture	70.8%	
Sugar		
Reducing	0.3%	1.0%
Non-reducing	0.3%	1.0%
Hemicellulose	4.4%	15.1%

Citation(s): Yanovsky 1938.

336 *Cirsium scariosum* var. *congdonii*

Synonym(s): *Cirsium congdonii*
Common Name(s): Leafy or Dwarf Thistle.

Usage:　Food:　　**Kawaiisu (Tehachapi):** peeled stalk eaten raw in Spring.
Citation(s): Zigmond 1981.

Terminology: **Kawaiisu (Tehachapi):** paziiyavɨ, "water"
Čiiyavɨ "thistle."
Citation(s): Zigmond 1981.

337 *Cirsium vulgare*

Synonym(s): *Cirsium lanceolatum*
Common Name(s): Bull Thistle.

Usage: Misc: **Pomo (Southwestern), Coast Yuki, Kawaiisu (Tehachapi):** named but not used.
Citation(s): Gifford 1939, 1967; Zigmond1981; Goodrich et al 1980.
Terminology: **Yuki:** xalet. **Pomo (Southwestern):** pôtlo hiti, "thistle," "gonorrhea (pôtlo) "thorn" (hiti). **Kawaiisu (Tehachapi):** tuquziiyavɨ.
Citation(s): Gifford 1939, 1967; Zigmond 1981; Goodrich et al 1980.
Note(s): naturalized from Europe.
Citation(s): Munz and Keck 1965.

338 *Citrullus lanatus*

Family: *Cucurbitaceae*
Common Name(s): Watermelom.

Usage: Food: **Cahuilla:** ate fresh or cut peel into strips and dried them for Winter use; may have buried them in sand for short-term storage.
Citation(s): Bean and Saubel 1972.
Terminology: **Cahuilla:** istochen, estuish (?).
Citation(s): Bean and Saubel 1972.
Note(s): native of Africa; the plants seems to have either preceded, or followed, the advance of the Spanish into the Southwest.

Citation(s): Munz and Keck 1965; Munz 1968; Bean and Saubel 1972.

339 *Cladium californicum*

Family: *Cyperaceae*
Synonym(s): *Cladium maricus, Cladium mariscus* var. *californicum*
Common Name(s): Saw Grass, Swamp Grass,

Usage: Material: **Yokuts:** used root, which grew up to 3 feet long, as wrapping in baskets. **Miwok:** used root as foundation material in basketry. **Salinan, Yokuts:** used root for stitching coiled baskets.
Citation(s): Merrill 1923; Strike 1994.

340 *Clarkia amoena*

Family: *Onagraceae*
Synonym(s): *Godetia amoena*
Common Name(s): Summer's Darling.

Usage: Food: **Miwok:** entire plant was pulled up as soon as the flowering was over and dried to retrieve the seeds; seeds parched, pulverized, and eaten dry. **Shelter Cove Sinkyone:** seeds used for making pinole.
Citation(s): Barrett and Gifford 1933; Baumhoff 1958.
Terminology: **Miwok (Central Sierra):** sipsibe.
Citation(s): Barrett and Gifford 1933.

341　*Clarkia biloba*

Synonym(s): *Godetia biloba*
Common Name(s): Farewell To Spring.

Usage:　Food:　　　**Miwok:** seeds gathered in June; parched and pulverized before eating.
Citation(s): Barrett and Gifford 1933.
Terminology: **Miwok (Central Sierra):** witala.
Citation(s): Barrett and Gifford 1933.

342　*Clarkia purpurea* ssp. *quadrivulnera*

Synonym(s): *Godetia albescens*
Common Name(s): Farewell-to-Spring.

Usage:　Food:　　　**Yuki:** seeds used in pinole. **Miwok:** considered seeds as well as the acorn from the Black Oak (*Quercus Kelloggii*) as their most prized foods.
　　　　Medicine:　**Yuki:** decoction of the leaves used as a wash for sore eyes; plant mashed, used a salve on irritated skin.
Citation(s): Chesnut 1902; Strike 1994.

343　*Clarkia purpurea* ssp. *viminea*

Synonym(s): *Godetia viminea*
Common Name(s): Farewell To Spring.

Usage:　Food:　　　**Miwok:** seeds used; dried, pulverized, and eaten dry and uncooked.
　　　　Misc:　　　**Tubatulabal:** considered weeds.
Citation(s): Barrett and Gifford 1933; Voegelin 1938.
Terminology: **Miwok (Central Sierra):** nuwati; nō'wasī.
Citation(s): Barrett and Gifford 1933.

344 *Clarkia rhomboidea*

Usage: Food: **Yana:** the seeds eaten raw, or, parched
and pounded up fine and then eaten.
Citation(s): Sapir and Spier 1943.
Terminology: **Yana:** ga$^{r\text{"}}$na` (seeds).
Citation(s): Sapir and Spier 1943.

345 *Clarkia unguiculata*
Synonym(s): *Clarkia elegans*

Usage: Food: **Miwok:** seeds were dried, parched, and
pulverized, and eaten dry with acorn
(*Quercus* spp.) mush.
Citatiuon(s): Barrett and Gifford 1933.
Terminology: **Miwok (Central Sierra):** tcikali, sokowila.
Citation(s): Barrett and Gifford 1933.

346 *Clavipes purpurea*
Family: *Hypocreaceae*
Common Name(s): Ergot.

Usage: Medicine: **Yuki:** considered to have medicinal
uses; gathered with/on *Elymus
triticoides.*
Citation(s): Chesnut 1902; Strike 1994.
Note(s): parasitic in the ovaries of grains; principle components
in this fungus are d-lysergic acid amide acid, and d-
lysergic acid methycarbinolanide, ergoline
derivations closely allied to lysergic acid
diethylamide; ergot alkaloids in small daily does
cause vasoconstriction and predispose to thromboses
with occlusion of circllation in the extremities (dry
gangrene); first signs of convulsive ergotism is
hyperexcitability; considered ecbolic, haemostatic,

emmenagigue, parturient.

Citation(s): Bailey 1950; Schultes 1969; Kingsbury 1964; Stuhr 1933.

347 *Claytonia sibirica*

Family: *Montiaceae*
Synonym(s): *Montia sibirica*
Common Name(s): Indian Lettuce, Candy Flower.

Usage: Misc: **Karok:** children play a game of hooking each other's flowers until one breaks and is the loser; a score is kept to see who wins.

Citation(s): Schenck and Gifford 1952.
Terminology: **Karok:** chishihiich, "make-believe [imitation] dog."
Citation(s): Schenck and Gifford 1952.

348 *Clematis lasiantha*

Family: *Ranunculaceae*
Common Name(s): Coyote's Rope, Pipe-Stem.

Usage: Medicine: **Shasta:** the whole stem or just the bark was pounded and boiled and the face steamed in it for a cold. **Miwok:** pulverized charcoal was dusted on running sores and burns; tea made from roots used as an emetic, as a general health tonic. **Chumash:** leaves, rubbed on skin to treat ringworm, skin eruptions, perhaps venereal disease.

Misc: **Karok:** nut not used but figures in coyote stories.

Citation(s): Barrett and Gifford 1933; Holt 1946; Schenck and Gifford 1952; Strike 1994; Timbrook 2007.

Terminology: **Miwok (Central Sierra):** wakilwakilu.
Karok: pineeftatapuwa, "coyote's sting," "coyote's
trap." **Shasta:** gaw'tagatupúkiras, "coyote's rope."
Citation(s): Barrett and Gifford 1933; Schenck and
Gifford 1952; Holt 1946.

349 *Clematis ligusticifolia*
Common Name(s): Virgin's Bower.

Usage: Medicine: **Costanoan:** poultice of foliage used to
relieve chest pains. **Yuki:** stems and
leaves chewed for sore throats, colds.
Maidu: stem and bark steam inhaled to
cure colds. **Chumash:** leaves, rubbed on
skin to treat ringworm, skin eruptions,
colds, sore throats, perhaps venereal
disease.
Misc: **Kawaiisu (Tehachapi):** named but not used.
Citation(s): Zigmond 1981; Strike 1994; Rayburn 2012;
Timbrook 2007.
Terminology: **Kawaiisu (Tehachapi):** pog^{w}itɨna hiapiina,
"Grizzly Bear's Trap." **Chumash (Barbareño,
Purisemeño):** makhsik'; **(Ynezeño):** 'alamakhwak'ay;
(Ventureño): makhsik.
Citation(s): Zigmond 1981; Timbrook 2007.

350 *Clematis pauciflora*

Usage: Medicine: **Diegueño:** bark tea used to reduce
fevers. **Salinan:** used on wounds,
stopped coughs, cured bladder
problems.
Citation(s): Strike 1994.
Note(s): plant considered as poisonous.
Citation(s): Hardin and Arena 1974.

351 *Cleomella lutea*
    Family: *Cleomaceae*

Usage: Misc: **Owens Valley Paiute:** named but not used.
Citation(s): Steward 1933.
Terminology: **Owens Valley Paiute:** cuyuhunava.
Citation(s): Steward 1933.

352 *Clinopodium douglasiii*
    Family: *Lamiaceae*
    Synonym(s): *Micromeria chamissonis, Micromeria Douglasii, Satureja douglasii, Satureja chamissonia*
    Common Name(s): Yerba Buena.

Usage: Food: **Maidu:** leafy vines made into tea. **Luiseño:** tea is used as a beverage. **Pomo (Southwestern):** crawling stems and leaves boiled into beverage tea.
    Material: **Hupa:** stems were sometimes tied up with the hair to impart their perfume. **Karok:** leaves put into hats and clothes as perfume; the vine is sometimes hung around the neck for the same reason.
    Medicine: **Luiseño:** a tea is used as a medicine. **Maidu:** tea is taken to relieve colic or "to purify the blood." **Pomo (Southwestern):** used as a tea to purify the blood; was boiled and drunk for upset stomach or when a person was getting thin. **Coast Yuki:** green or dried leaves steeped to make tea drink by the sick. **Cahuilla:** plants parts boiled, infusion for reducing fever and curing

colds; quantity of plant bound around head as headache remedy. **Costanoan:** strong decoction held in mouth for toothache, used for pinworms, used for fever, used for menstruation; leaves, warmed, wrapped on jaw to remedy toothache; used for gas. **Esselen:** used medicinally for a number of purposes.

Citation(s): Chesnut 1902; Gifford 1939, 1967; Goddard 1903; Schenck and Gifford 1952; Sparkman 1908; Goodrich et al 1980; Bean and Saubel 1972; Rayburn 2012, Breschini and Haversat 2004.

Terminology: **Luiseño:** huvaumal. **Pomo (Southwestern):** mishekale, "to smell" [refers to the odor of the plant]. **Yuki:** milmaktam. **Karok:** champinnishhich, "meadow plant."

Citation(s): Sparkman 1908; Gifford 1939, 1967; Schenck and Gifford 1952.

353 *Clintonia andrewsiana*
Family: *Liliaceae*

Usage: Misc: **Pomo (Southwestern):** considered plant as poisonous.

Citation(s): Gifford 1967.

Terminology: **Pomo (Southwestern):** silom.

Citation(s): Gifford 1967.

Note(s): plant reputed to be diuretic, demulcent, mild tonic.

Citation(s): Stuhr 1933.

354 *Clintonia uniflora*

> Common Name(s): Bride's Bonnet.

> Usage: Medicine: **Maidu:** roasted, or mashed leaves, poultice used for sore eyes, cuts.
> Citation(s): Strike 1994.

355 *Cneoridium dumosum*

> Family: *Rutaceae*
> Common Name(s): Bushrue.

> Usage: Medicine: **Luiseño:** used for medicine. **Diegueño:** decoction used to ease toothache, as a mouthwash, or gargle.
> Citation(s): Sparkman 1908; Strike 1994.
> Terminology: **Luiseño:** navish.
> Citation(s): Sparkman 1908.

356 *Coleogyne ramosissima*

> Family: *Rosaceae*
> Common Name(s): Blackbrush, Blackbush.

> Usage: Misc: **Kawaiisu (Tehachapi):** named but not used.
> Citation(s): Zigmond 1981.
> Terminology: **Kawaiisu (Tehachapi):** tociyavɨ (toci) "head."
> Citation(s): Zigmond 1981.

357 *Collinsia heterophylla*

> Family: *Plantaginaceae*
> Common Name(s): Chinese Houses, Innocence.

> Usage: Medicine: **Maidu:** leaves used to poultice insect or snake bites.
> Citation(s): Strike 1994.

358 *Collinsia tinctoria*

Usage: Misc: **Miwok:** named but not used.
Citation(s): Barrett and Gifford 1933.
Terminology: **Miwok (Central Sierra):** tcata lawati,
    "rattlesnake's rattle," istamü.
Citation(s): Barrett and Gifford 1933.

359 *Collinsia torreyi*

Usage: Misc: **Miwok:** named but not used.
Citation(s): Barrett and Gifford 1933.
Terminology: **Miwok (Central Sierra):** osta.
Citation(s): Barrett and Gifford 1933.

360 *Collomia grandiflora*
Family: *Polemoniaceae*
Synonym(s): *Gilia grandiflora*

Usage: Misc: **Yuki, Miwok:** named but not used.
Citation(s): Barrett and Gifford 9133; Chesnut 1902.
Terminology: **Miwok (Central Sierra):** witcima.
Citation(s): Barrett and Gifford 1933.

361 *Comandra umbellata* ssp. *californica*
Family: *Comandraceae*
Synonym(s): *Comandra pallida*
Common Name(s): Bastard, Sandalwood, Toad Flax.

Usage: Food: **Maidu:** fruit eaten, if too many, nausea
    would result.
    Medicine: **Maidu:** root preparation used to sooth
    sore, inflamed eyes.
Citation(s): Strike 1994.

362　*Convolvulus spp.*

Family: *Convolvulaceae*
Common Name(s): Morning-Glory,
Bindweed.

Usage:　Food:　　**Miwok (Coast):** seeds used in pinole.
Citation(s): Strike 1994.

363　*Convolvulus arvensis*

Common Name(s): Morning-Glory,
Bindweed.

Usage:　Medicine:　**Pomo (Southwestern):** stem with leaves
boiled, tea drunk to stop excessive
menstruation.
Citation(s): Goodrich et al.
Terminology: **Pomo (Southwestern):** daʈˊiʈím qʰale,
"tangled plant."
Citation(s): Goodrich et al 1980.

364　*Coprinus comatus*

Family: *Agaricaceae*
Common Name(s): Shaggy Mane, Shaggy Ink Cap.

Usage:　Food:　　**Cahuilla, Miwok:** eaten.
Citation(s): Strike 1994.

365　*Corallorhiza maculata*

Family: *Orchidaceae*
Common Name(s): Coral Root, Spotted Coral Root.

Usage:　Medicine:　**Maidu:** used to reduce fevers, or as a
sedative.
Citation(s): Strike 1994.

366 *Cordylanthus* spp.

Family: *Orobanchaceae*
Synonym(s): *Adenostegia* spp.
Common Name(s): Bird's-beak.

Usage: Medicine: **Luiseño:** as an emetic.
Citation(s): Sparkman 1908.
Terminology: **Luiseño:** yumayut.
Citation(s): Sparkman 1908.

367 *Corethrogyne filaginifolia*

Family: *Asteraceae*
Common Name(s): Cudweed, Sand Aster, California Aster.

Usage: Medicine: **Kawaiisu (Tehachapi):** twigs and leaves are placed in water which is heated with hot rocks; head is covered, vapor inhaled for relief of colds; procedure induces sweating. **Diegueño:** drank infusion of boiled flowers to relieve chest pains.
Citation(s): Zigmond 1981; Strike 1994.

368 *Cornus* spp.

Family: *Cornaceae*
Common Name(s): Dogwood.

Usage: Material: **Yuki:** made bows, arrows from shoots.
Citation(s): Foster 1944.

369   *Cornus glabrata*
    Common Name(s): Creek Dogwood.

Usage: Material: **Achomawi:** long shoots used for baskets. **Yuki:** used in making baskets. **Chumash:** made arrows, bows.
Citation(s): Curtin 1957; Merriam 1967; Timbrook 2007.
Terminology: **Yuki:** chá-me.
Citation(s): Curtis 1957.

370   *Cornus nuttallii*
    Common Name(s): Mountain Dogwood.

Usage: Material: **Pomo (Southwestern):** long slender branches used in making baby baskets. **Gabrielino:** made throwing sticks for rrabbits, other small game.
         Medicine: **Maidu:** bark, or root, decoction used as tonic, laxative, emetic, cathartic.
         Misc: **Karok:** good luck charm used by men only.
Citation(s): Schenck and Gifford 1952; Goodrich et al 1980; Strike 1994.
Terminology: **Karok:** oya'amma. **Pomo (Southwestern):** hayu g^{h}ale, "dog tree."
Citation(s): Schenck and Gifford 1952; Goordich et al 1980.

371   *Cornus occidentalis* ssp. *occidentalis*
    Common Name(s): Creek Dogwood.

Usage: Material: **Karok:** arrows tipped with foreshafts of Serviceberry (*Amelanchier spp.*) Wood instead of flint or obsidian.
         Medicine: **Maidu:** bark, or root, decoction used as

tonic, laxative, emetic, cathartic, by women after childbirth.
Citation(s): Strike 1994.

372 *Cornus sericea* ssp. *californica*
Synonym(s): *Cornus californica, Cornus x californica*
Common Name(s): Creek Dogwood.

Usage: Material: **Karok:** branches used for arrows, with tip of *Amelanchier pallida*.

Medicine: **Costanoan:** inner bark, used for fever remedy.
Citation(s): Schenck and Gifford 1952; Rayburn 2012.
Terminology: **Karok:** furah'puum.
Citation(s): Schenck and Gifford 1952.

373 *Cornus sericea* ssp. *occidentalis*
Synonym(s): *Cornus stolonifera*
Common Name(s): Red Osier Dogwood, Creek Dogwood.

Usage: Material: **Karok, Maidu:** branches used as thick cordage, or twisted together as rope. **Chumash:** made fishing poles, canoe ribs, bows, arrows, cradles, hoop for hoop-and-pole game, parts of baby cradles

Medicine: **Maidu:** bark, or root, decoction used as tonic, laxative, emetic, cathartic.
Citation(s): Strike 1994; Timbrook 2007.
Terminology: **Chumash (Barbareño):** wiliq'ap; **(Ynezeño):** wiqap; **(Ventureño):** wiliqap.
Citation(s): Timbrook 2007.

374 *Corylus cornuta* var. *californica*

Family: *Betulaceae*
Synonym(s): *Corylus californica, Corylus rostrata, Corylus rostrata californica.*
Common Name(s): Hazel-Nut, California Hazel, Hazel, Hazel-Brush.

Usage: Food:

Yuki: nuts gathered in Autumn; sun-dried, and roasted and used as occasion demanded. **Yokuts (Northern Hill):** gathered in August at same time as Sugar Pine (*Pinus lambertiana*) nuts. **Shasta, Maidu, Southern (Nisenan):** nuts gathered in quantities, **(Northern):** nuts used. **Wintun:** gathered nuts in the hills during July and August, hulled at leisure by hand. **Sinkyone:** nuts husked, broken, cooked into lumpy mush, or dried and stored. **Wappo, Wiyot:** nuts eaten. **Karok:** nuts are eaten in season; gathered and stored. **Miwok:** nuts used to a limited extent. **Pomo (Southwestern):** nuts eaten fresh, dried and stored for winter.

Material:

Bear River: made loose basket warps of fiber. **Wappo:** made arrow shafts from wood. **Hupa:** the shoots were used as foundation for baskets. **Tolowa, Hupa, Wiyot:** used stem as warp, woof, and dyed as black pattern in baskets. **Lassik:** used stem as warp, woof, and rim hoop in baskets. **Wailaki:** used stem as warp, woof, rim hoop, and dyed black as pattern in baskets. **Yana:** used stem as warp, woof, and foundation in baskets. **Miwok:** used stem as warp and foundation in baskets. **Whilkut (Chilula), Sinkyone, Achomawi, Atsugewi, Pomo, Yuki:** used stem as warp and woof in baskets. **Chimariko, Karok, Yokuts:** used stem as warp in baskets. **Yurok:** used stem dyed black as pattern in baskets; vertical rods in baskets. **Shasta:** used for ribs in basketry. **Karok:** withes are twisted to make rope; poles are used on the fish-trigger or set net; bent wood is used as heavy part of the frame for snowshoes. **Karok, Yurok:** gathered sticks in the Spring, April or May at the latest; pelled, cured in the sun, graded as to sizes; used in basketry. **Yuki:** slender twigs used to make fish traps. **Maidu (Northern):** shoots used for the radial elements in burden baskets. **Pomo (Southwestern):** switches used in basketry; straight branches used for arrows.

Misc: **Pomo (Southwestern):** named but not used (?).

Citation(s): Barrett and Gifford 1933; Beals 1933; Curtin 1957; Dixon 1905, 1907; Chesnut 1902; Driver 1936; Du Bois 1935; Foster 1944; Gayton 1948; Gifford 1967; Goddard 1903; Holt 1946; Merriam 1967; Merrill 1923; Nomland 1935, 1938; Loud 1918; O'Neale 1932; Schenck and Gifford 1952; Goodrich et al 1980.

Terminology: **Miwok (Northern Sierra):** so'lōkŌ; so'llogū, **(Central Sierra):** mūla', so'lokū, lī'ma, **(Southern Sierra):** mü'la. **Yurok:** hali L (stick). **Wiyot:** logoLès-wèl. **Hupa:** mûkaikitLoi, "on it one makes a basket." **Yuki**: Ōlmäm. **Pomo (Southwestern)**: dish, "hazel nut." **Karok**: assis, assis huntapan (nuts), sar'ip (sticks, prepared for baskets).

Citation(s): Barrett and Gifford 1933; O'Neale 1932; Loud 1918; Goddard 1903; Curetin 1957; Gifford 1967; Schenck and Gifford 1952; Goodrich et al 1980.

Note(s): note the differences in the usage given for **Pomo (Southwestern)**; perhaps this is a temporal problem or one with different informants (?) [author].

375 *Cotula coronopifolia*
Family: *Ateraceae*

Usage: Misc: **Tubatulabal**: considered weeds.
Citation(s): Voegelin 1938.
Note(s): naturalized from South Africa.
Citation(s): Munz and Keck 1965.

376 *Crataegus douglasii*
Family: *Rosaceae*
Synonym(s): *Crataegus rivularis*
Common Name(s): Black Haw.

Usage: Food: **Northern Paiute (Paviotso)**: the berry was eaten fresh, or dried.

Material: **Yuki**: wood used sparingly for fuel; spines considered poisonous.

Medicine: **Maidu**: bark and root decoctions used to relieve stomach ache; spines used to open boils although spines were considered poisonous.

Citation(s): Chesnut 1902; Kelly 1932; Strike 1994.
Terminology: **Northern Paiute (Paviotso)**: kwinü´pc.
Citation(s): Kelly 1932.
Note(S): the berries ripen and fall in August and September.

Food values – berries:

	(wet)	(dry)
Moisture	6.9%	
Sugar		
Reducing	3.2%	3.4%
Non-reducing	0.0%	0.0%
Hemicellulose	14.3%	15.4%
Protein	0.3%	
Ash	3.2%	

Citation(s): Sargent 1922; Yanovsky 1938.

377 *Crepis acuminata*
Family: *Asteraceae*

Usage: Food: **Karok:** the stems are peeled and eaten as greens.
Citation(s): Schenck and Gifford 1952.
Terminology: **Karok:** axarashpuuf.
Citation(s): Schenck and Gifford 1952.

378 *Crepis occidentalis*

Usage: Food: **Northern Paiute (Paviotso):** the leaves were eaten raw; the stem and root were not used.
Citation(s): Kelly 1932.

379 *Croton californicus*
Family: *Euphorbiaceae*
Common Name(s): Croton.

Usage: Material: **Chumash:** burned to obtain ashes as lye in soap making
Medicine: **Diegueño:** stem and leaf tea said to

produce abortion; plant boiled, infusion used to wash sore eyes. **Luiseño:** stem and leave tea said to produce abortion. **Cahuilla:** mashed and cooked stems and leaves; decoction placed in ear for earache; thimbleful of decoction taken very hot to relieve congestion; used only in small dosages as plant is toxic. **Chumash:** tea, drunk hot at bedtime for colds.

Citation(s): Sparkman 1908; Bean And Saubel 1972; Timbrook 2007.

Terminology: **Luiseño:** shuikawut. **Cahuilla:** te'ayal. **Chumash (Barbareño):** 'i'iaq', 'ilaq; **(Ventureño):** mo', smakhna'atl.

Citation(s): Sparkman 1908; Bean and Saubel 1972; Timbrook 2007.

Note(s): plant has purgative properties; sometimes listed as poisonous.

Citation(s): Stuhr 1933; Kingsbury1964.

380 *Croton setiger*

Family: *Euphorbiaceae*
Synonym(s): *Eremocarpus setigerus*
Common Name(s): Little Mullen, Turkey Mullen.

Usage: Material: **Pomo, Maidu:** bruised leaves used to poison fish. **Wintun, Yokuts, Monache (Western Mono):** used plant to poison fish. **Pomo (Southwestern):** knew of its use as fish poison but did not use it as such. **Costanoan:** roots, pounded, thrown in freshwater, dammed streams, to stupify fish.

Medicine: **Maidu:** fresh leaves are bruised and

applied to chest as a counter-irritant poultice for internal pain; a decoction of the leaves taken internally to cure chills and fever; decoction of the leaves put in warm water for typhoid and other fevers. **Maidu, Southern (Nisenan):** ague is believed to be cured by decoction of the plant. **Maidu, Northwestern (Konkow):** fresh leaves used as a counter-irritant for internal pains, as a wash to relieve fevers. **Achomawi:** when picked at the right time (about August 23 or 34) and dried for a years, it takes on great power and is the best medicine for dropsy, treating edema. **Pomo (Southwestern)** plant mashed, boiled, and the solution used for treating bleeding diarrhea, chills, fevers. **Kawaiisu (Tehachapi):** effective remedy for headache; various parts, but not root, boiled; decoction used as a head wash to relieve diarrhea, chills, fevers. **Coastanoan:** decoction used for dysentery. **Bear River:** decoction used to treat chills, fevers.

Misc: **Tubatulabal:** considered weeds. **Karok:** named but not used.

Citation(s): Chesnut 1902; Gifford 1967; Heizer 1953; Merriam 1967; Powers 1877; Schenck and Gifford 1952; Voegelin 1938; Zigmond 1981; Goodrich et al 1980; Strike 1994; Rayburn 2012.

Terminology: **Pomo (Southwestern):** ashapashi, "fish" (asha) "poison" (pashi). **Karok:** isyarukpihriv munevxat, "Across-Water Widower's stinking armpit." [Across-Water Widower is a mythological character].

Citation(s): Gifford 1967; Schenck and Gifford 1952; Goodrich et al 1980.

Note(s): the plant is carminative, febrifuge, poisonous, counter-
irritant.

Citation(s): Stuhr 1933.

381 *Cryptantha* spp.

Family: *Boraginaceae*
Synonym(s): *Cryptanthe* spp.

Usage: Food: **Chumash:** seeds may have been eaten
        Misc: **Miwok:** named but not used.
Citation(s): Barrett and Gifford 1933; Strike 1994.
Terminology: **Miwok:** susu.
Citation(s): Barrett and Gifford 1933.

382 *Cryptantha affinis*

Synonym(s): *Cryptanthe affinis*

Usage: Misc: **Miwok:** named but not used.
Citation(s): Barrett and Gifford 1933.
Terminology: **Miwok (Central Sierra):** sosolina.
Citation(s): Barrett and Gifford 1933.

383 *Cryptantha circumscissa*

Synonym(s): *Greenocharis circumscissa*

Usage: Misc: **Tubatulabal:** considered weeds.
Citation(s): Voegelin 1938.

384 *Cryptantha muricata*

Synonym(s): *Cryptantha miriculata*

Usage: Misc: **Tubatulabal:** considered weeds.
Citation(s): Voegelin 1938.

385   *Cucumis melo* var. *dudaim*
        Family: *Cucurbitaceae*
        Common Name(s): Muskmelon.

Usage: Food: **Cahuilla:** grown and eaten.
Citation(s): Bean and Saubel 1972.
Note(s): among the cultivated crops grown early in post-
        contact period; noted in 1853 by Don José Maria
        Estudillo of the Romero party.
Citation(s): Bean and Saubel 1972.

386   *Cucurbita foetidissima*
        Family: *Cucurbitaceae*
        Synonym(s): *Cucurbita perennis*
        Common Name(s): Wild Squash, Mock Orange.

Usage: Food: **Luiseño:** seeds are eaten. **Cahuilla:**
                        seeds collected in Spring, ground into
                        flour for mush.
          Material: **Luiseño:** the fruit is used when ripe as
                        a substitute for soap. **Tubatulabal:**
                        inside of the fruit used as soap.
                        Kawaiisu (Tehachapi): gourd broken
                        open and clothes to be washed are
                        rubbed against inner surface; piece of
                        gourd may be put in boiling water with
                        clothes. **Cahuilla:** root and squash were
                        cut into small fine pieces, used as hand
                        and laundry soap; used as bleach,
                        plants parts and material to be bleached
                        soaked in water for a long time; shell of
                        fruit ground up, used as hair shampoo;
                        squash and roots gathered in quantity
                        and stored until needed; yellow
                        blossoms used as dye; dried gourds

used to make ladles, occasionally as rattles. **Chumash:** root, scraped, sliced, to use as soap; gourds, made containers, drinking cups, dippers.

Medicine: **Cahuilla:** root, crushed and mixed with sugar, and applied to saddle sores; root was macerated, applied to ulcers; pulp of squash used as medication on open sores; dried root, boiled in water, used as an emetic or physic. **Chumash:** tendrils, crushed, or pounded roots, put in water, drunk as a purgative

Citation(s): Barrows 1900; Sparkman 1908; Voegelin 1938; Zigmond 1981; Bean and Saubel 1972; Timbrook 2007.

Terminology: **Cahuilla:** ne-kish, nekhish. **Chumash (Barbareño, Ynezeño):** mo'kh; **(Ventureño):** mo'pkh.

Citation(s): Barrows 1900, Bean and Saubel 1972; Timbrook 2007.

Note(s): seeds contain:
protein   33.8%
oils   33.9%.

Citation(s): Bean and Saubel 1972.

387 *Cuscuta californica*

Family: *Convolvulaceae*
Common Name(s): California Dodder, Witch's Hair.

Usage: Food: **Maidu:** seeds parched, eaten
Material: **Cahuilla:** handfuls of the plant used as scouring pads for cleaning. **Maidu:** obtained a orange/yellow dye using plant.

Medicine: **Kawaiisu (Tehachapi):** mass of stems chewed; resultant juice snuffed into

nostrils for nose bleed; dried powder
used the same way.

Misc: **Tubatulabal:** considered weeds.

Citation(s): Voegelin 1938; Zigmond 1981; Bean and Saubel;
Strike 1994.

Terminology: **Kawaiisu (Tehachapi):** Čigɨpiiwanavɨ, "lizard's
net." **Cahuilla:** wikat.

Citation(s): Zigmond 1981; Bean and Saubel 1972.

388 *Cylindropuntia acanthocarpa*

Synonym(s): *Opuntia acanthocarpa*
Common Name(s): Buckthorn Cholla.

Usage: Food: **Cahuilla:** fruit gathered in Spring,
eaten fresh, or dried for storage.

Medicine: **Cahuilla:** ashes of the stems applied to
cuts and burn to facilitate healing.

Citation(s): Bean and Saubel 1972.

Terminology: **Cahuilla:** mutal.

Citation(s): Bean and Saubel 1972.

389 *Cylindropuntia bigelovii*

Synonym(s): *Opuntia bigelovii*
Common Name(s): Jumping Cholla, Bull Cholla.

Usage: Food: **Cahuilla:** buds available from late April
to June; prepared and preserved similar
to # 887 (below).

Citation(s): Bean and Saubel 1972.

Terminology: **Cahuilla:** chukal.

Citation(s): Bean and Saubel 1972.

390 *Cylindropuntia echinocarpa*

Synonym(s): *Opuntia echinocarpa*
Common Name(s): Silver Cholla.

Usage: Misc: **Kawaiisu (Tehachapi):** named but not used.
Citation(s): Zigmond 1981.
Terminology: **Kawaiisu (Tehachapi):** wiarɨbɨ (wiya "awl").
Citation(s): Zigmond 1981.

391 *Cylindropuntia ramosissima*

Synonym(s): *Opuntia ramosissima*
Common Name(s): Pencil Cactus.

Usage: Food: **Cahuilla:** edible fruits gathered, April and May, eaten fresh, or dried for storage; stalk boiled into soup, or dried for later use.
Citation(s): Bean and Saubel 1972.
Terminology: **Cahuilla:** wival.
Citation(s): Bean and Saubel 1972.

392 *Cynoglossum grande*

Family: *Boraginaceae*
Common Name(s): Hound's Tongue.

Usage: Food: **Yuki, Maidu:** ate cooked roots.
Medicine: **Maidu:** grated roots used to draw out the inflammation from burns and scabs, to relieve stomach aches. **Pomo:** used grated roots to relieve stomach aches.
Misc: **Pomo (Southwestern):** named but not used.
Citation(s): Chesnut 1902; Goodrich et al 1980; Strike 1994.
Terminology: **Pomo (Southwestern):** duwi šíma·ta.

Citation(s): Goodrich et al 1980.

Note(s): some obvious conflict in different ethnographic accounts for the **Pomo**. [Author]

393 *Cyperus* spp.
Family: *Cyperaceae*

Usage: Material: **Miwok:** leaves used as a seat outdoors as well as indoors.
Citation(s): Barrett and Gifford 1933.
Terminology: **Miwok (Central Sierra):** kisti.
Citation(s): Barrett and Gifford 1933.

394 *Cyperus erythrorhizos*

Usage: Food: **Kamia:** the fine seeds collected in a pot, pulverized, and cooked as mush.
Citation(s): Gifford 1931.

395 *Cyperus esculentus*
Common Name(s): Nut Grass.

Usage: Food: **Pomo (Southwestern):** tubers eaten raw, or baked, or boiled. **Owens Valley Paiute:** seeds eaten. **Chumash:** tubers, charred, found in two pit-ovens in archaeological excavations.
Citation(s): Goodrich et al 1980; Strike 1994; Timbrook 2007.
Terminology: **Pomo (Southwestern):** ?ac^{h}im?.
Citation(s): Goodrich et al 1980.
Note(s):   seeds contain:

Ash	2.4%
Protein	6.1%
Oil	27.4%

Citation(s): Earle and Jones 1962.

396 *Cyperus rotundus*

        Synonym(s): *Cyperus virens*

Usage: Material: **Miwok:** leaves bundled together to form a container for carrying fishing worms.

Citation(s): Strike 1994.

Note(s): native to Eurasia.

Citation(s): Baldwin et al 2012.

397 *Dalea* spp.

        Family: *Fabaceae*

        Synonym(s): *Parosela spp.*

Usage: Food: **Owens Valley Paiute:** used as food.

Citation(s): Steward 1933.

Terminology: **Owens Valley Paiute:** cututsi[va].

Citation(s): Steward 1933.

398 *Darmera peltata*

        Family: *Saxifragaceae*

        Synonym(s): *Peltiphyllum peltatum*

Usage: Food: **Karok:** the young shoots are eaten raw as greens. **Miwok:** pulverized root mixed with acorn meal to whiten it. **Maidu:** root eaten.

        Medicine: **Karok:** roots cut up and soaked in water; a pregnant woman drinks the infusion so that the baby will not be too large.

Citation(s): Barrett and Gifford 1933; Schenck and Gifford 1952; Strike 1994.

Terminology: **Karok:** kaaf. **Miwok (Central Sierra):** senseteko.

Citation(s): Schenck and Gifford 1952; Barrett and Gifford 1933.

399 *Datisca glomerata*

Family: *Datiscaceae*
Common Name(s): Durango Root.

Usage: Material: **Pomo:** leaves and roots pulverized and then thrown into streams as a means of taking trout. **Karok:** used a dye for basketry materials. **Chumash:** root, used for yellow dye for basketry.

Medicine: **Miwok:** root was pulverized and a decoction made as a wash for sores and rheumatism. **Costanoan:** made decoction to relieve sore throats, swollen tonsils. **Chumash:** root used for sore throat.

Misc: **Tubatulabal:** considered weeds.

Citation(s): Barrett 1952; Barrett and Gifford 1933; Chesnut 1902; Schenck and Gifford 1952; Voegelin 1938; Strike 1994; Rayburn 2012; Timbrook 2007.

Terminology: **Miwok (Central Sierra): insiňotayi.**
Karok: ihyivkanva, "shout across." **Chumash (Barbareño):** 'ansiwa'wu'y; **(Ventureño):** 'aluqchahay 'isakhpilil.

Citation(s): Barrett and Gifford 1933; Schenck and Gifford 1952; Timbrook 2007.

Note(s): leaves and stems; bitter tonic; seeds, leaves, seed capsules are highly toxic to sheep and cattle.

Citation(s): Stuhr 1933; Kingsbury 1964.

400 *Datura wrightii*

Family: *Solanaceae*
Synonym(s): *Datura metaloides*
Common Name(s): Jimson Weed; Thorn Apple; Toloache, Tolguacha.

Usage: Medicine: **Yokuts:** the roots are pounded up, "good for anything;" good for a cut, a gunshot wound, a bruise, etc.; **(Southern Valley):** used for drinking ritual; **(Northern Hills):** often taken by shaman. **Luiseño, Tubatulabal:** juice of the roots used at boy's puberty ceremony to induce stupefaction of the novices. **Cahuilla:** whole plant pounded up, mixed with water and drunk in small quantities to produce stimulus and eventual unconsciousness; roots most commonly used in drink served at rituals; used in puberty ceremonies for boys; leaves were generally smoked; leaves and roots crushed with other parts, mixed into a medicinal paste; recognized as a pain-killer in setting bones, alleviating pain in specific area of the body, relieving toothaches and swellings; salve used as a hot poultice; paste used to cure bites of tarantulas, snakes, spiders, and various insects, alleviate saddle sores on horses; leaves steamed, vapor inhaled for severe bronchial or nasal congestion; hunters used plant on long treks to increaser their strength, allay hunger, and acquire power to capture game. **Monache (Western Mono):** drunk ritually by both men and women; often used as poultice for severe wounds or fractures; also drunk as an anaesthetic. **Miwok:** shamans ate the root or drank a decoction to induce a delirium, during which they ran about wildly and saw strange visions; gave supernatural power and the ability to look into the future. **Chumash:** in the past probably most important medicinal plant; used to establish contact with supernatural guardians, contacting the dead, curing the effects of injury; leaves or roots taken internally as a tea and/or applied as a poultice for broken bones and wounds; taken as a painkiller for serious injuries, broken bones, for ridding body of tapeworms; leaves, crushed, used as poultice for hemorrhoids, put on skin over any aching place; root, roasted, used as poultice for wounds; used for vision-inducing properties. **Costanoan:** leaves, heated, used in compresses for chest pain, respiratory problems; leaves, ground into salve, used to treat boils; leaves, dried, smoked, used as purgative, as a hallucinogen (visions were believed to reflect one's future); seeds, when mixed with tobacco, smoked, used as aphrodisiac; dew, collected from inside flowers, used as eyewash. **Kawaiisu (Tehachapi):** served medicinal and hallucinatory purposes; mashed or shredded root was soaked and rheumatic, arthritic limbs bathed in infusion; a year or two after attaining puberty, boys and girls underwent a Datura Ceremony; limited to the Winter season

Citation(s): Barrett and Gifford 1933; Barrows 1900; Bean and Saubel 1972; Curtis 1924; Gayton 1948; Powers 1877;

Sparkman 1908; Timbrook 1987, 2007; Rayburn 2012; Zigmond 1981.

Terminology: **Yokuts (Northern Hill):** t'anai, ta´ñai; **(Southern Valley):** ta´nai, ma´nai. **Luiseño:** naktomush. **Owens Valley Paiute:** taɳ aniva. **Miwok (Central Sierra):** monayu; **(Southern Sierra):** mō´nūya. **Cahuilla:** ki-ki-sow-il, kiksawva'al. **Chumash (Barbareño, Island):** mo'moy; **(Yenezeño, Ventureño):** momoy.

Citation(s): Gayton 1948; Sparkman 1908; Steward 1933; Barrett and Gifford 1933; Barrows 1900; Bean and Saubel 1972; Timbrook 2007.

Note(s): plant contains alkaloids in its leaves, roots, stems: hyoscyamine, atropine, hyoscine; causes headache, nausea, vertigo; extreme thirst, dry burning sensation in the skin, dilated pupils, loss of sight and voluntary motion; symptoms are thirst, pupil dilation, dry mouth, redness of the skin, headache, hallucinations, nausea, rapid pulse, temperature elevation, high blood pressure, delirium, convulsions, coma, and in extreme cases mania, convulsions, death; plant possibly introduced from Mexico.

Citation(s): Muenscher 1951; Kingsbury 1964; Munz and Keck 1965; Hardin and Arena 1974.

401 *Daucus* spp.

Family: *Aplaceae*

Usage: Material: **Miwok:** children threw stems at each other, trying to make the seeds stick in each other's hair.

Medicine: **Tubatulabal:** crushed plants, boiled in a small amount of water, drunk to relieve pain of snakebite. **Diegueño:** used decoction to treat fevers, toothaches.

Citation(s): Strike 1994.

402 *Daucus carota*

Common Name(s): Wild Carrot.

Usage: Food: **Huchnom:** ate roots.
Citation(s): Foster 1944.
Note(s): a native of Europe, occasionally naturalized.
Citation(s): Munz and Keck 1965.

403 *Daucus pusillus*

Common Name(s): Rattlesnake Weed.

Usage: Medicine: **Miwok:** plant was chewed and placed on snake bite, mashed leaves applied to snakebite to prevent swelling.; decoction drunk to purify blood. **Costanoan:** decoction taken internally to clean blood, in early stages of colds, for itching, fever. **Chumash:** tea, drunk for sore throat; leaves, dried, smoked for paralysis caused by water on the brain; used in snake charming.

Misc: **Wailaki:** used as a talisman in gambling.

Citation(s): Barrett and Gifford 1933; Chesnut 1902; Strike 1994; Rayburn 2012; Timbrook 2007.

Terminology: **Miwok (Central Sierra):** yotcitayu. **Chumash (Barbareño):** s'akhiyɨp 'ikhshap; **(Ynezeño):** shiyamsh 'ikhshap; **(Ventureño):** ch'atɨshwɨ 'ikhshap.

Citation(s): Barret and Gifford 1933; Timbrook 2007.

404 *Deinandra fasciculata*
>   Family: *Asteraceae*
>   Synonym(s): *Hemizonia fasciculata, Hemizonia ramosissima*
>   Common Name(s): Tarweed.

Usage: Food: **Cahuilla:** the entire plant, including the seeds, was used as a famine plant; boiled down into a thick tarry consistency and eaten. **Chumash:** seeds, dried, rubbed between hands, winnowed the chaff, pounded in mortar, mixed with water, made balls, eaten raw; sometimes seeds were roasted first.

Medicine: **Diegueño:** boiled, steam inhaled to relieve headaches.

Citation(s): Bean and Saubel 1972; Timbrook 2007; Strike 1994.
Terminology: **Chumash (Barbareño, Ynezeño, Purisimeño, Ventureño):** swey.
Citation(s): Timbrook 2007.

405 *Delphinium* spp.
>   Family: *Ranunculaceae*

Usage: Food **Miwok:** boiled in March when young.
Citation(s): Barrett and Gifford 1933; Strike 1994.
Terminology: **Miwok (Central Sierra):** witilima.
Citation(s): Barrett and Gifford 1933.
Note(s): contains alkaloids: delphinine, delphinoidine, delphisine, staphisagroine; causes loss of appetite, general uneasiness, and staggering gait, constipated and nauseated; generally most toxic in its youngest growth.
Citation(s): Nuenscher 1951; Kingsbury 1964.

406 *Delphinium decorum*

Common Name(s): Larkspur.

Usage: Material: **Karok:** the flowers are pounded in a small stone mortar, mixed with salmon glue and fresh berries of *Berberis aquifolium*, and used to paint arrows and bows.

Medicine: **Miwok (Coast):** used to prepare a strong emetic to cure stomach aches.

Citation(s) Schenck and Gifford 1952.

Terminology: **Karok:** kuniharekxurikkar, "thing for decorating arrows."

Citation(s): Schenck and Gifford 1952.

407 *Delphinium hansenii*

Usage: Misc: **Tubatulabal:** considered weeds.

Citation(s): Voegelin 1938.

408 *Delphinium hesperium*

Common Name(s): Western Larkspur.

Usage: Food: **Miwok:** leaves and flowers boiled and eaten as greens.

Misc: **Yuki:** named but not used.

Citation(s): Barrett and Gifford 1933; Chesnut 1902.

Terminology: **Miwok (Central Sierra):** kowe.

Citation(s): Barrett and Gifford 133.

409 *Delphinium nudicaule*

Usage: Misc: **Pomo:** root may be used to cause an opponent to become stupid while gambling.

Citation(s): Chesnut 1902.

410 *Delphinium parryi*

Usage: Medicine: **Kawaiisu (Tehachapi):** root, dried, ground, water added to make salve; applied to swollen limbs.

Citation(s): Zigmond 1981.
Terminology: **Kawaiisu (Tehachapi):** motoobɨ.
Citation(s): Zigmond 1981.

411 *Delphinium purpusii*

Usage: Misc: **Tubatulabal:** considered weeds.

Citation(s): Voegelin 1938.

412 *Dendroalsia abietina*
Family: *Cryphaaeceae*
Synonym(s): *Alsia abietina*
Common Name(s): Balsam-Tree Moss.

Usage: Material: **Pomo, Maidu:** used as bedding material, especially for babies.

Citation(s): Chesnut 1902; Strike 1994.

413 *Dendromecon rigida*

Family: *Papaveraceae*

Common Name(s): Bush Poppy, Chaparral Poppy, Tree Poppy.

Usage: Material: **Kawaiisu (Tehachapi):** leaves, a few, added to enhance the strength of tobacco (*Nicotiana*).

Citation(s): Strike 1994.

414 *Deschampsia cespitosa*

Family: *Poaceae*

Common Name(s): Tufted Hairgrass.

Usage: Food: **Maidu:** seeds eaten.

Citation(s): Strike 1994.

415 *Deschampsia danthonioides*

Common Name(s): Slender Hair Grass.

Usage: Food: **Kawaiisu (Tehachapi):** seeds pounded, cooked as mush.

Citation(s): Zigmond 1981.

Terminology: **Kawaiisu (Tehachapi):** pa(a)narrabɨ.

Citation(s): Zigmond 1981.

416 *Deschampsia elongata*

Synonym(s): *Aira elongata*

Usage: Food: **Karok, Maidu:** seeds eaten.

Citation(s): Schenck and Gifford 1952.

Terminology: **Karok:** ikravapuishnanich, "imitation ikravapu."

Citation(s): Schenck and Gifford 1952.

417   *Descurainia incisa* ssp. *incisa*

        Synonym(s): *Sisymbrium incisum, Descurainia richardsonii* var. *incisa*

Usage: Misc: **Tubatulabal:** considered weeds.
Citation(s): Voegelin 1938.

418   *Descurainia pinnata*

        Family: *Brassicaceae*
        Synonym(s): *Sisymbrium canescens, Sisymbrium pinnatum*

Usage: Food: **Atsugewi:** seeds gathered and prepared; pounded into flour, molded into cakes, eaten without cooking. **Cahuilla:** seed ground and cooked in large quantity of water and eaten with a little salt; leaves gathered in Spring as pot herbs. **Kawaiisu (Tehachapi):** when plant is turning brown in June, seeds collected, parched, pounded in bedrock mortar-hole, mixed with cold water; beverage considered nourishing.

      Medicine: **Cahuilla:** ground seeds used for stomach ailments. **Maidu:** tea made from seeds used as a diuretic, to ease coughs, to relieve "summer complaints."

Citation(s): Barrows 1900; Garth 1953; Zigmond 1981; Bean and Saubel 1972; Strike 1994.
Terminology: **Atsugewi:** kasaii. **Cahuilla:** ás-il, asily.
Citation(s): Garth 1953; Barrows 1900; Bean and Saubel 1972.

419 *Descurainia sophia*
Synonym(s): *Sisymbrium sophia*

Usage: Food: **Northern Paiute (Paviotso):** the seeds were eaten. **Kawaiisu (Tehachapi):** same as #418 (above).
Citation(s): Kelly 1932; Zigmond 1981.
Terminology: **Northern Paiute (Paviotso):** atsa′, "red" (seeds).
Citation(s): Kelly 1932.
Note(s): naturalized from Europe; introduced in Siskiyou and Modoc (Lutuami) counties.
Citation(s): Munz and Keck 1965; Jepson 1925.

420 *Dicentra formosa*
Common Name(s): Bleeding Heart.

Usage: Medicine: **Miwok (Coast):** leaves, dried, used to brew medication to relieve lung problems.
Citation(s): Strike 1994.
Note(s): alkaloids protopine and others, found throughout plant, cause trembling, staggering, convulsions, labored breathing. Large quantities can be fatal.
Citation(s): Hardin and Arena 1974.

421 *Dichelostemma capitatum*
Family: *Themidaceae*
Synonym(s: *Brodiaea capitata, Brodiaea pulchella*
Common Name(s): Wild Hyacinth, Grassnut, Blue Dicks.

Usage: Food: **Miwok:** bulbs steamed in earth oven and eaten. **Pomo:** bulbs eaten raw, but are far sweeter when cooked in ashes.

Pomo (Southwestern): bulbs cooks in hot ashes or boiled for food. **Yuki:** bulbs were relished for their sweetness. **Luiseño:** the bulb was eaten. **Owens Valley Paiute:** bulb used, harvested in the Fall; eaten raw when fresh; greater part dried and stored; roasted and ground into flour. **Chumash:** bulbs, dug in June after stalks died back, roasted in hot ashes of cooking fire, or, if large quantities, roasted in special earth oven.

Material: **Kawaiisu (Tehachapi):** corms crushed, used to line the insider of seed-gathering baskets to prevent small seeds slipping through.

Misc: **Karok:** Named but not used (?).

Citation(s): Barrett 1952; Barrett and Gifford 1933; Chesnut 1902; Curtin 1957; Schenck and Gifford 1952; Sparkman 1908; Steward 1933; Goodrich et al 1980; Strike 1994; Timbrook 2007.

Terminology: **Yuki:** lat. **Miwok (Central Sierra):** silüwü. **Luiseño:** tokapish. **Owens Valley Paiute:** tūpū'si'. **Karok:** tayiiθ. **Pomo (Southwestern):** hi?bú?la. **Chumash (Barbareño, Ynezeño):** shikh'ó'n; **(Ventureño):** shi'q'o.

Citation(s): Curtin 1957; Barrett and Gifford 1933; Sparkman 1908; Steward 1933; Schenck and Gifford 1952; Goodrich et al 1980; Timbrook 2007.

422 *Dichelostemma congesta*

Synonym(s): *Hookera congesta, Brodiaea congesta*
Common Name(s): Purple-Flowered Grass-Nut,
Long-Leafed Grass-Nut.

Usage: Food: **Hupa:** bulbs are roasted in ashes, or boiled in baskets. **Maidu, Southern (Nisenan):** eaten raw, or roasted in ashes, or boiled.
Citation(s): Goddard 1903; Powers 1877.

423 *Dichelostemma ida-maia*

Synonym(s): *Brodiaea ida-maia*
Common Name(s): Firecracker Plant.

Usage: Material: **Karok:** the seed pods are dried and hung up as ornaments.
Citation(s): Schenck and Gifford 1952; Strike 1994.
Terminology: **Karok:** ichyunihaa tayish, "throw-down potato."
Citation(s): Schenck and Gifford 1952.

424 *Dichelostemma multiflorum*

Synonym(s): *Brodiaea multiflora*

Usage: Food: **Hupa:** bulbs are roasted in ashes, or, boiled in baskets. **Atsugewi:** cook bulbs in earth oven all night, mashed and made into cakes, cooked again in pit, dried and stored; soaked for later use.
Citation(s): Garth 1953; Goddard 1903.
Terminology: **Atsugewi:** japwi.
Citation(s): Garth 1953.

425 *Dichelostemma volubile*

Synonym(s): *Hookera volubilis, Brodiaea volubilis*
Common Name(s): Climbing Grass-Nut.

Usage: Food: **Pomo:** bulbs eaten. **Maidu, Southern (Nisenan):** bulbs eaten raw, or roasted in ashes, or boiled.
Misc: **Miwok:** named but not used.
Citation(s): Barrett 1952; Barrett and Gifford 1933; Powers 1877.
Terminology: **Miwok (Central Sierra):** wakilwakilu.
Pomo (Central): hō'bū.
Citation(s): Barrett and Gifford 1933; Barrett 1952.

426 *Dirca occidentalis*

Family: *Theophrastaceae*
Common Name(s): Leatherwood, Wild Plum, Western Leatherwood.

Usage: Food: **Miwok (Coast):** fruits eaten.
Citation(S): Strike 1994.

427 *Distichlis spicata*

Family: *Poaceae*
Synonym(s): *Distichlis spicata* var. *divaricata*
Common Name(s): Salt Grass.

Usage: Food: **Tubatulabal:** rolled leaves and stems in hands with clover, then ate the clover; swished peeled willow (*Salix*) sticks the standing grass, later drew stick through the mouth to flavor food..
Cahuilla: plant was cut and beaten to detach salt; salt obtained from ashes after plant burned.
Material: **Cahuilla:** stiffness of plant made

excellent brushing material for cleaning implements or removing cactus thorns from objects.

Medicine: **Kawaiisu (Tehachapi):** dried seeds, mixed with sugar and water made into dried cakes; piece dissolved in water, drunk "when the heart beats fast" and as a laxative; sometimes the green plant immersed in cold water, strained and drunk as a laxative, or to relive heart palpitation, used to treat skin and venereal problems. **Diegueño:** boiled grass used as a mouth rinse to relieve soreness. **Yokuts:** heated until salt fell off, mixed with water to cure colds, revive poor appetites.

Citation(s): Voegelin 1938; Zigmond 1981; Bean and Saubel 1972; Strike 1994

Terminology: **Kawaiisu (Tehachapi):** yɨsuvibɨ. **Cahuilla:** simut(?). **Chumash (Ynezeño):** lit'on; **(Ventureño):** saha.

Citation(s): Zigmond 1981; Bean and Saubel 1972; Timbrook 2007.

428 *Dodecatheon clevelandii*

Family: *Primulaceae*
Common Name(s): Shooting Star.

Terminology: **Chumash (Barbareño):** stɨq' 'iwaq'aq'.
Citation(s): Timbrook 2007.

429 *Dodecatheon hendersoni*

Common Name(s): Shooting Star, Mosquito Bill.

Usage: Food: **Yuki:** roots and leaves roasted in ashes

and eaten.

Material: **Yuki:** flowers used by women to ornament themselves at dances. **Pomo (Southwestern):** flowers hung on baby baskets to make baby sleepy.

Medicine: **Maidu:** poultice used on swollen knees.

Citation(s): Chesnut 1902; Goodrich et al 1980; Strike 1994.

Terminology: **Pomo (Southwestern):** biŝiŝi.

Citation(s): Goodrich et al 1980.

430   *Draperia systyla*

Family: *Boraginaceae*

Usage: Misc: **Karok:** named but not used.

Citation(s): Schenck and Gifford 1952.

Terminology: **Karok:** pirish'axvaaharas, "Waxy-leaves," ahanatshinich, "gooseberry shine."

Citation(s): Schenck and Gifford 1952.

431   *Drosera rotundifolia*

Family: *Droseraceae*

Common Name(s): Dew Plant, Sundew.

Usage: Medicine: **Maidu:** sap and leaves used on warts; used as a diuretic, cure for bronchial problems.

Citation(s): Strike 1994.

432   *Drymocallis glandulosa*

Family: Rosaceae

Synonym(s): *Potentilla glandulosa*

Common Name(s): Sticky Cinquefoil.

Usage: Food: **Miwok (Coast):** beverage made. **Chumash:** root, boiled, tea drunk for fever, colds, good for the stomach; root, ground, or mashed, boiled, taken for as blood purifier, stimulant, Spanish flu.

Medicine: **Maidu:** used as tonic, or to relieve colic.

Misc: **Miwok:** named but not used.

Citation(s): Barrett and Gifford 1933; Strike 1994; Timbrook 2007.

Terminology: **Miwok (Central Sierra):** yatcatca. **Chumash (Barbareño):** chiqwi 'ikhakha'kh.

Citation(s): Barrett and Gifford 1933; Timbrook 2007.

Note(s): note different statements by the authors, perhaps different informants cited in the literature over time (?) [author]

433 *Dryopteris arguta*

Family: *Dryopteridaceae*

Synonym(s): *Aspidium rigidum arguta*

Common Name(s): California Wood Fern.

Usage: Medicine: **Maidu:** used to treat asthma, depression, to relieve pain; pulverized roots used to treat cuts. **Miwok:** root tea used to stop internal bleeding, vomiting. **Salinan:** root tea used to treat internal injuries. **Costanoan:** frond infusion used as hair wash.

Misc: **Karok:** named but not used.

Citation(s): Schenck and Gifford 1952; Strike 1994; Rayburn 2012.

Terminology: **Karok:** assak vaatzarakavruuk, "shouting down from a rock."

Citation(s): Schenck and Gifford 1952.

434 *Dudleya* spp.

Family: *Crassulaceae*
Common Name(s): Hen-and-Chickens.

Usage: Food: **Luiseño:** the juice of the leaves is used.
Cahuilla: fleshy leaves eaten raw; flowering stems are delicious and produce a lingering sweet taste.
Citation(s): Sparkman 1908; Bean and Saubel 1972.
Terminology: **Luiseño:** topnal. **Cahuilla:** ya'ish.
Citation(s): Sparkman 1908; Bean and Saubel 1972.

435 *Dudleya abramsii* ssp. *Setchelli*

Synonym(s): *Cotyledon laxa, Dudleya cymosa* ssp. *Setchelli*

Usage: Misc: **Tubatulabal:** considered weeds.
Citation(s): Voegelin 1938.

436 *Dudleya cymosa*

Common Name(s): Lax Dudleya.

Usage: Medicine: **Maidu:** leaf poultice used to hasten healing of sores, wounds; used as a diuretic.
Citation(s): Strike 1994.

437 *Dudlyea lanceolata*

Synonym(s): *Echerveris lanceolata*
Common Name(s): Rock-Lettuce.

Usage: Food: **Maidu, Southern (Nisenan):** eaten raw.
Citation(s): Powers 1877.

438   *Dudleya pulverulenta*
Common Name(s): Chalk Lettuce, Deer's Tongue.

Usage: Medicine: **Diegueño:** root boiled, decoction drunk as cure for breathing problems; heated leaves used as poultice, and to remove calluses.
Citation(s): Strike 1994.

439   *Dudleya saxosa*
Common Name(s): Live-forever.

Usage: Food: **Cahuilla:** leaves eaten raw.
Citation(s): Strike 1994.

440   *Dysphania ambrosioides*
Family: *Chenopodiaceae*
Synonym(s): *Chenopodium ambrosioides*
Common Name(s): Mexican Tea.

Usage: Medicine: **Miwok:** plant, either raw or boiled, was applied as a poultice to reduce swelling; used in mouth for toothache, or an ulcerated tooth; used as a wash for rheumatic parts.
Citation(s): Barrett and Gifford 1933.
Terminology: **Miwok (Central Sierra):** tistisu.
Citation(s): Barrett and Gifford 1933.
Note(s): weed in waste, especially damp, places; naturalized from tropical America; oil of *Chenopodium* used as an anthelmintic; overdoses of oil can poison; contains volatile oil of wormseed; in excess, symptoms are those of a narcotic, acid poison affecting the brain, spinal cord, and stomach; the plant is tonic, nervine, emmenagogue.

Citation(s): Munz and Keck 1965; Muenscher 1951; Pammel 1911; Trease and Evans 1966; Stuhr 1933.

441 *Dysphania pumilio*

Synonym(s): *Chenopodium carinatum, Chenopodium pumilio*

Usage: Food: **Atsugewi:** seeds gathered and prepared; pounded into flour, molded into cakes, and eaten without cooking.
Citation(s): Garth 1953.
Terminology: **Atsugewi:** pEru'we.
Citation(s): Garth 1953.
Note(s): occasional weed, mostly below 5,000 feet; naturalized from Australia.
Citation(s): Munz and Keck 1965.

442 *Echinocactus* spp.

Family: *Cactaceae*

Usage: Food: **Cahuilla:** fruit eaten.
Citation(s): Curtis 1924.

443 *Echinocactus polycephalus*

Common Name(s): Devil's Pincushion.

Usage: Food: **Cahuilla:** harvested between May and June.
Material: **Panamint Shoshone (Koso):** spines used as awls in making baskets. **Kawaiisu (Tehachapi):** spines used as awls in making coiled basketry.
Citation(s): Coville 1892; Merrill 1923; Zigmond 1981; Bean and Saubel 1972.
Terminology: **Kawaiisu (Tehachapi):** kuwavibɨ. **Cahuilla:**

u'ush.
Citation(s): Zigmond 1981; Bean and Saubel 1972.

444 *Echinochloa crus-galli*

Family: *Poaceae*
Common Name(s): Water Grass, Bunch Grass, Wild
Rice.

Usage: Food: **Owens Valley Paiute:** used as food.
Tubatulabal: used seeds.
Citation(s): Steward 1933; Voegelin 1938.
Terminology: **Owens Valley Paiute:** pā′wai.
Citation(s): Steward 1933.
Note(s): introduced from the Old World.
Citation(s): Munz and Keck 1965.

445 *Ehrendorferia chrysantha*

Family: *Papaveraceae*
Synonym(s): *Dicentra chrysantha*

Usage: Medicine: **Kawaiisu (Tehachapi):** roots, dried,
pounded, heated used as poultice on
chests to relieve pain. **Maidu:** roots
used as a diuretic.
Misc: **Tubatulabal:** considered weeds.
Citation(s): Voegelin 1932; Strike 1994.

446 *Eleocharis* spp.

Family: *Cyperaceae*
Synonym(s): *Heleocharis* spp.
Common Name(s): Spike Rush.

Usage: Food: **Mono Lake Paiute:** used bulb, eaten
raw when fresh; greater part dried and
stored; roasted and ground into flour.

Citation(s): Steward 1933.
Terminology: **Mono Lake Paiute:** ná'hāvīa, mā'hāvaita.
Citation(s): Steward 1933.

447 *Elymus* spp.

> Family: *Poaceae*
> Synonym(s): *Sitanion* spp.

Usage: Food: **Pomo (Southwestern):** grain is ground into fine powder, used in pinole.

Medicine: **Modoc (Lutuami):** used to poultice bruises, swellings. **Maidu:** decoction used to bathe sore eyes. **Miwok:** used dry or green by shaman to strike a patient with, before or after sucking.

Citation(s): Goodrich et al 1980; Strike 1994; Barrett and Gifford 1933.

Terminology: **Pomo (Southwestern):** kacáya. **Miwok (Central Sierra):** pokute.

Citation(s): Goodrich et al 1980; Barrett and Gifford 1933.

448 *Elymus condensatus*

> Synonym(s): *Leymus condensatus*
> Common Name(s): Rye-Grass, Giant Rye Grass.

Usage: Food: **Northern Paiute (Paviotso):** the seeds were eaten. **Owens Valley Paiute, Mono Lake Paiute:** grain a very important source of food.

Material: **Luiseño:** The mainshafts of arrows are made from the plant. **Owens Valley Paiute, Mono Lake Paiute:** bundles of the materials were used to cover cook house in Winter. **Cahuilla:** used as house thatching; for arrows, grass stems were fire hardened to serve as main shaft. **Chumash:** sometimes used as thatching, made doors, pipes, tubes to suck water from springs, ornaments in pierced nasal septum, newly pierced ears, knives for cutting meat, fish, gutting animals, cut umbilical cords of newborn, handles for paint brushes, counters in the walnut-shell game.

Medicine: **Luiseño:** roots pulverized, mixed with water, used to cure internal ailments, as a purgative. **Chumash:** new shoots, collected in March, boiled, drunk as a cure for gonorrhea

Citation(s): Barrows 1900; Sparkman 1908; Seward 1933; Kelly 1932; Iovin 1963; Bean and Saubel 1972; Strike 1994; Timbrook 2007.

Terminology: **Northern Paiute (Paviotos):** waiya′ (seeds). **Owens Valley Pauite, Mono Lake Paiute:** wai′ya. **Luiseño:** huikish. **Cahuilla:** pahankis., pahankish. **Chumash (Barbareño):** shtemelel; **(Island, Ynezeño, Ventureño):** shakh; **(Obispeño):** tqmimu′; **(Purisimeño):** shtemele.

Citation(s): Kelly 1932; Steward 1933; Sparkman 1908; Bean and Saubel 1972; Timbrook 2007.

449 *Elymus elymoides*

Synonym(s): *Sitanion elymoides, Sitanion hystrix*
Common Name(s): Squirreltail, Foxtail.

Usage: Food: **Northern Paiute (Paviotso), Maidu:** seeds were eaten.

Misc: **Yuki:** named but not used (?).

Citation(s): Chesnut 1902; Kelly 1932; Strike 1994.

Terminology: **Northern Paiute (Paviotso):** mono′pü (seeds).

Citation(s): Kelly 1932.

450   *Elymus glaucus*
Common Name(s): Western Rye Grass.

Usage: Food: **Karok:** seeds parched and pounded into flour; flour mixed with water and eaten as paste.

        Misc: **Karok:** used as "medicine" to settle quarrels between families or individuals' "can only be made once, but it works."

Citation(s): Schenck and Gifford 1952; Strike 1994.
Terminology: **Karok:** purukuri.
Citation(s): Schenck and Gifford 1952.

451   *Elymus multisetus*
Synonym(s): *Sitanion jubatum*
Common Name(s): Squirrel Tail.

Usage: Food: **Kawaiisu (Tehachapi):** seeds parched, pounded, cooked into mush.

Citation(s): Zigmond 1981.
Terminology: **Kawaiisu (Tehachapi):** kʷasiyavɨ, "tail."
Citation(s): Zigmond 1981.

452   *Elymus triticoides*
Common Name(s): Squaw Grass, Wild Wheat.

Usage: Food: **Yuki:** used seeds in pinole. **Kawaiisu (Tehachapi):** seeds, pounded, cooked, eaten as mush.

Citation(s): Chesnut 1902.
Terminology: **Kawaiisu (Tehachapi):** aʔaasawunvɨ.
Citation(s): Zigmond 1981.

Note(s): see #346 - *Clavipes purpurea*.

453 *Emmenanthe penduliflora*
Family: *Boraginaceae*

Usage: Misc: **Tubatulabal:** considered weeds.
Citation(s): Voegelin 1938.

454 *Encelia farinosa*
Family: *Asteraceae*
Common Name(s): Brittle-Bush, Incienso.

Usage: Medicine: **Cahuilla, Mohave:** gum from plant
heated, applied to chest to relieve pain;
boiled blossoms, leaves, stems;
decoction held in mouth to relieve
toothache.
Citation(s): Bean and Saubel 1972.
Terminology: **Cahuilla:** pa'akal.
Citation(s): Bean and Saubel 1972.

455 *Encelia virginensis*
Family: *Compositae*
Synonym(s): *Encelia actoni, Encelia virginensis* ssp.
Actoni

Usage: Medicine: **Kawaiisu (Tehachapi), Tubatulabal:**
leaves boiled, decoction used as a wash
on cuts and bruises on horses.
Misc: **Tubatulabal:** considered weeds.
Citation(s): Voegelin 1938; Zigmond 1981; Strike 1994.
Terminology: **Kawaiisu (Tehachapi):** ka?miČarabɨ.
Citation(s): Zigmond 1981.
Note(s) some difference in reporting between Voegelin (1938)
and Strike (1994) for Tubatulabal [author].

456 *Ephedra* spp.

Family: *Ephedraceae*

Note(s): various species of *Ephedra* are used for a source of the alkaloid ephedrine; the plants when dry has little odor, the taste is slightly bitter; ephedrine is used for relief of asthma and hay fever, its action being more prolonged than that of adrenaline, and it has the further advantage that it need not be given by injection, but may be administered by mouth.

Citation(s): Trease And Evans 1966.

457 *Ephedra californica*

Common Name(s): Mountain Tea.

Usage: Food: **Kawaiisu (Tehachapi):** brewed into tea drunk as nonmedicinal beverage although considered a remedy for backache.

Material: **Kawaiisu (Tehachapi):** made best charcoal for tattooing.

Medicine: **Diegueño:** green, or dried, leaf tea used to treat colds, coughs, to purify the blood. **Salinan:** tea used to treat bladder problems, nasal complaints. **Chumash:** stems, root, decoction drunk cold for kidney, bladder disorders, to purify the blood, to treat unspecified "hidden illnesses" of men and women, applied externally to stop bleeding from wounds, good for washing cuts

Misc: **Owens Valley Paiute:** named but not used.

Citation(s): Steward 1933; Zigmond 1981; Strike 1994; Timbrook 2007.

Terminology: **Owens Valley Paiute:** tuduʹva. **Chumash (Barbareño):** woshkʹoʹloy; **(Ventureño):** kɨwɨkɨw.
Citation(s): Steward 1933; Timbrook 2007.

458   *Ephedra nevadensis*
        Common Name(s): Joint Pine, Mormon Tea, Miner's Tea, Mexican Tea.

Usage: Food: **Cahuilla:** bunches of the twigs steeped in water to form a tea; tea taken too long is "bad for the system;" seeds ground into meal, eaten as mush. **Panamint Shoshone (Koso):** seeds roasted, ground into flour and made into bitter cakes. **Kawaiisu (Tehachapi):** brewed into tea drunk as nonmedicinal beverage although considered a remedy for backache.
          Material: **Kawaiisu (Tehachapi):** made best charcoal for tattooing.
Citation(s): Barrows 1900; Coville 1892; Zigmond 1981; Bean and Saubel 1972.
Terminology: **Cahuilla:** tú-tut, tutut.
Citation(s): Barrows 1900; Bean and Saubel 1972.

459   *Ephedra viridis*
        Common Name(s): Mexican Tea, Joint Pine, Mountain Tea.

Usage: Food: **Owens Valley Paiute:** leafless needles boiled a short time to make drink. **Tubatulaba:** used leaves and stalks. **Kawaiisu (Tehachapi):** brewed into tea drunk as nonmedicinal beverage although considered a remedy for backache.

Medicine: **Washo:** a tea is taken for delayed or difficult menstruation.

Material: **Kawaiisu (Tehachapi):** made best charcoal for tattooing.

Misc: **Chumash:** reported from general region in collections of materials found in cave sites.

Citation(s): Grant 1964; Steward 1933; Train et al 1941; Voegelin 1938; Zigmond 1981.

Terminology: **Owens Valley Paiute:** to'yatudu'va. **Washo:** mag-gel, mah-gah.

Citation(s): Steward 1933; Train et al 1941.

460 *Epilobium* spp.

Family: *Onagarceae*

Usage: Misc: **Miwok:** named but not used.
Citation(s): Barrett and Gifford 1933.
Terminology: **Miwok:** samlili.
Citation(s): Barrett and Gifford 1933.

461 *Epilobium canum*

Family: *Onagraceae*
Synonym(s): *Zauschneria californica*
Common Name(s): California Fushia

Usage: Medicine: **Chumash:** leaves, dried, powdered, sprinkled on wound, or boiled, as a wash; used on cuts, sores, sprains, particularly in livestock.

Citation(s): Timbrook 2007.

Terminology: **Chumash (Barbareño):** s'akht'utun 'iyukhnuts.

Citation(s): Timbrook 2007.

462 *Epilobium canum* ssp. *canum*

Common Name(s): California Fushia, Mexican Balsmea.

Usage: Medicine: **Costanoan:** leaf infusion drunk for kidney, bladder ailment; decoction used to cure infected sores, for infants' fever; as a general remedy. **Miwok:** leaf decoction used to cure person who vomited blood; leaf infusion drunk for kidney, bladder ailment. **Maidu:** leaf infusion drunk for kidney, bladder ailment; leaves, dried, pulverized, or infusion, used on sores, cuts, wounds, rashes caused by Poison Oak.

Citation(s): Strike 1994; Rayburn 2012.

463 *Epilobium canum* ssp. *latifolium*

Synonym(s): *Zauschneria californica* spp. *latifolia*

Usage: Food: **Karok:** sometimes pick the blossoms and suck the nectar from the head of the flower.

Misc: **Kawaiisu (Tehachapi):** named but not used.

Citation(s): Schenck and Gifford 1952; Zigmond 1981.

Terminology: **Miwok:** husī´k<u>u</u>. **Karok:** punichi banich, "make-believe-huckleberry." **Kawaiisu (Tehachapi):** agakidɨbɨ.

Citation(s): Barrett and Gifford 1933; Schenck and Gifford 1952; Zigmond 1981.

464　*Epilobium ciliatum* ssp. *ciliatum*

> Synonym(s): *Epilobium adenocaulon, Epilobium californicum, Epilobium adenocaulon* var. *Parishii*
> Common Name(s): Willowherb.

Usage:　Medicine:　**Maidu:** infusion, or poultice, used to treat pain, muscle cramps, bruises, sores; leaf decoction used for diarrhea, sore throats.

　　　　Misc:　　　**Miwok:** named but not used. **Tubatulabal:** considered weeds

Citation(s): Barrett and Gifford 1933; Strike 1994; Voegelin 1938.

Terminology: **Miwok:** tcêpise.

Citation(s): Barrett and Gifford 1933.

465　*Epilobium densiflorum*

> Family: *Onagraceae*
> Synonym(s): *Boisduvalia densiflora*
> Common Name(s): Dense-Flowered Evening Primrose, Upright Evening Primrose.

Usage:　Food:　　　**Pomo:** used seeds in pinole and bread. **Miwok:** seeds parched, pulverized, and eaten dry.

　　　　Medicine:　**Maidu:** rubbed flowers on heads to relieve headaches.

Citation(s): Barrett 1952; Barrett and Gifford 1933; Chesnut 1902; Strike 1994.

Terminology: **Pomo (Central):** msī'pal, tsī'pal. **Miwok (Central Sierra):** winiwayu, **(Southern Sierra):** waw'na.

Citation(s): Barrett 1952; Barrett and Gifford 1933.

466 *Epilobium torreyi*

Synonym(s): *Boisduvalia stricta*
Common Name(s): Upright Evening
Primrose.

Usage: Food: **Miwok:** seeds parched, pulverized, and
eaten dry.
Citation(s): Barrett and Gifford 1933.

467 *Epipactis gigantea*

Family: *Orchidaceae*
Common Name(s): Stream Orchids.

Usage: Medicine: **Maidu:** root decoction used for
illnesses manifested by overall soreness,
paralysis.
Misc: **Tubatulabal:** considered weeds. **Karok:**
have no use for the plant except for its
"pretty flowers."
Citation(s): Schenck and Gifford 1932; Voegelin 1938; Strike
1994.
Terminology: **Karok:** pinef yukuku, "coyote shoes."
Citation(s): Schenck and Gifford 1952.

468 *Equisetum* spp.

Family: *Equisetaceae*

Usage: Material: **Achomawi:** used for polishing arrows.
Cahuilla: used in the manner of a
cleaning pad. **Chumash:** stems, used to
polish bowls.
Medicine: **Miwok:** stems used for medicine.
Achomawi: made into tea for coughs
and bladders problems. **Chumash:**
considered suitable for treating "hidden

illnesses" of men and women.

Citation(s): Barrett and Gifford 1933; Merriam 1967; Bean and Saubel 1972; Timbrook 2007.

Terminology: **Northern Paiute (Paviotso):** pazoi'winup. **Chumash (Barbareño):** woshk'o'loy; **(Ventureño):** kïwïkïw.

Citation(s): Kelly 1932; Timbrook 2007.

Note(s): stems are diuretic, astringent; a strong decoction acts as an emmenagogue.

Citation(s): Grieve 1931.

469 *Equisetum arvense*

Common Name(s): Common Horsetail.

Usage: Medicine: **Pomo (Southwestern):** plant boiled; tea used for washing itching and open sores.

  Material: **Miwok:** used to finish smoothing surface of bows as well as arrow shafts.

  Misc: **Karok:** named but not used. **Costanoan:** split rhizomes used as stitching on coiled baskets.

Citation(s): Barrett and Gifford 1933; Schenck and Gifford 1952; Goodrich et al 1980; Strike 1994.

Terminology: **Miwok (Central Sierra):** sakayu. **Karok:** chimchikara tunuweich, "immitation chimchikara." **Pomo (Southwestern):** šima·yu.

Citation(s): Barrett and Gifford 1933; Schenck and Gifford 1952; Goodrich et al 1980.

470 *Equisetum hymale* ssp. *affine*

Synonym(s): *Equisetum hymale robustum, Equisetum robustum, Equisetum hymale*

Common Name(s): Common Scouring-Rush, Horsetail Rush.

Usage: Material: **Hupa:** used to polish the stems of tobacco pipes during manufacture. **Karok:** the dried stalk served as an abrasive to sharpen the edge of mussel shells used to scrape and prepare Iris (*Iris* spp.) fibers for cordage; also used to polish arrows. **Coast Yuki:** stems used to "sandpaper" arrow shafts. **Yuki:** used as sandpaper in finishing off arrows and other woodwork; thrown into fire by medicine men so that they explode and by virtue of their continued crackling stimulate their patients to renewed vigor.

Medicine: **Karok:** stalk is soaked in water and applied as a remedy for sore "bad" eyes; also a boiled decoction of the plant is used to wash the eyes; used for ceremonial cleansing of shamans who participated in the First Salmon Ceremony. **Maidu:** decoction used to rid themselves of lice, fleas.

Citation(s): Gifford 1939; Schenck and Gifford 1952; Goddard 1903; Chesnut 1902; Strike 1994.

Terminology: **Karok:** chimchikara. **Yuki:** shetet, chiis.

Citation(s): Schenck and Gifford 1952; Gifford 1939.

Note(s): identified by Chesnut (1902) as *Equisetum variegatum*. All modern distributions of *E. Variegatum* do not show it as being present in California. It is sometimes confused with *E. hymale* [author].

471 *Equisetum laevigatum*

Synonym(s): *Equisetum Funstoni*
Common Name(s): River Horsetail, Horsetail Rush.

Usage: Medicine: **Pomo (Southwestern):** whole plant boiled; tea drunk for kidney trouble and associated back trouble. **Costanoan:** decoction used for bladder problems and as contraceptive and to stimulate a delayed menstruation; used as hair wash. **Esselen:** stem, decoction, used for various internal disorders, including kidney and bladder problems.

Citation(s): Goodrich et al 1980; Strike 1994; Rayburn 2012; Breschini and Haversat 2004.

Terminology: **Pomo (Southwestern):** šimaʻyu.

Citation(s): Goodrich et al 1980.

472 *Equisetum telmateia*

Synonym(s): *Equisetum telmateia, Equisetum telmateia* var. *braunii*
Common Name(s): Horsetail, Great Horsetail.

Usage: Material: **Yuki:** dry stalks used to smooth the stems of Indian Hemp (*Apocynum cannabinum*) and were also used for polishing arrows. **Pomo (Southwestern):** stems used as sandpaper to smooth arrows, drill shafts; used as a binding to fasten feathers to coat.

Medicine: Yuki: a decoction of the plant is used as a diuretic. **Pomo (Southwestern):** stem boiled; tea used for menstrual cramps. **Washo:** decoction used to wash sore eyes, taken as a diuretic.

Misc: **Pomo (Southwestern):** named but not used.

Citation(s): Curtin 1957; Gifford 1967; Goodrich et al 1980; Strike 1994.

Terminology: **Pomo (Southwestern):** shimayu, šima·yu. **Yuki:** shân-tum'.

Citation(s): Gifford 1967; Curtin 1957; Goodrich et al 1980.

Note(s): stems are diuretic, astringent; a strong decoction acts as an emmenagogue. Also note the different statements regarding the **Pomo (Southwestern)** probably indicating ethnographers visiting at different points in time, or, perhaps different informants [author].

Citation(s): Grieve 1931.

473 *Eragrostis mexicana* ssp. *virescens*

Family: *Poaceae*
Synonym(s): *Eragrostis secundiflora, Eragrostis oxylepis*
Common Name(s): Love Grass.

Usage: Food: **Owens Valley Paiute:** seeds, gathered in late Summer, pounded to separate chaff, winnowed, parched, ground into flour.

Citation(s): Steward 1933; Strike 1994.

Terminology: **Owens Valley Paiute:** mō'nō.

Citation(s): Steward 1933.

474 *Eremalche* spp.
Family: *Malvaceae*

Usage: Medicine: **Luiseño:** a decoction of the leaves was used as an emetic.
Citation(s): Sparkman 1908.
Terminology: **Luiseño:** kaukat.
Citation(s): Sparkman 1908.

475 *Eremalche exilis*
Synonym(s): *Sphaeralcea exile, Malvastrum exile*

Usage: Misc: **Tubatulabal:** considered weeds.
Citation(s): Voegelin 1938.

476 *Eremogone aculeta*
Family: *Caryophyllaceae*

Usage: Medicine: **Maidu:** root decoction used as eyewash.
Citation(s): Strike 1994.

477 *Eremogone macradenia*
Common Name(s): Mojave Sandwort.

Usage: Medicine: **Kawaiisu (Tehachapi):** root used as medicine; smoke used to clear sinuses, relieve headache; root pounded, mixed with water, applied externally to relieve pain; paste applied to face to eliminate pimples.
Citation(s): Zigmond 1981; Strike 1994.
Terminology: **Kawaiisu (Tehachapi):** arigĭdɨbɨ.
Citation(s): Zigmond 1981.

478 *Eremothera boothhii*
> Synonym(s): *Oenothera alyssoides, Oenothera alyssiodes* var, *villosa, Oenothera boothhii* ssp. *villosa*

Usage: Misc: **Tubatulabal:** considered weeds.
Citation(s): Voegelin 1938.

479 *Eriastrum densifolium*
> Family: *Polemoniaceae*
> Common Name(s): Starflower.

Usage: Medicine: **Kawaiisu (Tehachapi):** dried flowers, or root, pounded, made into salve, applied to venereal sores.
Citation(s): Zigmond 1981.

480 *Eriastrum* eremicum ssp. eremicum
> Common Name(s): Starflower.

Usage: Medicine: **Tubatulabal:** used whole plant as poultice for ulcerated sores.
Citation(s): Strike 1994.

481 *Eriastrum filifolium*
> Synonym(s): *Huegelia filifolia*

Usage: Misc: **Tubatulabal:** considered weeds.
Citation(s): Voegelin 1938.

482 *Eriastrum virgatum*
> Synonym(s): *Huegelia virgata*

Usage: Medicine: **Owens Valley Paiute:** entire plant boiled into a very strong gargle for sore throat, used as an emetic.

Citation(s): Steward 1933; Strike 1994.
Terminology: **Owens Valley Paiute:** tūma'nāva. **Mono Lake Paiute:** tü'manava.
Citation(s): Steward 1933.

483 *Ericameria* spp.
        Family: *Asteraceae*

Usage: Medicine: **Owens Valley Paiute:** tea for stomach trouble, diarrhea, etc.
Citation(s): Steward 1933.
Terminology: **Owens Valley Paiute:** pawatsiva, sigū'pᵃ.
Citation(s): Steward 1933.

484 *Ericameria arborescens*
        Common Name(s): Tree Haplopappus.

Usage: Medicine: **Miwok:** boiled decoction of leaves drunk hot to cure stomach trouble; applied to rheumatic parts; during menstruation and after parturition women sometimes drank it to relieve pain; considered too strong a remedy to use during parturition; twigs and leaves bound on rheumatic parts; leaves applied to boils to bring them to a head; also bound over a sore on one's foot when traveling. **Esselen:** considered one of the most important plants.
Citation(s): Barrett and Gifford 1933; Breschini and Haversat 2004.
Terminology: **Miwok:** tce'ktceka.
Citation(s): Barrett and Gifford 1933.

485   *Ericameria cooperi*

Usage: Medicine: **Tubatulabal:** used externally to relieve rheumatic pain.

Citation(s): Strike 1994.

486   *Ericameria cuneata*
Synonym(s): *Ericameria cuneatus*

Usage: Medicine: **Miwok:** decoction of the stems drunk for colds.

Misc: **Tubatulabal:** considered weeds.

Citation(s): Barrett and Gifford 1933; Voegelin 1930.

487   *Ericameria eriocoidea*
Common Name(s): Mock Heather.

Usage: Medicine: **Miwok (Coast):** used to poultice sores.

Citation(s): Strike 1994.

488   *Ericameria laricifolia*
Common Name(s): Terpentine Bush.

Usage: Medicine: **Diegueño:** used to relieve body pains.

Citation(s): Strike 1994.

489   *Ericameria linearifolia*
Synonym(s): *Stenotopsis linearifolius*

Usage: Medicine: **Kawaiisu (Tehachapi):** decoction of leaves and flowers applied warm for relief of rheumatism, sores, bruises, cuts on men; sore backs on horses; washing

hair to make it grow; washing tired feet.
Tubatulabal: used externally to relieve
rheumatic pain. **Miwok:** leaves applied
to boils to bring them to a head, to feet
to alleviate pains from sores.

Misc: **Tubatulabal:** considered weeds.

Citation(s): Voegelin 1938; Zigmond 1981; Strike 1994.
Terminology: **Kawaiisu (Tehachapi):** sanaco?ovɨbɨ.
Citation(s): Zigmond 1981.
Note(s): note difference in reports by Voegelin and Strike - may
be a change through time? [author].

490 *Ericameria nauseosa*

Synonym(s): *Bigilovia graveolens, Chrysothamnus
nauseosus, Chrysothamnus nauseosus* var.
consimilis

Common Name(s): Rubber Rabbit Brush, Rabbit
Brush.

Usage: Food: **Northern Paiute (Paviotso):** the root at
surface level was chewed until gummy.
Kawaiisu (Tehachapi): pinyon nuts
threaded on a twig and kept until eaten.
Maidu: tips of new growth eaten.

Material: **Tubatulabal:** used to cover dwellings,
sweathouses; stems used for arrows,
wickerwork, windbreaks.

Medicine: **Cahuilla:** tea of the twigs is drunk for
coughs and pain in the chest; used to
relieve toothache. **Modoc (Lutuami)
(Lutuami):** used to poultice blisters.
Maidu: tonic made from plant used to
relieve stomach aches, treat bronchial
problems.

Citation(s): Barrows 1900; Kelly 1932; Zigmond 1981; Bean and

Saubel 1972; Strike 1994.

Terminology: **Cahuilla:** tĕs-i-nit (?), mocee (?). **Kawaiisu (Tehachapi):** tivaposoni(m)bɨ. **Northern Paiute (Paviotso):** sigu'p'.

Citation(s): Barrows 1900; Zigmond 1981; Bean and Saubel 1972; Kelly 1932.

491 *Ericameria palmeri*

Synonym(s): *Haplopappus palmeri, Ericameria venetus*

Usage: Material: **Cahuilla:** straight stem used as arrow foreshafts. **Diegueño:** stalks used to make brooms.

Medicine: **Cahuilla:** leaves and twigs are bound upon the feet, or arms, together with stones, to relieve swelling and pain, or to reduce possibility of infection from cuts; horses occasionally washed with infusion to keep away insects. **Diegueño:** tea used to relieve bad colds.

Citation(s): Barrows 1900; Zigmond 1981; Bean and Saubel 1972; Strike 1994.

Terminology: **Cahuilla:** Le-nil. **Kawaiisu (Tehachapi):** paahoovɨ.

Citation(s): Barrows 1900; Zigmond 1981.

492 *Ericameria palmeri* var. *pachylepsis*

Synonym(s): *Ericameria acradenius*

Usage: Material: **Cahuilla:** plant occasionally used to build fences as protection for cold winds.

Medicine: **Cahuilla:** for colds, root cut into small pieces, boiled in water, infusion drank at hourly intervals; fore sore throats,

leave soaked in pan of hot water,
patient knelt over pan, blanket over
head, steam inhaled until symptoms
disappeared; leaves used as poultice for
sores.
Citation(s): Bean and Saubel 1972.

493 *Ericameria parishii* var. *parishii*
Synonym(s): *Chrysoma parishii*

Usage: Food: **Luiseño:** the seeds are eaten.
         Medicine: **Luiseño:** the plant is used medicinally.
Citation(s): Sparkman 1908.
Terminology: **Luiseño:** sanmikut.
Citation(s): Sparkman 1908.

494 *Ericameria teretifolia*
Synonym(s): *Chrysothamnus teretifolus*

Usage: Misc: **Tubatulabal:** considered weeds.
Citation(s): Voegelin 1938.

495 *Erigeron breweri*
Family: *Asteraceae*

Usage: Medicine: **Kawaiisu (Tehachapi):** piece of root
                      held between teeth for relief of
                      toothache; root boiled and used as a
                      head wash for lice.
Citation(s): Zigmond 1981.

496 *Erigeron canadensis*
Family: *Compositae*
Synonym(s): *Erigeron canadensis, Conyza canadensis*
Common Name(s): Horseweed.

Usage: Food: **Miwok:** leaves and tender tops pounded in bedrock mortar; eaten pulverized, but uncooked; flavor like onions.

Medicine: **Cahuilla:** boiled to make infusion for curing diarrhea. **Chumash:** ground, rubbed on any aching part; tea, drunk for kidney problem.

Citation(s): Barrett and Gifford 1933; Bean and Saubel 1972; Timbrook 2007.

Terminology: **Miwok (Central Sierra):** mututa. **Chumash (Barbareño):** wili'lik'; **(Ynezeño):** wililik'; **(Ventureño):** wililik.

Citation(s): Barrett and Gifford 1933; Timbrook 2007.

497 *Erigeron divergens*

Usage: Misc: **Tubatulabal:** considered weeds.
Citation(s): Voegelin 1938.

498 *Erigeron foliosus* var. *foliosus*

Synonym(s): *Erigeron foliosus* var. *stenophyllus*
Common Name(s): Fleabane.

Usage: Medicine: **Miwok:** boiled and cooled decoction of the washed and pounded root drunk to abate fever and ague; root chewed and placed in cavity for toothache. **Kawaiisu (Tehachapi):** piece of root held between teeth for relief of toothache; root boiled and used as a head wash for lice.

Misc: **Miwok:** plant traded from the north for beads, shells, and baskets.

Citation(s): Barrett and Gifford 1933; Zigmond 1981.

Terminology: **Miwok (Central Sierra):** we'ne.
Citation(s): Barrett and Gifford 1933.

499 *Erigeron glacialis* var. *glacialis*
Synonym(s): *Erigeron peregrinus*
Common Name(s): Wandering Daisy.

Usage: Medicine: **Maidu:** pulverized, dried plant infusion
used internally and externally for aches,
dizziness, drowsiness.
Citation(s): Strike 1994.

500 *Erigeron miser*
Common Name(s): Starved Daisy.

Usage: Medicine: **Maidu:** used as a tonic, a diuretic, for
diarrhea.
Citation(s): Strike 1994.

501 *Erigeron philadelphicus* var. *philadelphicus*
Synonym(s): *Erigeron philadelphicus*
Common Name(s): Canadian Fleabane.

Usage: Medicine: **Maidu:** medication for fevers, stomach
aches, menstrual problems; sniffed into
nostrils to promote sneezing to relieve
head-cold congestion.
Citation(s): Strike 1994.

502 *Eriodictyon californicum*
Family: *Boraginaceae*
Common Name(s): Yerba Santa, Mountain Balm.

Usage: Medicine: **Karok:** good remedy for colds, pleurisy,
and tuberculous; leaves boiled in basket

with hot rocks. **Yuki:** leaves used as a cure for colds and asthma; leaf placed on scabby sore helps it heal; generally used as a cough syrup. **Pomo:** used as a cure for grippe. **Pomo (Southwestern):** leaves boiled, tea used as cough medicine and a blood purifier; tea will bring down a fever; leaves boiled to make a wash for sores. **Miwok:** leaves and flowers steeped in hot water and drunk for coughs, colds, stomach ache, and rheumatism; leaves sometimes chewed for same purpose; leaves warmed and used as plasters on aching or sore spots; leaves smoked to relieve coughs and colds; mashed leaves applied to cuts, wounds, and abrasions, over fractured bones to keep down swelling, aid knitting, and relieve pain. **Cahuilla:** used leaves for a poultice or liniment; the leaves are pounded up and bound upon the sores of both men and beasts and a strong decoction is used for bathing sore parts or the limbs when painful or fatigued. **Atsugewi:** person steamed in moisture from branches and leaves for rheumatism; chewed and juice swallowed for colds and whooping cough. **Kawaiisu (Tehachapi):** leaves brewed into tea "good for stomach" and for relief of gonorrhea. **Costanoan:** decoction used to purify the blood, used to treat tuberculosis, rheumatism; combined with other herbs to treat infected sores, used for asthma; leaves, heated, stuck on forehead for headache; tea, used as eyewash, used to treat new colds; chewed, or smoked, to relieve asthma. **Esselen:** leaves, fresh, or dried, boiled to make bitter tea, taken to cure colds, sore throat, asthma, tuberculosis, rheumatism, as a blood purifier; as a wash to reduce fever.

Citation(s): Barrett and Gifford 1933; Barrows 1900; Chesnut 1902; Curtin 1957; Garth 1953; Schenck and Gifford 1952; Zigmond 1981; Goodrich et al 1980; Rayburn 2012; Breschini and Haversat 2004.

Terminology: **Karok:** pirishaxwaharash, "plant with lots of pitch." **Yuki:** tel-at'-mo, "stick catch." **Atsugewi:** natatsĭrop. **Miwok (Central Sierra):** pa'ssalu. **Cahuilla:** tán-wi-vel. **Kawaiisu (Tehachapi):** wupigidɨbɨ. **Pomo (Southwestern):** du?ċán? qʰale, "sticky plant."

Citation(s): Schenck and Gifford 1952; Curtin 1957; Garth 1953; Barrett and Gifford 1933; Barrows 1900; Zigmond 1981; Goodrich et al 1980.

Note(s): the plant has been used pharmaceutically for coughs, colds, asthma, and inflammations of the urogenital organs.

Citation(s): Tease and Evans 1966.

503 *Eriodictyon crassifolium*

Usage: Medicine: **Luiseño:** used medicinally. **Chumash:** leaves, simmered in water, warm liquid used for coughs, colds, asthma, to expel mucus, as a general tonic; considered to have conferred some degree of supernatural power for healing, games.

Citation(s): Strike 1994; Timbrook 2007.

Terminology: **Chumash (Barbareño):** wishap'; **(Ynezeño, Ventureño):** wishap.
Citation(s): Timbrook 2007.

504 *Eriodictyon parryi*
Synonym(s): *Turricula parryi*
Common Name(s): Poodle-Dog Bush.

Usage: Medicine: **Luiseño:** used for medicinal purposes. **Kawaiisu (Tehachapi):** leaves brewed; used liquid as a wash for relief of swelling and rheumatism.
Citation(s): Sparkman 1908; Zigmond 1981.
Terminology: **Luiseño:** atovikut. **Kawaiisu (Tehachapi):** tihiyago?opi, "deer tobacco."
Citation(s): Sparkman 1908; Zigmond 1981.

505 *Eriodictyon tomenstosum*
Common Name(s: Yerba Santa.

Usage: Medicine: **Liuseño:** Much valued for medicinal purposes. **Yokuts:** a decoction used for fever and for bad blood.
Citation(s): Sparkman 1908; Powers 1877.
Terminology: **Liuseño:** palwut.
Citation(s): Sparkman 1908.

506 *Eriodictyon trichocalyx*
Common Name(s): Yerba Santa.

Usage: Food: **Cahuilla:** leaves sometimes chewed as thirst quencher.
Medicine: **Cahuilla:** thick, sticky leaves, fresh, or dried, boiled, mixed with sweetening agents, drunk, said to be a blood

purifier, a cure for coughs, colds, sore throats, asthma, catarrh, tuberculosis, rheumatism; fresh leaves pounded, placed on sore or fatigued limbs, or heated and placed on body to cure rheumatism.

Citation(s): Bean and Saubel 1972.

Terminology: **Cahuilla:** tanwivel.

Citation(s): Bean and Saubel 1972.

507   *Eriogonum* spp.
Family: *Polygonaceae*

Usage: Food: **Tubatulabal:** used seeds. **Cahuilla:** edible shoots gathered from February to may; seeds gathered June until September.

Medicine: **Kawaiisu (Tehachapi):** plant is boiled, drunk for relief of diarrhea. **Cahuilla:** strong, black decoction made from leaves drunk as cure for headaches, stomach disorders; white flowers steeped to make eye wash, or drink to clean out intestines; leaves growing near root used as physic; tea made from plant said to cause uterus to shrink, inhibiting dysmenorrhea; oldest plants said to most efficacious. **Costanoan:** berries, decoction, used for stomach ache; leaves, tea, used for headache.

Citation(s): Voegelin 1938; Zigmond 1981; Bean and Saubel 1972; Rayburn 2012.

Terminology: **Cahuilla:** hulaqal.

Citation(s): Bean and Saubel 1972.

508 *Eriogonum angulosum*

Usage: Food: **Kawaiisu (Tehachapi):** seeds pounded and eaten.

Misc: **Tubatulabal:** considered weeds

Citation(s): Voegelin 1938; Zigmond 1981.

509 *Eriogonum argillosum*

Usage: Misc: **Tubatulabal:** considered weeds.

Citation(s): Voegelin 1938.

510 *Eriogonum baileyi*

Common Name(s): Bailey, Buckwheat.

Usage: Food: **Kawaiisu (Tehachapi):** seeds, when ripe in August, gathered, pounded into meal, eaten dry, or mixed with water and drunk.

Medicine: **Tubatulabal:** used lotion for skin eruptions.

Citation(s): Zigmond 1981.

511 *Eriogonum deflexum* var. *deflexum*

Synonym(s): *Eriogonum insigne*

Usage: Material: **Kawaiisu (Tehachapi):** canes could substitute for *Phragmites australis* for making pipes.

Citation(s): Zigmond 1981.

512 *Eriogonum elongatum* var. *elongatum*

Synonym(s): *Eriogonum elongatum gramineum*

Usage: Material: **Tubatulabal:** used 4-inch sections of tubular stalks to collect juice of *Ascelpias* *erosa* in order to make chewing gum.

Citation(s): Voegelin 1938.

513 *Eriogonum fasciculatum*

Common Name(s): Wild Buckwheat, Gray California Buckwheat.

Usage: Material: **Kawaiisu (Tehachapi):** hard wood employed for piercing ears; leaves utilized in lining acorn granary.

Medicine: **Cahuilla:** strong, black decoction of the leaves is drunk for pain in the stomach or pain in the head, used to relieve back pain during pregnancy, used as a laxative; the white flour is steeped to make an eye-wash and as a purgative. **Costanoan:** plant, boiled, used as poultice(?) for urinary problems. **Chumash:** plant, top, boiled, tea drank, also bathed in it, for rheumatism, irregular menstruation; tea combined with sage, used to suppress menstruation; plant, or flowers, simmered, decoction drunk hot for stomach trouble.

Citation(s): Barrows 1900; Zigmond 1981; Strike 1994; Rayburn 2012; Timbrook 2007.

Terminology: **Cahuilla:** hú-la-kal. **Kawaiisu (Tehachapi):** sag^wi?avɨ. **Chumash (Ventureño):** tswana'atl 'ishup.

Citation(s): Barrows 1900; Zigmond 1981; Timbrook 2007.

514   *Eriogonum gracillimum*

Usage: Medicine: **Tubatulabal:** used lotion for skin eruptions.
Citation(s): Strike 1994.

515   *Eriogonum inflatum*
Common Name(s): Desert Trumpet.

Usage: Food: **Kawaiisu (Tehachapi):** seeds, ripe in August, are eaten.
Citation(s): Zigmond 1981.
Terminology: **Kawaiisu (Tehachapi):** tɨniporobɨ.
Citation(s): Zigmond 1981.

516   *Eriogonum latifolium*
Common Name(s): Sour Grass.

Usage: Food: **Yuki:** the young stems have a very agreeable acid taste and are eagerly sought after by children in May or June before the flowers have developed.
        Medicine: **Yuki:** leaves, stem, and woody root are used in the form of a decoction for pain in the stomach, for headache, and for female complaints; a decoction of the root is used for sore eyes. **Maidu:** decoction relieved colds, coughs. **Costanoan:** root, stems, leaves, decoction, taken internally for colds, coughs.
Citation(s): Chesnut 1902; Strike 1994; Rayburn 2012.

517 *Eriogonum mohavense*

Usage: Misc: **Tubatulabal:** considered weeds.
Citation(s): Voegelin 1938.

518 *Eriogonum nudum*

Synonym(s): *Eriogonum latifolium* ssp. *Nudum*
Common Name(s): Tibinaqua.

Usage: Food: **Karok:** the sour-tasting stems are eaten raw as greens. **Miwok:** eaten raw, sour taste.

Material: **Miwok:** twigs and leaves used to clear ground around Manzanita (*Arctostaphylos* spp.) bushes before knocking off the berries. **Kawaiisu (Tehachapi):** hollow stems used as drinking tubes, pipes for smoking.

Medicine: **Kawaiisu (Tehachapi):** roots boiled; infusion drunk hot for relief of coughs and colds. **Chumash:** tea, used to stop hemorrhage; root, tea, remedy for fever; leaves, decoction for thirst.

Misc: **Karok:** the children play a game with the stems by hooking each other's plant; the first to let go is the loser.

Citation(s): Barrett and Gifford 1933; Schenck and Gifford 1952; Zigmond 1981; Timbrook 2007.

Terminology: **Karok:** tahukannaich, "imitation hook." **Miwok (Central Sierra):** sapu'la, sapasu. **Kawaiisu (Tehachapi):** paako?orɨbɨ. **Chumash (Ventureño):** 'an.

Citation(s): Schenck and Gifford 1952; Barrett and Gifford 1933; Zigmond 1981; Timbrook 2007.

519   *Eriogonum nudum* var. *oblongifolium*

Synonym(s): *Eriogonum latifolium* ssp. *Sulphureum*

Usage: Food: **Karok:** the sour-tasting stems are eaten raw as greens.

Material: **Kawaiisu (Tehachapi):** hollow stems used as drinking tubes, pipes for smoking.

Medicine: **Kawaiisu (Tehachapi):** roots boiled; infusion drunk hot for relief of coughs and colds.

Citation(s): Schenck and Gifford 1952; Zigmond 1981.

Terminology: **Karok:** tahukannaich, "imitation hook."

Citation(s): Schenck and Gifford 1952.

520   *Eriogonum plumatella*

Common Name(s): Flat Top.

Usage: Food: **Kawaiisu (Tehachapi):** seeds pounded, cooked into mush.

Citation(s): Zigmond 1981.

521   *Eriogonum pusillum*

Common Name(s): Yellow Turban, Wild Buckwheat.

Usage: Food: **Kawaiisu (Tehachapi):** seeds pounded, eaten dry; flowers mixed with *Mentzelia* spp. seeds and eaten.

Citation(s): Zigmond 1981.

Terminology: **Kawaiisu (Tehachapi):** tiniporobɨ.

Citation(s): Zigmond 1981.

523 *Eriogonum roseum*

Synonym(s): *Eriogonum virgatum*

Usage: Medicine: **Tubatulabal:** used lotion for skin eruptions.
Citation(s): Strike 1994.

524 *Eriogonum umbellatum*

Common Name(s): Sulphur-flowered Eriogonum.

Usage: Medicine: **Mono Lake Paiute, Owens Valley Paiute:** tea made from the roots; used for colds. **Kawaiisu (Tehachapi), Modoc (Lutuami):** flowers mashed to make salve for gonorrhea sores, and wounds. **Maidu:** used to treat colds, stomach aches, colic, headaches, sore eyes.
Citation(s): Steward 1933; Zigmond 1981; Strike 1994.
Terminology: **Mono Lake Paiute:** naka'donup.
Citation(s): Steward 1933.

525 *Eriogonum wrightii*

Common Name(s): Wright Buckwheat.

Usage: Medicine: **Kawaiisu (Tehachapi):** flowers mashed to make salve for gonorrhea sores.
Citation(s): Zigmond 1981.

526 *Eriophyllum ambiguum*

Family: *Asteraceae*
Common Name(s): Wooly Daisy.

Usage: Food: **Kawaiisu (Tehachapi):** seeds parched, pounded, eaten dry.

Misc: **Tubatulabal:** considered weeds.
Citation(s): Voegelin 1938; Zigmond 1981.

527   *Eriophyllum confertiflorum*
Common Name(s): Golden Yarrow.

Usage: Food: **Cahuilla:** seeds, gathered June to
                              November, parched, ground into flour.
            Medicine: **Miwok (Coast):** boiled, used steam to
                              relieve rheumatic pains in legs.
            Misc: **Tubatulabal:** considered weeds.
Citation(s): Voegelin 1938; Bean and Saubel 1972; Strike 1994.

528   *Eriophyllum lanatum* var. *achilleoides*
Synonym(s): *Eriophyllum caespitosum, Eriophyllum
                      lanatum* var. *leucuphyllum*
Common Name(s): Golden Yarrow.

Usage: Medicine: **Miwok:** leaves bound on the body over
                              aching parts.
Citation(s): Barrett and Gifford 1933.
Terminology: **Miwok (Central Sierra):** lakma, pusukele.
Citation(s): Barrett and Gifford 1933.

529   *Eriophyllum pringlei*

Usage: Misc: **Tubatulabal:** considered weeds.
Citation(s): Voegelin 1938.

530   *Erodium* spp.
Family: *Geraniaceae*
Common Name(s): Storkbill, Filaree, Clocks.

Usage: Food: **Cahuilla:** gathered prior to blossom
                              between January and April, cooked

while fresh, or eaten uncooked;
preserved for a brief time. **Huchnom,
Maidu, Southern (Nisenan), Yuki:**
leaves, fresh, or cooked, eaten.
Citation(s): Bean and Saubel 1972; Strike 1994.
Terminology: **Cahuilla:** pakhanat.
Citation(s): Bean and Sabel 1972.

531 *Erodium cicutarium*

Common Name(s): Red-Stem Filaree.

Usage: Food: **Chumash:** esteemed as livestock fodder.

Medicine: **Costanoan:** leaves, tea used to treat typhoid fever.

Misc: **Karok:** named but not used (?).

Citation(s): Schenck and Gifford 1952; Rayburn 2012; Timbrook 2007.
Terminology: **Karok:** pinhiich, "imitation pin."
Citation(s): Schenck and Gifford 1952.
Note(s): naturalized very early from the Mediterranean region.
Citation(s): Munz and Keck 1965.

532 *Erodium moschatum*

Common Name(s): Alfilerilla, Mountain Fillery, Filaree.

Usage: Food: **Yokuts:** eaten with Salt Grass (*Distichlis spicata*) while in the tender stage. **Yuki, Huchnom:** eaten as greens. **Maidu, Southern (Nisenan):** eaten when young and tender, or boiled for greens.

Medicine: **Chumash:** whole plant, boiled, used tonic for the blood.

Citation(s): Foster 1944; Gayton 1948; Powers 1877; Timbrook
2007.
Terminology: **Chumash (Barbareño):** chikwɨ', s'u'wlima';
(Ynezeño): s'u'wlima; **(Ventureño):** kwɨ'ɨn.
Citation(s): Timbrook 2007.
Note(s): a native of Eurasia; native of the Mediterranean region;
plants is astringent, diaphoretic.
Citation(s): Bailey 1950; Munz and Keck 1965; Stuhr 1933.

533 *Erysimum capitatum*
Family: *Brassicaceae*
Common Name(s): Wallflower.

Usage: Medicine: **Maidu:** dried, pulverized, infusion
rubbed on face, head to prevent
sunburn, to alleviate heat exposure;
chewed root rubbed on back for
pneumonia.
Citation(s): Strike 1994.

534 *Erythronium grandiflorum* ssp. *grandiflorum*
Family: *Liliaceae*
Synonym(s): *Erythronium grandiflorum* var. *pallidum*
Common Name(s): Adder's Tongue.

Usage: Medicine: **Wailaki:** bulbs, crushed, used to
poultice boils, rubbed on breasts of
nursing mothers.
Citation(s): Strike 1994.

535 *Erythronium oregonum*
Synonym(s): *Erythronium giganteum, Erythronium
oregonum* ssp. *leucandrum*

Usage: Food: **Wailaki:** eaten but not in large quantity.

Medicine: **Wailaki:** used crushed corms as a poultice for boils.
Citation(s): Chesnut 1902.

536 *Eschscholzia* spp.

Family: *Papaveraceae*
Common Name(s): California Poppy

Usage: Material: **Cahuilla:** women used pollen as a facial cosmetic.

Medicine; **Cahuilla:** plant said to provide a sedative for babies.
Citation(s): Bean and Saubel 1972.
Terminology: **Cahuilla:** tesinat.
Citation(s): Bean and Saubel 1972.

537 *Eschscholzia californica*

Family: *Papaveraceae*
Synonym(s): *Eschscholzia douglasii, Eschscholzia californica* var. *douglasii*
Common Name(s): California Poppy.

Usage: Food: **Maidu, Southern (Nisenan):** either boiled or roast with hot stones and then laid in water. **Luiseño:** leaves used for greens; the flowers chewed with chewing gum.

Medicine: **Costanoan:** flowers, decocted, liquid rubbed in hair to kill lice; 1-2 flowers placed beneath bed to put child to sleep; plant avoided by pregnant, or lactating, women as smell was believed to be poisonous. **Chumash:** used for colic. **Yuki:** fresh root placed in cavity of tooth to stop toothache; extract of root used as a wash or liniment for headache, suppurating sores, or to stop secretion of milk in women; internally, to cause vomiting, cure stomach ache, and to cure constipation; leaves eaten as green after boiling, the water being thrown away. **Pomo (Southwestern):** seed pod mashed, rubbed on a nursing mother's breast to dry up her milk, or can be boiled and rubbed on her breasts.

Misc: **Tubatulabal, Pomo (Southwestern):** named but not used. **Pomo, Yuki:** a nursing mother should not touch the plant because "her milk would dry up;" to purify mother's milk, a small quantity was placed on a hot stone as a symbolic rite. **Chumash:** believed that if boys and girls gathered the flowers together, the girls would yield to the boys because the beauty of the flowers would overcome the girls.

Citation(s): Curtin 1957; Gifford 1967; Powers 1877; Schenck and Gifford 1952; Sparkman 1908; Voegelin 1938; Rayburn 2012; Timbrook, Chesnut 1902; Goodrich et al 1980.

Terminology: **Luiseño:** ataushanut. **Yuki:** hui-con'-il. **Pomo:** si'dohe'q'ale, "milk disappear plant." **Pomo (Southwestern):** ši?dóhcʰo?, shidocho kale, "poppy flower," "breast" (shido) "die away" (cho); silom?. **Karok:** sinvanahich. **Chumash (Barbareño, Ynezeño, Ventureño):** qupe.

Citation(s): Saprkman 1908; Schenck and Gifford 1952; Goodrich et al 1980; Strike 1994; Timbrook 2007.

Note(s): plant reputed soporific, analgesic.

Citation(s): Stuhr 1933.

538   *Eschscholzia minutiflora*

Usage: Misc: **Tubatulabal:** named but not used.
Citation(s): Voegelin 1938.

539   *Eschscholzia parishii*

Usage: Meicine: **Kawaiisu (Tehachapi):** root dried,
                         ground, and applied as a dry powder
                         on moistened venereal sores.
Citation(s): Zigmond 1981.
Terminology: **Kawaiisu (Tehachapi):** hiyogʷivɨ.
Citation(s): Zigmond 1981.

540   *Eucalyptus* spp.
                Family: *Myrataceae*
                Common Name(s): Eucalyptus, Gum Tree.

Usage: Medicine: **Cahuilla:** leaves boiled in water, patient
                         held head over bowl, blanket placed
                         over head, inhaled steam to relieve
                         sinus congestion.
Citation(s): Bean and Saubel 1972
Terminology: **Cahuilla:** qahich'a waavu'it, "tall thing."
Citation(s): Bean and Saubel 1972.
Note(s): native of Australia, introduced into California in the
         19th century.
Citation(s): Munz and Keck 1965; Bean and Saubel 1972.

541 *Eucephalus* spp.
       Family: Asterace*ae*
       Synonym(s): *Aster* spp.

Usage: Misc: **Mono Lake Paiute:** named but not
                      used.
Citation(s): Steward 1933.
Terminology: **Mono Lake Paiute:** tonikup.
Citation(s): plant acts as secondary selenium absorbers.
Citation(s): Kingsbury 1964.

542 *Euphorbia* spp.
       Family: *Euphorbiaceae*
       Synonym(s): *Chamaesyce* spp.
       Common Name(s): Spurge.

Usage: Medicine: **Cahuilla:** plant boiled; decoction drunk
                      to cure sores in moth, reducing fever, as
                      a cure for chicken pox and smallpox; a
                      paste made from plant, applied as
                      poultice for snakebite; sap used to cure
                      earaches, a remedy for bee stings,
                      placed on sores to facilitate healing.
                      Chumash: used for skin diseases, warts,
                      cataracts.
Citation(s): Bean and Saubel 1972; Timbrook 2007.
Terminology: **Cahuilla:** temal hepi', "earth's milk."
Citation(s): Bean and Saubel 1972.
Note(s): the milky juice of many species of spurge are
       suspected of causing dermatitis.
Citation(s): Hardin and Arena 1974.

543   *Euphorbia albomargininata*

Usage: Medicine: **Kawaiisu (Tehachapi):** leaves and
                             flowers ground into salve, applied to
                             rattle snake bite.
Citation(s): Zigmond 1981.
Terminology: **Kawaiisu (Tehachapi):** tɨvikagivɨ.
Citation(s): Zigmond 1981.

544   *Euphorbia crenulata*
              Common Name(s): Milkweed.

Usage: Misc: **Pomo (Southwestern):** named but not
                             used.
Citation(s): Goodrich et al 1980.
Terminology: **Pomo (Southwestern):** ši?do qʰále, "milkplant."
Citation(s): Goodrich et al 1980.

545   *Euphorbia maculata*
              Common Name(s): Golondrina.

Usage: Medicine: **Costanoan:** tea drunk to purify the
                             blood; leaf, infusion, used as wash for
                             facial blemishes (milky juice applied to
                             pimples); decoction, used as eyewash,
                             to wash cuts.
Citation(s): Rayburn 2012.

546   *Euphorbia ocellata*
              Common Name(s): Rattlesnake Weed.

Usage: Medicine: **Miwok:** leaves were mashed and
                             rubbed into snake bite; a decoction was
                             drunk as a blood purifier.
Citation(s): Barrett and Gifford 1933.

Terminology: **Miwok (Northern Sierra):** pē'sippēsa, **(Central Sierra):** pē'sippesa.
Citation(s): Barrett and Gifford 1933.

547 *Euphorbia polycarpa*
Common Name(s): Yerba Golondrina.

Usage: Medicine: **Luiseño:** reputed to be beneficial in the case of snake bite. **Cahuilla, Tubatulabal:** used on rattlesnake bites. **Diegueño:** used on rattlesnake bites.
Citation(s): Sparkman 1908; Strike 1994.
Terminology: **Luiseño:** kenhamal.
Citation(s): Sparkman 1908.

548 *Euphorbia serpyllifolia*
Common Name(s): Thyme-Leaf Spurge.

Usage: Medicine: **Miwok:** running sores were washed with a decoction of the leaves, after which a green powder made from the leaves of *Solidago californica* was dusted on the sores; said to cure rattlesnake bites if applied immediately.

Misc: **Tubatulabal:** considered weeds.
Citation(s): Barrett and Gifford 1933; Voegelin 1938.
Note(s): the plant is disphoretic, emetic.
Citation(s): Stuhr 1933.

549 *Evernia* spp.
Family: *Parmeliaceae*

Usage: Material: **Achomawi:** the principal ingredient for the poison used for the stone arrow tips; the arrow points were embedded

im masses of the wet lichen and allowed to remain for an entire year, rattlesnake venom was sometimes added. **Yuki:** dwellings of wood, or bark, sometimes caulked with lichen.
Citation(s): Merriam 1967; Nash et al 2001; Strike 1994.

550   *Evernia prunastri*

Common Name(s): Oak Lichen, Flabby Lichen.

Usage: Material: **Maidu:** obtained a lavender dye.
Citation(s): Strike 1994.

551   *Evernia vulpina*

Common Name(s): Lichen, Wolf Moss, Tree Lichen, Yellow Lichen.

Usage: Material: **Hupa:** used to dye the leaves of *Xerophyllum tenax* a bright yellow; porcupine quills are sometimes dyed the same way. **Modoc (Lutuami) (Lutuami):** porcupine quills for basketry decoration are dyed yellow. **Yurok, Wintun, Northern Paiute (Paviotso):** used as a yellow dye. **Karok:** yellow dyes used for porcupine quills which are worked into the design of some basket caps; not used in other kinds of baskets. **Yuki:** used as a bedding material; thick decoction was used as a kind of paint, not as a dye.

Medicine: **Yuki, Wailaki:** used to dry up running sores.
Citation(s): Barrett 1910; Chesnut 1902; Goddard 1903; Merrill 1923; O'Neale 1932; Schenck

and Gifford 1952.
Terminology: **Yurok:** mece'n. **Karok:** manilmaashaxaeme,
"mountain moss."
Citation(s): O'Neale 1932; Schenck and Gifford 1952.

552 *Ferocactus cylindraceus*
Family: *Cactaceae*
Synonym(s): *Echinocactus acanthodes*
Common Name(s): Barrel Cactus.

Usage: Food: **Cahuilla:** buds gathered by women, plucked from cactus with a pair of short stick, placed in a gathering basket, bud usually parboiled several time to remove bitter taste; eaten or sun dried fore storage; dried buds recooked in water, usually salted to enhance taste; sometimes steam cooked in a fire pit for several hours ,could also be sun dried until dehydrated and stored; buds also used in strew, or with mountain sheep or jackrabbit meat; mature flower cooked, prepared, and preserved same manner as bud; to obtain water, top sliced off, pulp squeezed by hand to release water; body of plant could be used for cooking, heated by hot stones dropped into depression in the interior.

Citation(s): Bean and Saubel 1972.
Terminology: **Cahuilla:** kupash.
Citation(s): Bean and Saubel 1972.

553 *Festuca* spp.

Family: *Poaceae*
Common Name(s): Fescue.

Usage: Food: **Maidu:** seeds eaten.
Citation(s): Strike 1994.

554 *Festuca microstachys*

Synonym(s): *Festuca pacifica*
Usage: Misc: **Kawaiisu (Tehachapi):** named but not
used.
Citation(s): Zigmond 1981.
Terminology: **Kawaiisu (Tehahchapi):** takʷičiyavɨ.
Citation(s): Zigmond 1981.

555 *Festuca myuros*

Synonym(s): *Festuca megalura*

Usage: Misc: **Kawaiisu (Tehachapi):** named but not
used.
Citation(s): Zigmond 1981.
Terminology: **Kawaiisu (Tehachapi):** takʷičiyavɨ.
Citation(s): Zigmond 1981.

556 *Festuca temulentum*

Family: *Poaceae*
Synonym(s): *Lolium temulentum*

Usage: Food: **Yuki, Pomo:** used seeds in pinole.
Citation(s): Chesnut 1902.
Note(s): seeds frequently infected with a fungus; only
occasionally fatal; symptoms consist of apathy,
giddiness, or a feeling of intoxication, accompanied
by ataxa, various abnormal sensations, mydriasis,
nausea, vomiting, gastric pain, and diarrhea; the

seeds contain 30-50% starch; plant introduced from Europe.
Citation(s): Kingsbury 1964; Stille and Maisch 1880; Munz and Keck 1965.

557 *Filago californica*
Common Name(s): California Cotton Rose.

Usage: Medicine: **Maidu:** used to treat fevers, chills, promote healing of sores, wounds.
Citation(s): Strike 1994.

558 *Foeniculum vulgare*
Family: *Apiaceae*
Common Name(s): Sweet Fennel.

Usage: Medicine: **Pomo (Southwestern):** seeds gathered, cleaned; a pinch chewed for stomach upset, indigestion, heart burn; seeds boiled for eye wash.
Citation(s): Goodrich et al 1980.
Note(s): naturalized from Europe.
Citation(s): Munz and Keck 1965.

559 *Fomitopsis pinicola*
Family: *Fomitopsidaceae*
Synonym(s): *Fomes pinicola*
Common Name(s): Shelf Fungus.

Usage: Material: **Karok:** peel the white underside of this fir (*Abies* spp.) log dwelling fungus and rub it on buckskin to "polish", or smooth the skin; not eaten.

Citation(s): Schenck and Gifford 1952.

Terminology: **Karok:** tuxuwai ixbahakaiwish.

Citation(s): Schenck and Gifford 1952.

560 *Fomitopsis rosea*

Synonym(s): *Trametes subrosa*

Family: *Polyporaceae*

Usage: Material: **Karok:** used to smooth buckskin much
the same as *Fomitopsis pinicola* was used.

Citation(s): Schenck and Gifford 1952.

Terminology: **Karok:** tuxuwai ixbanakaiwish.

Citation(s): Schenck and Gifford 1952.

Note(s): occurs on conifers.

Citation(s): Krieger 1936; Kim et al 2005.

561 *Fontinalis* spp.

Family: *Fontinalaceae*

Common Name(s): Water Moss.

Usage: Misc: **Miwok:** named but not used.

Citation(s): Barrett and Gifford 1933.

Terminology: **Miwok:** tcepekula.

Citation(s): Barrett and Gifford 1933; Kim et al 2005.

562 *Forestiera pubescens*

Family: *Oleaceae*

Synonym(s): *Foresteria neomexicana*

Usage: Misc: **Owens Valley Paiute, Kawaiisu
(Tehachapi):** named but not used.

Citation(s): Steward 1933; Zigmond 1981.

Terminology: **Owens Valley Paiute:** pa'tsanāva. **Kawaiisu
(Tehachapi):** to(m)bovɨ.

Citation(s): Steward 1933; Zigmond 1981.

563 *Fouquieria splendens*

Family: *Fouquieriaceae*
Common Name(s): Ochotillo, Candlewood.

Usage: Food: **Cahuilla:** the blossoms were eaten fresh
                   or soaked in water to make a Summer
                   drink; seeds were parched and ground
                   into flour and made into cakes or mush.

   Material: **Cahuilla:** utilized for firewood; built
                   fences around garments as prevention
                   against rodents.

Citation(s): Barrows 1900; Bean and Saubel 1972.
Terminology: **Cahuilla:** o-tos, utush.
Citation(s): Barrows 1900; Bean and Saubel 1972.
Note(s): seeds contain:
   protein   28.8%
   oil   18.6%.
Citation(s): Bean and Saubel 1972.

564 *Fragaria spp.*

Family: *Rosaceae*
Common Name(s): Strawberry, Wood Strawberry.

Usage: Food: **Bear River:** fruit eaten in season.
                   Salinan, Maidu (Northern): berries
                   eaten. **Sinkyone:** berries eaten raw.
                   Pomo (Southwestern): berries eaten
                   fresh.

   Material: **Pomo (Southwestern):** flowers used in
                   wreaths in flower dance at Strawberry
                   Festival, which is danced by young
                   girls; wild strawberries only eaten after

they are danced and blessed.
Citation(s): Dixon 1905; Mason 1912; Nomland 1935; 1938; Goodrich et al 1980.

565 *Fragaria chiloensis*

Synonym(s): *Fragaria chilensis, Fragaria chilensis* ssp. *pacifica*
Common Name(s): Sand Strawberry.

Usage: Food: **Wiyot:** the berries were eaten. **Pomo (Southwestern):** berries eaten fresh.
Citation(s): Loud 1918; Goodrich et al 1980.
Terminology: **Pomo (Southwestern):** qʰám?šudu.
Citation(s): Goodrich et al 1980.

566 *Fragaria vesca*

Synonym(s): *Fragaria californica, Fragaria vesca* ssp. *californica*
Common Name(s): Wood Strawberry.

Usage: Food: **Karok:** fresh berries eaten when ripe but are not preserved. **Yuki:** eaten, especially by children, direct from the vines. **Coast Yuki:** berries eaten raw. **Pomo (Southwestern):** berries eaten raw when they ripen in June. **Cahuilla:** fruit always eaten fresh.
Citation(s): Chesnut 1902; Gifford 1939, 1967; Schenck and Gifford 1952; Bean and Saubel 1972.
Terminology: **Yuki:** tululmen. **Karok:** uxnahich, "little berry." **Cahuilla:** piklyam.
Citation(s): Gifford 1939; Schenck and Gifford 1952; Bean and Saubel 1972.

567 *Frangula californica* ssp. *occidentalis*

Family: *Rhamnaceae*

Synonym(s): *Rhamnus californica* ssp. *occidentalis*

Common Name(s): Coffee Berry.

Usage: Medicine: **Cahuilla:** berries steeped in water, drunk as laxative or tonic; bark, dried, ground into powder in stone mortar, used as cure for constipation.

Citation(s): Bean and Saubel 1972.

568 *Frangula purshiana*

Synonym(s): *Rhamnus purshiana*

Common Name(s): Cascara Sagrada, California Buckthorn.

Usage: Food: **Coast Yuki:** berries eaten raw.

Medicine: **Pomo:** bark was made into an infusion which was drunk as a very strong cathartic. **Karok:** an infusion of the bark is used as a physic. **Yurok:** bark was boiled and used as a cathartic. **Modoc (Lutuami) (Lutuami):** bark, leaves used as a laxative; berries as an emetic. **Costanoan:** bark, berries, used as a cathartic..

Citation(s): Gifford 1939; Merriam 1967; Schenck and Gifford 1952; Strike 1994; Rayburn 2012.

Terminology: **Yuki:** kal′a, cheel. **Pomo (Northern):** tī′ta cēwa; **(Central):** hō′mtsat kale; **(Southwestern):** xōwa′l, batsa′sa. **Karok:** xoutyeupin.

Citation(s): Gifford 1939; Barrett 1952; Schenck and Gifford 1952.

Note(s): a source of cascara, a commonly used laxative; if used in excess: nausea, dizziness, vomiting, abdominal

pain, watery bloody diarrhea.
Citation(s): Hardin and Arena 1974; Kingsbury 1964; Fuller and McClintok 1986.

569 *Frangula rubra*

Synonym(s): *Rhamnus rubra*
Common Name(s): Cascara, Buckthorn Berry.

Usage: Food: **Atsugewi:** berries gathered and eaten when fresh.

         Medicine: **Miwok:** decoction of the bark drunk as a cathartic, as a laxative. **Maidu:** bark decoction used asa a cathartic.

         Misc: **Tubatulabal:** when berries ripen in the Fall, it was time to gather *Pinus* seeds.

Citation(s): Barrett and Gifford 1933; Garth 1953; Strike 1994.
Terminology: **Miwok (Northern Sierra):** lo'o. **Atsugewi:** yuhaiup.
Citation(s): Barrett and Gifford 1933; Garth 1953.

570 *Frankenia salina*

Family: *Frankeniaceae*
Synonym(s): *Frankenia grandifolia*

Usage: Medicine: **Diegueño:** tea brewed to treat colic.
Citation(s): Strike 1994.

571 *Frasera albicaulis*

Family: *Gentianaceae*
Common Name(s): White-stemmed Frasera.

Usage: Medicine: **Maidu:** used to treat infected sores.
Citation(s): Strike 1994.

572   *Frasera albicaulis var. nitida*
        Synonym(s): *Swertia nitida*
        Common Name(s): Frasera.

        Usage: Misc: **Karok:** named but not used.
        Citation(s): Schenck and Gifford 1952.
        Terminology: **Karok:** mahkkapafich.
        Citation(s): Schenck and Gifford 1952.

573   *Frangula californica*
        Family: *Rhamnaceae*
        Synonym(s): *Rhamnus californica*
        Common Name(s): Coffee Berry, Pigeon Berry, Wood
                Strawberry,   California Buckthorn,
                Buckthorn.

        Usage: Food: **Owens Valley Paiute:** eaten fresh or
                                boiled in pots. **Kawaiisu (Tehachapi):**
                                berries eaten. **Chumash:** berries, felt
                                that fruit was bear's food and that
                                eating them would make a person
                                crazy.
                Medicine: **Maidu, Southern (Nisenan):** for
                                toothache it is heated as hot as can be
                                borne, placed in the mouth against the
                                offending member and tightly gripped
                                between the teeth. **Maidu:** leaf
                                decoction used to soothe rashes caused
                                by Poison Oak. **Costanoan:** leaf
                                decoction used to soothe rashes caused
                                by Poison Oak; bark, inner, decoction
                                used as a purgative; dried, ground,
                                used as a laxative preparation. **Pomo:**
                                bark used as a cathartic and as a kidney
                                remedy; bark boiled in water as a cure

for mania; **(Southwestern):** bark boiled and the decoction used for consumption, constipation; bark has to be stored for a year before usage. **Achomawi:** used as a cathartic; as a medicine for rheumatism. **Kawaiisu (Tehachapi):** berries had a laxative effect; crushed berries used to stop flow of blood; on sores to stop infections, and to counteract poisoning; crushed leaves and buds plus immature berries ground slightly, rubbed into burns – heals the wound. **Atsugewi, Miwok:** bark tea used to relieve constipation. **Esselen:** used as a cathartic. **Chumash:** leaves, rubbed on skin to treat rheumatism; leaves boiled, bathe in decoction to treat Poison Oak rash; bark, boiled, drunk for stomach gas.

Misc: **Tubatulabal:** when berries ripe it is time to gather piñon; seeds rejected as food as "no good, bitter." **Karok, Miwok:** named but not used. **Pomo (Southwestern):** considered berries as poisonous.

Citation(s): Barrett and Gifford 1933; Chesnut 1902; Gifford 1967; Merriam 1967; Powers 1877; Schenck and Gifford 1952; Steward 1933; Voegelin 1938; Zigmond 1981; Goodrich et al 1980; Strike 1994; Rayburn 2012; Breschini and Haversat 2004; Timbrook 2007.

Terminology: **Pomo (Southwestern):** kamsudu, basata, ba Śaśa. **Owens Valley Paiute:** pōgōpū. **Miwok (Central Sierra):** lo'o. **Karok:** akrapuk afishi, "rig-tailed cat berries." **Chumash (Barbareño, Ynezeño):** puq'; **(Ventureño):** chatɨshwɨ 'ikhus.

Citation(s): Gifford 1967; Steward 1933; Barrett and Gifford

1933; Schenck and Gifford 1952; Goodrich et al 1980; Timbrook 2007.

Note(s): the plants contain glycosides which are fairly strong laxatives. For the **Pomo (Southwestern)** there are conflicting reports from the ethnographers, possibly due to the passage of time, or perhaps due to different informants [author].

Citation(s): Hardin and Arena 1974.

574 *Fraxinus dipetala*

Family: *Oleaceae*
Common Name(s): Foothill Ash.

Usage: Material: **Chumash:** wood, burned, charcoal ground, mixed with water to make black paint; twigs, steeped until water turned bluish, particularly good medicine for sick people, very cooling.

Citation(s): Timbrook 2007.

Terminology: **Chumash (Barbareño, Ynezeño):** wɨntɨ'y; **(Ventureño):** wɨltɨ'y.

Citation(s): Timbrook 2007.

575 *Fraxinus latifolia*

Family: *Oleacaea*
Synonym(s): *Fraxinus oregana*
Common Name(s): Fresno Tree, Oregon Ash.

Usage: Food: **Maidu:** fruit, sap eaten. **Pomo (Eastern):** Army Worm (*Homoncocnemis fortis*) gathered in late Spring from tree, parched, eaten.

      Material: **Tubatulabal:** branches used for firewood. **Yuki:** used for fuel and tobacco pipes. **Pomo:** make canoes, small tools. **Miwok:** made two-pronged harpoon from long poles; used for taking salmon. **Kawaiisu (Tehachapi):** a peeled pole, ten or more feet in length, is employed to knock down pinyon cones. **Costanoan:** leaves placed in sandals as snake repellant.

Medicine: **Pomo:** fresh roots mashed; valued as cure for wounds. **Maidu:** root, seed, leaves used to treat many illnesses; root poultice used to treat bear wounds. **Karok:** bark used to prepare special medicine to avert an impure person's harmful influence. **Costanoan:** twigs, stepped in cold water, drank for fever remedy.

Citation(s): Barrett and Gifford 1933; Chesnut 1902; Curtin 1957; Voegelin 1938; Zigmond 1981; Strike 1994; Rayburn 2012.

Terminology: **Karok:** akravshiip. **Miwok:** pa'ñasu̲. **Yuki:** pök. **Kawaiisu (Tehachapi):** edɨvɨ.

Citation(s): Schenck and Gifford 1952; Barrett and Gifford 1933; Curtin 1957; Zigmond 1981.

576 *Fremontodendron californicum*

Family: *Malvaceae*
Synonym(s): *Fremontica californica, Fremontia californica*
Common Name(s): Flannel Bush, Fremontia.

Usage: Material: **Owens Valley Paiute:** made carrying straps of braided fibers. **Tubatulabal:** peeled, forked branches used for frames of cradles; rope made from outer bark. **Miwok:** made string from fibers; made hoop of bark wrapped with buckskin for the hoop and pole game. **Kawaiisu (Tehachapi):** bark fiber used to make cordage. **Yokuts:** bark twisted into good bow string; large branches split to make bows; small branches used for arrows.

Medicine: **Kawaiisu (Tehachapi):** inner bark soaked in water and drunk as physic.

Citation(s): Barrett and Gifford 1933; Steward 1933; Voegelin 1938; Zigmond 1981; Strike 1994.

Terminology: **Kwaiisu:** unparabɨ.

Citation(s): Zigmond 1981.

577 *Fritillaria* spp.

Family: *Liliaceae*

Usage: Food: **Yana, Maidu, Shasta:** roots were roasted in earth oven all day and eaten.

Misc: **Wailaki:** feared digging up all bulbs would causes all acorns to drop off Oak Trees (*Quercus spp.*).

Citation(s): Sapir and Spier 1942; Strike 1994.

Terminology: **Yana:** t'a'ka.

Citation(s): Sapir and Spier 1943.

578 *Frittaria affinis*

Synonym(s): *Fritillaria mutica, Frittaria lanceolata*

Usage: Misc: **Pomo:** named but not used. **Wailaki:** believed that if roots are dug up the acorns will drop off the oak.

Citation(s): Chesnut 1902.

579 *Fritillaria recurva*

Usage: Food: **Shasta:** the bulbs were eaten.
Citation(s): Holt 1946.
Terminology: **Shasta:** chwahú'.
Citation(s): Holt 1946.

580 *Fucus* spp.
Family: *Fucaceae*
Common Name(s): Edible Seaweed.

Usage: Food: **Yurok:** dried and eaten without cooking.
Citation(s): Merriam 1967.

581 *Funaria hygrometrica*
Family: *Funariaceae*

Usage: Misc: **Miwok:** named but not used.
Citation(s): Barrett and Gifford 1933.
Terminology: **Miwok (Central Sierra):** tcepekupa.
Citation(s): Barrett and Gifford 1933.

582 *Funastrum cyanchoides* spp. *hartwegii*
Family: *Asclepiadaceae*
Synonym(s): *Philibertia heterophylla, Sarcostemma cyanchoides* spp. *hartwegii*

Usage: Food: **Luiseño:** eaten raw with salt.
Citation(s): Sparkman 1908.
Terminology: **Luiseño:** towunla.
Citation(s): Sparkman 1908.

583 *Galium* spp.

Family: *Rubiaceae*
Common Name(s): Parasite Vine.

Usage: Material: **Maidu:** root made red or purple dye.

Medicine: **Maidu, Southern (Nisenan):** for rheumatism the leaves and stems pf the vine, which grows in the middle of the Chaparral (*Ceanothus spp.*) bush, are heated, or burned, and then clapped hot on the place where the affectation is located. **Miwok:** drank infusion to reduce inflammation. **Costanoan:** decoction given for dysentery, used externally for rheumatism, wounds.

Citation(s): Powers 1877; Strike 1994; Rayburn 2012.

584 *Galium angustifolium*

Usage: Medicine: **Diegueño:** leaves used to treat diarrhea.

Misc: **Tubatulabal:** considered weeds.

Citation(s): Voegelin 1938; Strike 1994.

585 *Galium triflorum*

Common Name(s): Sweet Bedstraw, Sweet-Scented Bedstraw.

Usage: Medicine: **Karok:** made into "love medicine" by women. **Miwok:** boiled and drunk as tea for dropsy. **Costanoan:** decoction used to stop diarrhea; used as skin lotion for wounds, rheumatic pains.

Citation(s): Barrett and Gifford 1933; Schenck and Gifford 1952; Strike 1994.

Terminology: **Miwok (Central Sierra):** tutumkalali. **Karok:**

akkwanwawup.

Citation(s): Barrett and Gifford 1933; Schenck and Gifford 1952.

586   *Garrya* spp.

Family: *Garryaceae*

Common Name(s): Fever Bush, Skunk Bush, Sil-tassel Bush.

Usage: Material: **Wiyot, Yurok:** fire-hardened wood used as pry bars in mussel gathering.

Medicine: **Kawaiisu (Tehachapi), Yuki, Mohave:** leaves boiled, infusion used for stomach problems, diarrhea.

Citation(s): Strike 1994.

587   *Garrya elliptica*

Usage: Material: **Yurok:** used for mussel bar to pry mussels off the rocks; first hardened by fire.

Medicine: **Coast Yuki:** boiled decoction of leaves drunk for stomach trouble. **Pomo (Southwestern):** leaves made into tea to bring on women's period when it is late; very strong and could cause an abortion.

Citation(s): Gifford 1939; Merriam 1967; Gooodrich et al 1980.

Terminology: **Yuki:** wesika. **Pomo (Southwestern):** du?čán?, "sticky plant."

Citation(s): Gifford 1939; Goodrich et al 1980.

588 *Garrya flavescens*
Common Name(s): Silk-Tassel Bush.

Usage: Medicine: **Kawaiisu (Tehachapi):** decoction from leaves drunk hot for relief of colds.
Citation(s): Zigmond 1981.

589 *Garrya fremontii*
Common Name(s): Bear Bush.

Usage: Misc: **Karok:** named but not used.
Citation(s): Schenck and Gifford 1952.
Terminology: **Karok:** oshoxurip, "bitter sugar pine."
Citation(s): Schenck and Gifford 1952.

590 *Gastridium phleoides*
Family: *Poaceae*
Synonym(s): *Gastridium vantricosum*
Common Name(s): Nit Grass.

Usage: Misc: **Pomo (Southwestern):** ƛ'a?šu qá?di.
Citation(s): Goodrich et al 1980.
Note(s): introduced from Europe.
Citation(s): Munz and Keck 1965.

591 *Gaultheria ovatifolia*
Common Name(s): Oregon Wintergreen.

Usage: Food: **Maidu:** berries eaten fresh, or cooked.
Citation(s): Strike 1994.

592 *Gaultheria shallon*

Family: *Ericaceae*

Common Name(s): Salal.

Usage: Food: **Karok:** the berries are gathered and eaten when ripe but are not preserved. **Pomo (Southwestern):** berries are eaten raw or cooked. **Coast Yuki:** berries eaten raw.

Material: **Karok:** berries rubbed on basket caps to stain black.

Misc: **Pomo (Southwestern):** pregnant women or fathers-to-be not to eat berries or baby will come out dark when born.

Citation(s): Gifford 1939, 1967; Schenck and Gifford 1952; Goodrich et al 1980.

Terminology: **Pomo (Southwestern):** koishosho. **Wiyot:** mĪkwel. **Yuki:** hisimel, sala. **Karok:** purisukams, "big huckleberry."

Citation(s): Gifford 1939, 1967; Loud 1918; Schenck and Gifford 1952; Goodrich et al 1980.

593 *Gayophytum ramosissimum*

Family: *Onagraceae*

Usage: Medicine: **Maidu:** infusion used to soothe irritated skin.

Citation(s): Strike 1994.

594 *Gentiana* spp.

Family: *Gentianaceae*

Usage: Medicine: **Maidu:** used as tonic and to reduce fevers.

Citation(s): Strike 1994.

595 *Geranium carolinianum*
Family: *Geraniaceae*
Common Name(s): Carolina Geranium.

Usage: Medicine: **Maidu:** used to reduce fevers, as a diuretic; infusion to sooth children's rashes.
Citation(s): Strike 1994.

596 *Geranium dissectum*
Common Name(s): Common Geranium.

Usage: Misc: **Karok:** named but not used (?).
Citation(s): Schenck and Gifford 1952.
Terminology: **Karok:** `atahvichkiinach.
Citation(s): Schenck and Gifford 1952.
Note(s): naturalized from Europe.
Citation(s): Munz and Keck 1965.

597 *Geranium oreganum*
Synonym(s): *Geranium incisum*

Usage: Medicine: **Miwok:** root was pulverized and steeped; the decoction was rubbed on aching joints but not on open sores.
Citation(s): Barrett and Gifford 1933.
Terminology: **Miwok (Central Sierra):** olosena.
Citation(s): Barrett and Gifford 1933.seeds eaten.
Note(s): *Geranium* ssp. rhizomes are astringent, styptic.
Citation(s): Stuhr 1933.

598   *Gilia achilleifolia*
        Family: *Polemoniaceae*

Usage: Misc: **Miwok:** named but not used.
Citation(s): Barrett and Gifford 1933.
Terminology: **Miwok (Central Sierra):** Lelema.
Citation(s): Barrett and Gifford 1933.

599   *Gilia capitata*

Usage: Misc: **Tubatulabal:** considered weeds.
Citation(s): Voegelin 1938.

600   *Gilia capitata* ssp. *Staminea*

Usage: Food: **Luiseño:** seeds eaten.
Citation(s): Sparkman 1908.
Terminology: **Luiseño:** chachwomal.
Citation(s): Sparkman 1908.

601   *Gilia latiflora*

Usage: Misc: **Tubatulabal:** considered   weeds.
                         Kawaiisu (Tehachapi): named but not
                         used.
Citation(s): Voegelin 1938; Zigmond 1981.
Terminology: **Kwaiisu:** sanawagadɨbɨ.
Citation(s): Zigmond 1981.

602   *Glyceria borealis*
        Family: *Poaceae*
        Common Name(s): Manna Grass.

Usage: Food: **Northern Paiute (Paviotso):** seeds were
                         eaten.

Citation(s): Kelly 1932; Strike 1994.
Terminology: **Northern Paiute (Paviotso):** so·pi′.
Citation(s): Kelly 1932.

603 *Glycyrrhiza lepidota*
Family: *Fabaceae*
Common Name(s): Liquoria.

Usage: Material: **Panamint Shoshone (Koso):** stems made into cordage used to make woven bag or sack used by women to carry a child on her back.

Medicine: **Owens Valley Paiute:** made tea for sickness.

Citation(s): Steward 1933; Strike 1994.
Terminology: **Owens Valley Paiute:** atsia′ɦ dava.
Citation(s): Steward 1933.

604 *Gnaphalium* spp.
Family: *Poaceae*

Usage: Misc: **Chumash:** reported in area from collections from in caves. **Wiyot:** girls must stay away from plant and on no account touch it; if they do they are likely to have a baby.

Citation(s): Grant 1964; Merriam 1967.

605 *Gnaphalium palustre*
Common Name(s): Cudweed.

Usage: Medicine: **Maidu:** leaf decoction used to relieve throat inflammation; crushed plant used as poultice on bruises.

Misc: **Kawaiisu (Tehachapi):** named but not

used.
Citation(s): Zigmond 1981; Strike 1994.
Terminology: **Kawaiisu (Tehachapi):** piiyazigad<del>i</del>.
Citation(s): Zigmond 1981.

606 *Goodyera oblongifolia*
>>> Family: *Orchidaceae*
>>> Synonym(s): *Peramium decipens*
>>> Common Name(s): Rattlesnake Plantain.

Usage: Misc: **Karok:** named but not used.
Citation(s): Schenck and Gifford 1952.
Terminology: **Karok:** achnapuichti, "ring-tailed cat's ears."
Citation(s): Schenck and Gifford 1952.

607 *Gravia spinosa*
>>> Family: *Chenopodiaceae*

Usage: Misc: **Mono Lake Paiute:** named but not used.
Citation(s): Steward 1933.
Terminology: **Mono Lake Paiute:** poxo'pi.
Citation(s): Steward 1933.

608 *Grayia spinosa*
>>> Family: *Chenopodiaceae*
>>> Common Name(s): Spiny Hop-Sage.

Usage: Misc: **Kawaiisu (Tehachapi):** named but not used.
Citation(s): Zigmond 1981.
Terminology: **Kawaiisu (Tehachapi):** murunav<del>i</del>.
Citation(s): Zigmond 1981.

609 *Grindelia* spp.

Family: *Asteraceae*

Usage: Food: **Pomo:** used leaves for tea. **Karok:** eaten, in the Spring, as greens.

Medicine: **Pomo:** decoction of the whole plant used as a blood purifier, to open bowels, and to cure colds and colic, especially in children.

Citation(s): Chesnut 1902; Strike 1994.

Note(s): several species are secondary or facultative selenium absorbers; the plant is expectorant and sedative, with an action resembling atropine.

Citation(s): Kingsbury 1964; Grieve 1931.

610 *Grindelia camporum*

Synonym(s): *Grindelia robusta*
Common Name(s): Gumweed, Gum Plant.

Usage: Medicine: **Kawaiisu (Tehachapi):** leaves and flowers brewed to make decoction applied to sore parts of body. **Cahuilla:** boiled, infusion used as a remedy for rashes caused by Poison Oak, skin disease, coughs, throat and bronchial problems, cure colds. **Maidu:** used as a diuretic, a tonic, as a poultice for burns, sores, wounds, including rashes, sores caused by Poison Oak. **Miwok:** used for treating blood disorders. **Costanoan:** decoction used for dermatitis, especially from Poison Oak, also for boils, wounds. **Chumash:** fresh plant, decoction, used to treat Poison Oak rash, skin diseases, pulmonary

problems. **Miwok:** the leaves are steeped and the decoction used as a wash for running sores; steeped material was pulverized and applied to sores.
Citation(s): Zigmond 1981; Strike 1994; Rayburn 2012; Timbrook 2007; Barrett and Gifford 1933.
Terminology: **Kawaiisu (Tehachapi):** sanawagadɨbɨ. **Chumash (Venureño):** stɨq shiʼshaʼw.
Citation(s): Zigmond 1981; Timbrook 2007.
Terminology: **Miwok (Central Sierra):** kalkala.
Citation(s): Barrett and Gifford 1933.

611 *Grindelia hissutula*

Synonym(s): *Grindelia robusta patens, Grindelia hissutula* ssp. *Rubricaulis*
Common Name(s): Gum Plant.

Usage: Food: **Karok:** plants grows on flats and is gathered in the Spring and eaten raw as greens.

Material: **Karok:** the roots are boiled in a cooking basket and are used as shampoo to kill lice in the hair.
Citation(s): Schenck and Gifford 1952.
Terminology: **Karok:** offid.
Citation(s): Schenck and Gifford 1952.

612 *Grindelia squarrosa* var. *serrulata*

Synonym(s): *Grindelia squarrosa*
Common Name(s): Gum Plant.

Usage: Medicine: **Cahuilla:** decoction from plant used internally to cure colds.
Citation(s): Bean and Saubel 1972.

613   *Grindelia stricta*
> Synonym(s): *Grindelia latifolia*
> Common Name(s): Coastal Gum Plant.

> Usage: Medicine: **Costanoan:** may have been used for Poison Oak remedy.
> Citation(s): Rayburn 2012.

614 *Grindelia stricta* var. *platyphylla*
> Synonym(s): *Grindelia stricta* ssp. *venulosa*
> Common Name(s): Gum Plant.

> Usage: Material: **Pomo (Southwestern):** sticky sap used like glue.
> Citation(s): Goodrich et al 1980.
> Terminology: **Pomo (Southwestern):** qʰaqáhwe.
> Citation(s): Goodrich et al 1980.

615   *Gutierrezia californica*
> Family: *Asteraceae*
> Common Name(s): San Joaquin Matchweed.

> Usage: Medicine: **Kawaiisu (Tehachapi):** upper part of plant mashed and placed on hot rock; aching back or limbs placed directly on hot material.
>
> Material: **Kawaiisu (Tehachapi):** used as wall filler in the construction of the winter house.
> Citation(s): Zigmond 1981.
> Terminology: **Kawaiisu (Tehachapi):** šivapɨ.
> Citation(s); Zigmond 1981.

616 *Gutierrezia microcephala*
>>>> Common Name(s): Matchweed, Snakeweed.

Usage: Medicine: **Cahuilla:** solution made from plant used as gargle to cure toothache, or, part of plant placed inside mouth periodically to ease pain.
Citation(s): Bean and Saubel 1972.

617 *Gutierrezia sarothrae*
>>>> Common Name(s): Common Matchweed, Snakeweed.

Usage: Misc: **Kawaiisu (Tehachapi):** named but not used.
Citation(s): Zigmond 1981.
Terminology: **Kawaiisu (Tehachapi):** pohniyavɨ.
Citation(s): Zigmond 1981.

618 *Habenaria* spp.
>>>> Family: *Orchidaceae*
>>>> Common Name(s): Rein Orchid.

Usage: Food: **Maidu:** roots eaten.
Citation(s): Strike 1994.

619 *Habenaria unalascensis*

Usage: Food: **Pomo (Southwestern):** bulb baked, eaten.
>>>> Misc: **Miwok:** named but not used; considered a poisonous plant which caused blood vomiting if eaten.
Citation(s): Barrett and Gifford 1933; Goodrich et al 1980.
Terminology: **Miwok (Central Sierra):** sasi. **Pomo**

(Southwestern): koyó?yo.
Citation(s): Barrett and Gifford 1933; Goodrich et al 1980.

620 *Hastingsia alba*

Family: *Agavaceae*
Synonym(s): *Schoenolirion album*

Usage: Food: **Maidu (Northern):** roots eaten.
        Misc: **Karok:** people amuse themselves by putting the large leaf over their teeth, sucking in their breath and breaking the leaf with a snapping sound.

Citation(s): Dixon 1905; Schenck and Gifford 1952.
Terminology: **Karok:** basrakupkam.
Citation(s): Schenck and Gifford 1952.

621 *Helianthus annus*

Family: *Asteraceae*
Synonym(s): *Helianthus annus* ssp. *lenticularis*
Common Name(s): Sunflower.

Usage: Food: **Cahuilla:** seed, gathered in quantity if Fall, dried, ground and mixed with flour of other seeds.

Citation(s): Bean and Saubel 1972.
Terminology: **Cahuilla:** pa'akal.
Citation(s): Bean and Saubel 1972.
Note(s): seeds contain:
         protein   26.2-43.8%
         ash   3.1-5.6%
         oil   28.2-54.5%
Citation(s): Bean and Saubel 1972.

622 *Helianthus Cusickii*

Usage: Medicine: **Shasta:** used to treat chills, fevers; burned root to purify a dwelling after a death had occurred.
Citation(s): Strike 1994.

623 *Helenium bigelovii*
Family: *Asteraceae*

Usage: Misc: **Miwok:** named but not used.
Citation(s): Barrett and Gifford 1933.
Terminology: **Miwok (Central Sierra):** hangu.
Citation(s): Barrett and Gifford 1933.

624 *Helenium puberulum*
Common Name(s): Sneezeweed.

Usage: Medicine: **Pomo:** three plants are boiled in a gallon of water and three tablespoons are taken before each meal for two or three days to cure certain venereal complaints. **Costanoan:** plant, dried, powdered, rubbed on forehead and nose of cold victims, sniffed to induce sneezing, sprinkled in wounds. **Chumash:** plant, as a tonic, treatment for scurvy; flowers, powdered, used a snuff to treat head colds, catarrh, influenza.

Misc: **Miwok:** named but not used.
Citation(s): Barrett and Gifford 1933; Chesnut 1902; Rayburn 2012; Timbrook 2007.
Terminology: **Miwok (Central Sierra):** polobia. **Chumash (Ventureño):** manakhshmu.

Citation(s): Barrett and Gifford 1933; Timbrook 2007.

625 *Helianthella californica*
Family: *Asteraceae*

Usage: Food: **Yana:** the yellow flowers were cooked and eaten.
Citation(s): Sapir and Spier 1943.
Terminology: **Yana:** ts'igā′naaua'I.
Citation(s): Sapir and Spier 1943.

626 *Helianthus annus*
Family: *Asteraceae*
Synonym(s): *Helianthus annus* ssp. *lenticularis*
Common Name(s): Wild Sunflower.

Usage: Food: **Huchnom:** used to make bread. **Luiseño, Northern Paiute (Paviotso):** seeds are eaten. **Kawaiisu (Tehachapi):** ripe seeds roasted lightly, pounded, ground into meal, eaten dry.
Citation(s): Kelly 1932; Powers 1877; Sparkman 1908; Zigmond 1981.
Terminology: **Northern Paiute (Paviotso):** pa^{hü} (seed), paukla. **Kawaiisu (Tehachapi):** pa?akatabɨ.
Citation(s): Kelly 1932; Sparkman 1908; Zigmond 1981.
Note(s): seeds contain:

Ash 4.4%

Protein 54.4%

Oil 48.8%

Citation(s): Earle and Jones 1962.

627 *Helianthus bolanderi*
>   Common Name(s): Sunflower.

Usage: Food: **Owens Valley Paiute, Mono Lake
                        Paiute:** flowers harvested in about
                        August, dried in the sun, threshed and
                        winnowed.
Citation(s): Steward 1933.
Terminology: **Owens Valley Paiute, Mono Lake Paiute:** pāk$^{\ddot{u}}$.
Citation(s): Steward 1933.

628 *Helianthus cusickii*

Usage: Medicine: **Shasta:** roots were burned in the house
                        after a death; also used for treatment of
                        a long, slow sickness characterized by
                        chills and fever.
Citation(s): Holt 1946.
Terminology: **Shasta:** garawihú'.
Citation(s): Holt 1946.

629 *Heliotropium curvassavicum* var. *oculatum*
>   Family: *Boraginaceae*
>   Synonym(s): *Heliotropium curvassavicum*
>   Common Name(s): Common Heliotrope.

Usage: Food: **Tubatulabal:** used seeds.
        Medicine: **Owens Valley Paiute, Tubatulabal:**
                        roots made into tea for diarrhea.
                        Diegueño: tea used to regulate
                        menstruation. **Maidu:** root decoction
                        used as gargle for sore throats, an as
                        emetic; sap dried and pulverized used
                        as a poultice on wounds, abrasions.
Citation(s): Steward 1933; Voegelin 1938; Strike 1994.

Terminology: **Cahuilla:** hoskos (?), hoshos (?).
Citation(s): Bean and Saubel 1972.

630 *Hemizonia congesta* ssp. *clevelandi*
Family: *Asteraceae*
Synonym(s): *Hemizonia clevelandi*

Usage: Food: **Pomo:** used seeds for pinole.
Citation(s): Barrett 1952.
Terminology: **Pomo (Central):** ala´.
Citation(s): Barrett 1952.

631 *Hemizonia congesta* ssp. *luzulaefolia*
Synonym(s): *Hemizonia luzulaefolia*

Usage: Food: **Pomo:** seeds gathered in the Fall and a small supply saved over winter and spring; used for pinole; slightly bitter but spicy when roasted.

Medicine: **Maidu:** treated urinary ailments.
Citation(s): Chesnut 1902; Strike 1994.

632 *Heracleum maximum*
Family: *Apiaceae*
Synonym(s): *Heracleum lanatum*
Common Name(s): Cow Parsnip.

Usage: Food: **Hupa:** the fresh shoots are eaten raw. **Karok:** ate the fresh shoots. **Yuki:** leaf and flower stalks eaten as greens in the Spring and early Summer before the flowers have expanded; the tender young stems are peeled and eaten raw. **Pomo (Southwestern):** new shoots peeled, eaten raw.

Material: **Karok:** used to dye porcupine quills yellow. **Pomo (Southwestern):** dried hollow stems used by boys as toy blowguns.

Medicine: **Pomo (Southwestern):** roots pounded raw, baked under ashes, cooked or warmed, and applied as a poultice for swellings and rheumatism. **Washo:** half-cupful of decoction of the root used as a tea to stop diarrhea. **Maidu:** leaves smoked to cure headaches, colds. **Miwok:** roots crushed to poultice swellings. **Miwok (Coast):** roots used to treat coughs. **Modoc (Lutuami):** infusion drunk to relieve headaches; root infusion, or roots chewed, to relieve colds, coughs, chest congestion; root compress used for eye ailments; plant mashed, rubbed on bruises, swellings.

Citation(s): Chesnut 1902; Curtin 1957; Gifford 1967; Goddard 1903; Schenck and Gifford 1952; Train et al 1941; Goodrich et al 1980; Strike 1994.

Terminology: **Yuki:** muń-shök. **Pomo (Southwestern):** shoshokale, butakashosho kale. **Northern Paiute (Pavitoso):** dotsi'toniga. **Hupa:** selkyō. **Karok:** ihyivkanva, "Shout across." **Washo:** comb-ho.

Note(s): the leaves can cause dermatitis; roots and leaves are reputed carminative, stimulant.

Food values – leaves and stems:

	(wet)	(dry)
Moisture	6.9%	
Sugar		
Reducing	3.0%	43.4%
Non-reducing	0.0%	0.0%

Starch	0.0%	0.05
Hemicellulose	0.8%	11.6%
Protein	0.0%	17.7%
Ash	0.0%	12.6%

Citation(s): Yanovsky 1938; Muenscher 1951; Stuhr 1933.

633 *Hericium coralloides*
>Family: *Hericiaceae*
>Common Name(s): Coral Mushroom.

Usage: Food: **Pomo (Southwestern):** baked.
Citation(s): Goodrich et al 1980.
Terminology: **Pomo (Southwestern):** s'ayiċe·.
Citation(s): Goodrich et al 1980.

634 *Hesperocallis undulata*
>Family: *Agavaceae*
>Common Name(s): Desert Lily, Creosote Bush.

Usage: Food: **Cahuilla:** bulb eaten raw, or baked in stone-lined pit, covered with hot ashes and leaves, baked for 12-24 hours; garlic-like flavor. **Luiseño:** eaten.
Citation(s): Bean and Saubel 1972; Strike 1994..

635 *Hesperocyparis* spp.
>Family: *Cupressaceae*
>Synonym(s): *Cupressus* spp.
>Common Name(s): Cedar.

Usage: Material: **Monache (Western Mono):** leaves used to line acorn leaching basket.
         Medicine: **Miwok:** decoction of the stems drunk as a remedy for colds and rheumatism.
Citation(s): Barrett and Gifford 1933; Gifford 1932.

636   *Hesperocyparis macrocarpa*
          Synonym(s): *Cupressus macrocarpa*
          Common Name(s): Monterey Cypress.

Usage: Medicine: **Costanoan:** foliage, decoction, used to
                       relieve rheumatic pain.
Citation(s): Strike 1994.

637   *Hesperocyparis nevadensis*
          Synonym(s): *Cupressus nevadensis*

Usage: Medicine: **Kawaiisu (Tehachapi):** cones boiled;
                       red infusion drunk hot or cold; remedy
                       for backache, kidney, menstruation
                       problems; seeds dried, made into tea
                       for colds, coughs (3 times a day for 1
                       month - no meat, fresh water at this
                       time).
Citation(s): Zigmond 1981; Strike 1993.

638   *Hesperocyparis sargentii*
          Synonym(s): *Cupressus sargentii*

Usage: Material: **Chumash:** may have used fibers for
                       cradle lining.
Citation(s): Grant 1964.

639   *Hesperolinum californicum*
          Family: *Linaceae*
          Synonym(s): *Linum californicum, Linum californica*
          Common Name(s): Blue Flax, California Dwarf Flax.

Usage: Medicine: **Maidu:** used to relieve inflamed
                       mucous membranes, to soothe skin

irritation. **Costanoan:** seeds, decoction, administered as a form of fomentation for fever.

Citation(s): Strike 1994; Rayburn 2012.

640 *Hesperoyucca whipplei*

Synonym(s): *Yucca whipplei*
Common Name(s): Spanish Dagger, Spanish Bayonet, Our Lord's Candle.

Usage: Food: **Luiseño:** the head is eaten; the flowers are boiled and eaten; the scape or stalk is also eaten, roasted and chewed for their sugar; leaf bases were cooked. **Tubatulabal:** stalks eaten in February to May. **Kamia:** the stem is good to eat when roasted green. **Cahuilla:** stalk is cut before flowering when sap is full; roasted in sections in a fire pit for one night, dried, ground, mixed with water to form cakes, or sliced, parboiled, cooked like squash; the dates or seeds bags are eaten as are the flowers, which, when in bloom, are picked and cooked in water in an olla; immature flowers parboiled, eaten; mature flowers boiled, up to three times, with salt before eating; blossoms dried and saved. **Kawaiisu (Tehachapi):** in the Spring the heart is removed, roasted for 2 days and eaten. **Chumash:** crowns, roasted in pits eaten . **Esselen:** stalk, fluid, was drunk; stalk, roasted, eaten.

Material: **Tubatulabal:** center stalk used in the manufacture of a bull-roarer. **Chumash:**

made cordage, and from that such items as netting, headbands, fishing lines, men's belts, sandals; needle, tattooing; flower stalks as tinder. **Kawaiisu (Tehachapi):** fibers from leaves used as cordage. **Diegueño:** leaves, buried in ground until fleshy part rotted away, the fine white fiber remaining often used as foundation material in basketry. **Esselen:** leaves, fiber, were used.

Citation(s): Barrows 1900; Curtis 1924; Merriam 1967; Sparkman 1908; Voegelin 1938; Zigmond 1981; Bean and Saubel 1972; Strike 1994; Landberg 1965; Breschini and Haversat 2004; Timbrook 2007.

Terminology: **Luiseño:** panal. **Cahuilla:** panuul (stalk). **Kawaiisu (Tehachapi):** kʷinuurɨbɨ. **Chumash (Barbareño, Ynezeño):** pokh; **(Island, Ventureño):** shtakuk; **(Obispeño):** ts'isuyi'.

Citation(s): Sparkman 1908; Zigmond 1981; Timbrook 2007.

Note(s): seeds contain:

Protein 24.4%

Oil 25.4%

Citation(s): Earle and Jones 1962.

641 *Heteromeles arbutifolia*

Family: *Rosaceae*

Synonym(s): *Photinea arbutifolia*

Common Name(s): Toyon, Christmas Berry, Little Madrona, California Holly.

Usage: Food: **Wappo, Salinan, Hupa:** berries eaten. **Maidu, Southern (Nisenan):** the bright-red berries eaten with relish. **Karok:** berries placed on basket in front of fire until they are wilted when they are eaten; they are not stored. **Miwok:** berries boiled, then baked in earth oven, or, stored for two months to soften, then parched and eaten. **Pomo (Southwestern):** berries wilted in hot ashes, then winnowed in a basket plate, eaten raw, usually without preparation; not stored. **Luiseño:** the berries are eaten parched(?). **Yuki:** occasionally eaten directly from tree; generally cooked first, most frequently by roasting, sometimes boiled. **Cahuilla:** ate berries raw, or cooked by roasting in fire, or tossing them on hot coals, which removed the slightly bitter taste. **Maidu:** berries made into cider. **Chumash:** berries eaten, or toasted over hot coals; dried in sun until black, mashed after standing "until it was good and ready."

Material: **Karok:** children throw the leaves into fire to hear them crack. **Chumash:** wood used for arrows, cooking implements, tools (awls, wedges, hide scrapers), composite fishhook, hairpin for hair buns (men) about 15 inches long with Woodpecker feathers attached to one end, thatching needles, digging sticks, pestles, bowls, drinking cups, walking sticks, war clubs, canoe pegs, cradle frames, flutes, gaming implements, fish smoking racks and fuel to smoke fish, headdress pins.

Medicine: **Yuki, Maidu:** a decoction of leaves and bark used for stomach ache and various aches and pains. **Diegueño:** leaves mashed to bathe sores. **Costanoan:** leaf tea used as blood purifier, to regulate girl's menses.

Citation(s): Barrett and Gifford 1933; Chesnut 1902; Curtis 1924; Driver 1936; Gifford 1967; Goddard 1903; Mason 1912; Powers 1877; Schenck and Gifford 1952; Sparkman 1908; Goodrich et al 1980; Bean and Saubel 1972; Strike 1994; Landberg 1965; Breschini and Haversat 2004; Timbrook 2007.

Terminology: **Pomo (Southwestern):** budu kale, budu (berry). **Miwok (Central Sierra):** koso. **Luiseño:** achawut. **Hupa:** isdewitc, "little madrona berries." **Karok:** pushiip, pusiyaa (berry). **Cahuilla:** ashwet. **Chumash (Barbareño, Island, Ynezeño):** qwe'; **(Obispeño):** ch'okoko, chmishɨ; **(Ventureño):** qwe.

Citation(s): Gifford 1967; Barrett and Gifford 1933; Sparkman 1908; Goddard 1903; Schenck and Gifford 1952; Goodrich et al 1980; Bean and Saubel 1972; Timbrook 2007.

Note(s): ripens in November and December.

Citation(s): Sargent 1922.

642 *Heterotheca grandiflora*

Family: *Asteraceae*

Common Name(s): Telegraph Weed.

Usage: Food: **Chumash:** seeds eaten (?).

Material: **Luiseño:** mainshafts of arrows were sometimes made from the stems of the plant. **Chumash:** plant, used to repel fleas.

Medicine: **Diegueño:** boiled, used to bathes sores; decoction drunk to treat infections.

Citation(s): Sparkman 1908; 1963; Strike 1994; Timbrook 2007.

Terminology: **Luiseño:** humut.

Citation(s): Sparkman 1908.

643 *Heuchera micrantha*
Family: *Saxifragaceae*
Common Name(s): Alum Root.

Usage: Food: **Miwok:** leaves first to be eaten in Spring; boiled or steamed; after steaming a certain quantity might be dried and stored.

Medicine: **Maidu:** root infusion used as eyewash, drunk as a general tonic.

Citation(s): Barrett and Gifford 1933.
Terminology: **Miwok (Central Sierra):** tcuyuma.
Citation(s): Barrett and Gifford 1933.

644 *Heuchera pilosissima*
Common Name(s): Alum Root.

Usage: Misc: **Karok:** named but not used (?).
Citation(s): Schenck and Gifford 1952.
Terminology: **Karok:** kafichtunveech, "little leaf."
Citation(s): Schenck and Gifford 1952.

645 *Hibiscus lasiocarpos var. occidentalis*
Family: *Malvaceae*
Synonym(s): *Hibiscus californicus*
Common Name(s): California Hibiscus, Rose Mallow, Hibiscus.

Usage: Medicine: **Maidu:** used as astringent to stop secretions or hemorrhages.
Citation(s): Strike 1994.

646 *Hieracium albiflorum*

        Family: *Asteraceae*
        Common Name(s): White Hawkweed.

Usage: Medicine: **Maidu:** used to ease toothaches, cure warts, as an astringent in treating hemorrhages, as a general tonic; plant and sap used as chewing gum.
Citation(s): Strike 1994.

647 *Hilaria* spp.

        Family: *Poaceae*
        Common Name(s): Galleta.

Usage: Material: **Serrano:** used in basketry.
Citation(s): Strike 1994.

648 *Hirschfeldia incana*

        Family: Brassicaceae
        Synonym(s): *Brassica geniculata*
        Common Name(s): Mustard.

Usage: Food: **Cahuilla:** an important Winter food plant; seed collected, ground into mush.
Citation(s): Bean and Saubel 1972.
Note(s): naturalized from Europe.
Citation(s): Munz and Keck 1965.

649 *Hoita macrostachya*

        Family: *Fabaceae*
        Synonym(s): *Psoralea macrostachya*
        Common Name(s): Leather Root.

Usage: Material: **Luiseño:** a yellow dye is made from the roots. **Pomo:** fiber for cordage is obtained from the bark and the root; the root fiber is said to be stronger. **Pomo (Southwestern:** probably used fiber from root. **Pomo, Maidu:** inner bark used for thread. **Salinan:** probably used root for yellow body paint. **Cahuilla:** roots boiled with basket materials to dyed them yellow.

Medicine: **Luiseño, Maidu:** root used for ulcers and sores.

Citation(s): Chesnut 1902; Barrett 1952; Gifford 1967; Mason 1912; Merrill 1923; Sparkman 1908; Strike 1994; Bean and Saubel 1972.

Terminology: **Luiseño:** pi'mukvul.

Citation(s): Sparkman 1908.

650 *Hoita orbicularis*

Synonym(s): Psoralea orbicularis
Common Name(s): Round-Leaved Hoita.

Usage: Food: **Luiseño Maidu:** used for greens. **Costanoan:** used as a tonic, and to reduce fever; decoction used for the blood.

Medicine: **Costanoan:** decoction used for fever.

Citation(s): Sparkman 1908; Strike 1994; Rayburn 2012.

Terminology: **Luiseño:** shi'kal.

Citation(s): Sparkman 1908; Strike 1994.

651 *Holcus lanatus*

Family: *Poaceae*
Common Name(s): Velvet Grass.

Usage: Material: **Miwok:** attached dried pods to

grasshopper legs left as bait for yellowjackets; when adults flew away with grasshopper leg, it was followed in order to find nest and get larvae which were considered as a food source.

Misc: **Pomo (Southwestern):** named but not used.

Citation(s): Barrett and Gifford 1933; Goodrich et al 1980.
Terminology: **Pomo (Southwestern):** qa?di boi, "grass soft."
Citation(s): Goodrich et al 1980.
Note(s): native of Europe.
Citation(s): Goodrich et al 1980; Munz and Keck 1965.

652 *Holocarpha virgata*

Family: *Asteraceae*
Synonym(s): *Hemizonia virgata, Holocarpus virgata*
Common Name(s): Tarweed.

Usage: Medicine: **Miwok:** bath with a decoction of the plant was used for measles and fever in general; must never be taken internally.

Citation(s): Barrett and Gifford 1933.
Terminology: **Miwok (Central Sierra):** kitimpa, tū'mō.
Citation(s): Barrett and Gifford 1933.

653 *Holodiscus discolor*

Family: *Rosaceae*
Synonym(s): *Holodiscus ariaefolius, Holodiscus discolor ariaefolius.*
Common Name(s): Cream Bush, Meadow Sweet, Indian Arrow-Root.

Usage: Food: **Cahuilla:** used small fruit.
        Material: **Karok, Hupa:** make sets of gambling

sticks from the shoots. **Pomo (Southwestern):** used for arrows; broad leaves used to make baby baskets. **Yurok:** used for arrows. **Hupa:** bundles of twigs made from the shrub are used in playing a game called *kin*; a form of armor was made consisting of a waistcoat of split rods placed vertically and bound together with a woof of native twine.

Citation(s): Barrows 1900; Gifford 1967; Goddard 1903; Merriam 1967; Schenck and Gifford 1952; Goodrich et al 1980.

Terminology: **Hupa:** kinLits. **Pomo (Southwestern):** chibuklan, cu?t'a, cʰi?búkʰlan. **Karok:** pitiri, "step on and flatten."

Citation(s): Goddard 1903; Gifford 1967; Schenck and Gifford 1952; Barrows 1900; Goodrich et al 1980.

654 *Holodiscus microphyllus*

Usage: Misc: **Cahuilla:** present day Cahuilla unfamiliar with its food use.

Citation(s): Bean and Saubel 1972.

Terminology: **Cahuilla:** tetnat.

Citation(s): Bean and Saubel 1972.

655 *Hordeum* spp.

Family: *Poaceae*
Common Name(s): Wild Barley.

Usage: Food: **Atsugewi:** seeds gathered and eaten; ground into flour and made into cakes. **Sinkyone:** seeds parched, made into pinole, stored for winter. **Pomo**

(**Southwestern**): seeds sometimes used in pinole.
Citation(s): Garth 1953; Nomland 1935; Goodrich et al 1980.
Terminology: **Pomo (Southwestern)**: na?qá qá?di.
Citation(s): Goodrich et al 1980.

656 *Hordeum brachyantherum*
  Synonym(s): *Hordeum nodosum*

Usage: Food: **Northern Paiute (Paviotso)**: the seeds were eaten.
Citation(s): Kelly 1932.
Terminology: **Northern Paiute (Paviotso)**: waha'bü.
Citation(s): Kelly 1932.

657 *Hordeum jubatum*
  Common Name(s): Foxtail.

Usage: Medicine: **Maidu:** root poultice used to treat sore eyes.
Citation(s): Strike 1994.

658 *Hordeum murinum* ssp. *glacum*
  Synonym(s): *Hordeum murinum, Hordeum Stebbinsi, Hordeum glaucum*
  Common Name(s): Wall Barley, Foxtail.

Usage: Food: **Karok:** used seeds (?). **Yuki:** used seeds in pinole. **Cahuilla:** seeds eaten occasionally when other foods were scarce.
  Medicine: **Costanoan:** plant, decoction, used to treat bladder ailments.
Citation(s): Garth 1953; Nomland 1935; Bean and Saubel 1972; Rayburn 2012.

Terminology: **Karok:** sitapvuuy, "mouse tail." **Cahilla:** pa'ish heqwas, "field mouse's tail."
Citation(s): Schenck and Gifford 1952; Bean and Saubel 1972.
Note(s): introduced from Europe.
Citation(s): Munz and Keck 1965, Munz 1968.

659 *Hordeum murinum* ssp. *leporinum*
Synonym(s): *Hordeum leporinum*
Common Name(s): Wild Barley.

Usage: Food: **Karok:** used seeds (?).
Citation(s): Schenck and Gifford 1952.
Terminology: **Karok:** `akϴiip.
Citation(s): Schenck and Gifford 1952.

660 *Hordeum vulgare*
Common Name(s): Common Barley.

Usage: Food: **Cahuilla:** grown since at least the mid-nineteenth century.
Citation(s): Bean and Saubel 1972.
Note(s): native of the Old World.
Citation(s): Munz and Keck 1965.

661 *Horkelia californica*
Family: *Rosaceae*

Usage: Medicine: **Pomo (Southwestern):** root boiled; tea used as a blood purifier.
Citation(s): Goodrich et al 1980.
Terminology: **Pomo (Southwestern):** balá·wenu qʰale.
Citation(s): Goodrich et al 1980.

662 *Hosackia oblongifolia*

Family: Fabaceae
Synonym(s): *Lotus torreyi, Lotus oblongifolius*

Usage: Misc: **Tubatulabal:** considered weeds.
Miwok, Kawaiisu (Tehachapi): named but not used.
Citation(s): Barrett and Gifford 1933; Voegelin 1938, Zigmond 1981.
Terminology: **Miwok (Central Sierra):** lulumati, susuyi.
Kawaiisu (Tehachapi): looko?ovaviži, paalaavaviži.
Citation(s): Barrett and Gifford 1933; Zigmond 1981.

663 *Hydnum repandum*

Family: *Hydnaceae*
Synonym(s): *Dentinum repandum*
Common Name(s): Hedgehog Mushroom.

Usage: Food: **Pomo (Southwestern):** baked.
Citation(s): Goodrich et al 1980.
Terminology: **Pomo (Southwestern):** ċala.
Citation(s): Goodrich et al 1980.

664 *Hydrocotyle* spp.

Family: *Araliaceae*
Common Name(s): Marsh Pennywort.

Usage: Food: **Cahuilla:** plant, gathered from top of water, used as greens
Citation(s): Bean and Saubel 1972.

665 *Hydrocotyle ranunculoides*
>   Common Name(s): Floating Pennywort.

>   Usage: Medicine: **Maidu:** used as a narcotic, emetic, to relieve skin irritation.
>   Citation(s): Strike 1994.

666 *Hydrocotyle verticillata*
>   Common Name(s): Whorled Pennywort.

>   Usage: Medicine: **Maidu:** used to treat venereal problems, to heal skin irritations.
>   Citation(s): Strike 1994.

667 *Hydrophyllum occidentale*
>   Family: *Boraginaceae*
>   Common Name(s): Waterleaf.

>   Usage: Food: **Maidu:** shoots, leaves, roots eaten raw, or cooked.
>   Citation(s): Strike 1994.

668 *Hypericum anagalloides*
>   Family: *Hypericaceae*
>   Common Name(s): Tinker's Penny.

>   Usage: Medicine: **Maidu:** used to accelerate healing of wounds, sores.
>   Citation(s): Strike 1994.

669 *Hypericum concinnum*
>   Common Name(s): Gold-Wire.

>   Usage: Medicine: **Miwok, Maidu:** boiled and used as a

wash for running sores.
Citation(s): Barrett and Gifford 1933; Strike 1994.
Terminology: **Miwok (Central Sierra):** hoyilü.
Citation(s): Barrett and Gifford 1933.

670 *Hypericum perforatum* ssp. *perforatum*
Synonym(s): *Hypericum perforatum*
Common Name(s): Klamath Weed.

Usage: Misc: **Karok:** named but not used.
Citation(s): Schenck and Gifford 1952.
Terminology: **Karok:** tsusinentaiwara, "spoil the ground."
Citation(s): Schenck and Gifford 1952.
Note(s): native of Europe; toxic to livestock.
Citation(s): Munz and Keck 1965; Baldwin et al 2012

671 *Hypericum scouleri*
Synonym(s): *Hypericum formosum, Hypericum formosum* var. *scouleri*
Common Name(s): St. John's Wort.

Usage: Food: **Miwok:** bulb eaten fresh as it came from the ground, or, dried, ground into flour, and used like acorn meal.

Medicine: **Maidu:** infusion used to bathe sores, swelling, to treat venereal ailments; powdered root used on wounds, cuts.

Citation(s): Barrett and Gifford 1933; Strike 1994.
Terminology: **Miwok (Central Sierra):** a′iisa.
Citation(s): Barrett and Gifford 1933.

672　*Hyptis emoryi*
>　Family: *Lamiaceae*
>　Common Name(s): Desert Lavender, Bee Sage.

>　Usage:　Medicine:　**Cahuilla:** boiled blossoms and leaves, infusion used to stop hemorrhages.
>　Citation(s): Bean and Saubel 1972; Strike 1994..

673　*Ipomopsis aggregata*
>　Family: *Polemoniaceae*
>　Common Name(s): Sky Rocket.

>　Usage:　Medicine:　**Maidu:** leaf decoction used as a purgative, eyewash, blood tonic, disinfectant, treat fevers; leaf infusion relieved stomach aches, used as laxative.
>　Citation(s): Strike 1994.

674　*Ipomopsis congesta*
>　Synonym(s): *Gilia congesta*

>　Usage:　Medicine:　**Washo:** for dropsy the tea was drunk and poultices of the crushed plant were applied.
>　Citation(s): Train et al 1941.
>　Terminology: **Washo:** wem-see.
>　Citation(s): Train et al 1941.

675　*Iris* spp.
>　Family: *Iridaceae*
>　Common Name(s): Wild Iris.

>　Usage:　Food:　**Monache (Western Mono), Yokuts:** seed made into flour.

Material:	**Wappo:** used fibers for cordage. **Miwok:** the leaf was used as foundation in making baskets. **Hupa:** fiber collected from plants growing under Oak (*Quercus*) considered superior to those from under Pine (*Pinus*). **Yokuts:** women gathering Manzanita (*Arctostaphylos*) berries in hot, dry areas wrapped babies in leaves to prevent dehydration. **Pomo:** covered corn meal in leaching pits with leaves before pouring water to leach out tannic acid.
Medicine:	**Yana:** roots chewed raw as cure for cough. **Modoc (Lutuami):** root decoction used to soothe sore eyes. **Maidu, Wintun:** root piece placed in tooth cavity to stop toothache. **Maidu:** root burned, smoke inhaled to alleviate dizziness; root decoction used as a cathartic, emetic, but large doses could cause severe digestive problems. **Pomo:** root medicine would accelerate the birthing process.
Misc:	**Northern Paiute (Paviotso), Mono Lake Paiute:** considered poisonous.

Citation(s): Driver 1936; Kelly 1932; Merrill 1923; Sapir and Spier 1943; Steward 1933; Strike 1994.

Terminology: **Mono Lake Paiute:** wasta'vü.

Citation(s): Steward 1933.

Note(s): leaves contain an irritant principle which produces gastroenteritis if ingested in sufficient, relatively large quantities.

Citation(s): Kingsbury 1964.

676   *Iris douglasiana*

Common Name(s): Mountain Iris.

Usage: Material: **Pomo:** leaf fiber used to make nets and ropes. **Pomo (Southwestern):** flowers used in wreaths for Strawberry Festival.

   Medicine: **Miwok (Coast):** weak tea used as an emetic.

   Misc: **Pomo (Southwestern):** named but not used.

Citation(s): Chesnut 1902; Gifford 1967; Goodrich et al 1980: Strike 1994.

Terminology: **Pomo (Southwestern):** siwitax.

Citation(s): Gifford 1967; Goodrich et al 1980.

Note(s): different ethnographers, different results for **Pomo (Southwestern)** in terms of plant usage [author].

677   *Iris hartwegii*

Usage: Material: **Miwok:** leaves used as starting knot and foundation in early stages of coiled basket.

Citation(s): Strike 1994.

678   *Iris macrosiphon*

Common Name(s): Iris, Ground Iris, Narrow-Leaf Iris.

Usage: Material: **Hupa:** nets and snares made from twine and rope made from the leaves gathered in the Fall. **Wiyot:** fish nets and rope were made from the fibers of the leaves. **Karok:** cord of various sizes, according to need, was made from dried leaves; scraped with mussel shell scraper. **Shelter Cove Sinkyone:** leaves

used for cord and nets. **Coast Yuki:**
cordage fibers made from leaves.
Citation(s): Baumhoff 1958; Gifford 1939; Goddard 1903; Loud
1918; Schenck and Gifford 1952.
Terminology: **Yuki:** chiwas. **Karok:** a`appakash, "rope
material;" aachirira aappakash, "bird aappakash."
Hupa: mestcelen.
Citation(s): Gifford 1939; Schenck and Gifford 1952; Goddard
1903.

679 *Iris purdyi*

Synonym(s): *Iris macrosiphon purdyi*

Usage: Material: **Karok:** used for rope.
Citation(s): Schenck and Gifford 1952.

680 *Juglans californica*

Family: *Juglandaceae*
Common Name(s): Black Walnut, California Walnut.

Usage: Food: **Chumash:** nuts, cracked with
hammerstone, eaten.

Material: **Chumash:** half shells, filled with tar,
used in gambling game called *pɨ*; bark
used in basketry.

Medicine: **Costanoan:** leaf infusion used a blood
medicine (for thin blood). **Cahuilla:**
hulls used to make dark brown dye for
basketry materials. **Chumash:** leaves,
tea good for the blood.
Citation(s): Strike 1994; Rayburn 2012; Timbrook 2007.
Terminology: **Chumash (Barbareño:** ktɨp; **(Ventureño):** tɨpk.
Citation(s): Timbrook 2007.

681 *Juncus* spp.

Family: *Juncaceae*

Common Name(s): Tule Grass-Nut, Rush, Tule-Grass, Small Bulrush.

Usage: Food: **Maidu, Southern (Nisenan):** eaten raw, or, roasted in ashes, or, boiled. **Luiseño:** fresh shoots eaten raw.

Material: **Maidu, Southern (Nisenan):** hatched with flints, or with fingernails, bleached and wove into breechcloths, **Luiseño:** The lower part of the rush furnishes the brown color for baskets; a mat is made in which articles used a religious ceremonies are kept by the religious chief of the clan; used in making open mesh baskets. **Cahuilla:** the leaf used as wrapping; the leaf dyed black or orange used for pattern in baskets. **Diegueño:** the leaf was used as wrapping in baskets. **Chumash:** used split stems to make brooms; hung abalone on split stems to dry in sun; used in basketry, mats, aprons for carrying game, Fox Dancer's headdress; stems, used to polish shell-bead money. **Maidu:** stems provided a brown dye. **Pomo:** used stem to hold drilled clamshell beads in place while rolling and shaping them on stone.

Medicine: **Maidu:** used as diuretic, cathartic.

Misc: **Miwok:** named but not used.

Citation(s): Barrett and Gifford 1933; Curtis 1924; Merrill 1923; Powers 1877; Sparkman 1908; Strike 1994; Timbrook 2007.

Terminology: **Miwok (Central Sierra):** kokyu. **Luiseño:** shoila.
Cahuilla: seil, seily.
Citation(s): Barrett and Gifford 1933; Sparkman 1908; Bean and
Saubel 1972.

682   *Juncus acuminatus*

Usage: Material: **Cahuilla:** the stalk was used as
wrapping in basketry.
Citation(s): Barrows 1900; Merrill 1923.
Note(s): was identified as *Juncus robustus* in the citations
although that label, or synonym, is not used in any of
the botanical literature searched [author].

683   *Juncus acutus* ssp. *leopoldii*

Synonym(s): *Juncus acutus, Juncus acutus* var.
sphaerocarpus

Usage: Material: **Chumash:, Diegueño:** used leaf as
wrapping, warp and woof in basketry;
leaf (dyed) as yellow and black pattern
in making baskets. **Luiseño, Cahuilla:**
used leaf as wrapping, warp and woof
in basketry; leaf (dyed) as yellow and
black pattern in making baskets; as
foundation in making baskets.
Citation(s): Merrill 1923.

684   *Juncus balticus*

Common Name(s): Wire Rush.

Usage: Food: **Owens Valley Paiute:** used seeds as
food; a minor source of sugar which
formed along the tops of plants,
gathered and eaten as candy.

Material: **Washo:** used leaf as foundation in baskets. **Pomo (Southwestern):** used to hold drilled clamshell beads in place when they are rolled on a stone slab to smooth them. **Kawaiisu (Tehachapi):** minor item for basketry.

Citation(s): Steward 1933; Gifford 1967; Zigmond 1981.

Terminology: **Owens Valley Paiute:** pa'wahapu'hia, sinā'va. **Pomo (Southwestern):** djiba.

Citation(s): Steward 1933; Gifford 1967.

685 *Juncus duranii*

Synonym(s): *Juncus mertensianus*

Usage: Material: **Luiseño:** An openwork basket is made and used for gathering acorns, cactus, etc.; another basket is made for cooking acorn meal; Another is made as a sieve; the leaf was used as foundation, wrapping, warp and woof in asketry.

Citation(s): Merrill 1923; Sparkman 1908.

Terminology(s): **Luiseño:** pivut.

Citation(s): Sparkman 1908.

686 *Juncus effusus* ssp. *pacificus*

Synonym(s): *Juncus effusus*

Usage: Material: **Yuki:** used leaf as warp and woof in practice baskets. **Pomo:** made fish trap 18 inches by 4 inches in diameter. **Chumash:** possibly used in basketry.

Misc: **Coast Yuki:** named but not used.

Citation(s): Chesnut 1902; Gifford 1939; Merrill 1923; Timbrook 2007.

Terminology: **Yuki:** lallem. **Chumash (Barbareño, Ynezeño):**

'ulat'; **(Ventureño):** 'esmu.
Citation(s): Gifford 1939; Timbrook 2007.

687 *Juncus ensifolius*

Usage: Material: **Karok:** has no value, but is sometimes used to teach little girls to make baskets.
Citation(s): Schenck and Gifford 1952.
Terminology: **Karok:** tapraratumnyuaich, "make-believe taprara."
Citation(s): Schenck and Gifford 1952.

688 *Juncus lescurii*

Usage: Material: **Cahuilla:** the stalk was used as wrapping in basketry.
Citation(s): Barrows 1900; Merrill 1923.
Terminology: **Cahuilla:** se-il (scape of plant), i-i-ul (red portion of scape), se-il-tul-iksh (scape dyed black) or black se-il.
Citation(s): Bean and Saubel 1972.
Note(s): this was listed as *Juncus lesenerii* which appears to have been a misspelling or typo [author].

689 *Juncus textilis*
Common Name(s): Basket Rush.

Usage: Material: **Cahuilla:** the stem was used as wrapping in baskets. **Chumash:** used for stitching and for foundation in basketry; stems used for 3-rod bundles or shredded to start coil in basketry.
Citation(s): Dawson and Deetz 1965; Grant 1964; Merrill 1923; Timbrook 2007.

Terminology: **Chumash (Ventureño):** tash, mekhme'y; **(Barbareño, Ynezeño):** mekhme'y; **(Island):** meqme'I; **(Ventureño):** syɨt (reddish base of stems only).
Citation(s): Timbrook 2007.

690 *Juncus xiphioides*

Usage: Misc: **Kawaiisu (Tehachapi):** named but not used.
Citation(s): Zigmond 1981.
Terminology: **Kawaiisu (Tehachapi):** nɨg?odɨbɨ.
Citation(s): Zigmond 1981.

691 *Juniperus* spp.
Family: *Cupressaceae*
Common Name(s): Juniper.

Usage: Food: **Wintun, Hill (Patwin):** berries were eaten, dried, or made into pinole. **Northern Paiute (Paviotso):** made a beverage from crushed berries. **Kawaiisu (Tehachapi):** berries, dried, stored in brush-lined pit covered with dirt.

Medicine: **Atsugewi, Achomawi:** berries were chewed raw to cure a cold; if no berries, leaves and bark were boiled to make a tea and the mixture was drunk. **Maidu:** berries chewed to relieve fever; bark used for colds, fever, constipation. **Maidu, Yuki:** twig inserted through septum as built-in inhaler for sinus problems. **Costanoan:** used to relieve pain, or to cause sweating.

Citation(s): Garth 1953; Kelly 1932; Kroeber 1932; Strike 1994.

Note(s): the berries contain:

volatile oil	½ - 2 ½%
sugar	30%
resins	10%
protein*	4%

* compounds, fat, wax, formic acids, acetic acids, malates.

Acts as a stimulant, stomachic, carminative, diuretic, and emmernagogue.

Citation(s): Stille and Maisch 1880.

692   *Juniperus californica*

Common Name(s): Juniper Berry.

Usage: Food: **Tubatulabal:** used berries. **Kawaiisu (Tehachapi):** berries important source of food; boiled fresh and eaten cold; dried, unseeded and stored; seeded, pounded into meal, eaten, or meal may be moistened and stored as cakes. **Cahuilla:** berries consumed in great quantities between June and August; eaten fresh or dried in the sun; dried berries ground into a flour, made into mush or bread. **Chumash:** seed cones, ground into bright yellow meal, eaten.

Material: **Kawaiisu (Tehachapi):** wood primary material for bows, arrows, acorn-mush stirrer, ladle; bark sometimes used as house covering. **Diegueño:** tea used to stop hiccups. **Luiseño:** berries chewed, or made into tea, to treat fevers, colds. **Chumash:** wood, made sinew-backed bows.

Medicine: **Costanoan:** leaf decoction taken to
cause sweating, for relief of pain.
Citation(s): Voegelin 1938; Zigmond 1981; Bean and Saubel
1972; Strike 1994; Rayburn 2012; Timbrook 2007.
Terminology: **Kawaiisu (Tehachapi):** wa?adabɨ. **Cahuilla:**
yuyily. **Chumash (Barbareño, Ynezeño, Ventureño):**
mulus (tree, fruit); **(Obispeño):** t'pi'nɨ.
Citation(s): Zigmond 1981; Bean and Saubel 1972; Timbrook
2007.
Note(s): Food values – berries:

	(wet)	(dry)
Moisture	16.1%	
Sugar		
Reducing	3.3%	3.9%
Non-reducing	0.0%	0.0%
Starch	0.0%	0.0%
Hemicellulose	11.0%	13.1%

Citation(s): Yanovsky 1938.

693   *Juniperus communis* var. *saxatilis*

Usage: Food: **Pomo:** the dry fruit is sometimes boiled.
Citation(s): Chesnut 1902.
Note(s): this plant was identified as *Juniperus californica* by
Chesnut (1902) but the range cited by Munz and Keck
(1965) would seem to indicate that it should be as
stated above [author].

694   *Juniperus occidentalis*
Common Name(s): Juniper.

Usage: Food: **Miwok:** nuts eaten when thoroughly
ripe. **Cahuilla:** fruit eaten. **Atsugewi:**
berries eaten fresh and also dried for
storage; pounded into flour.

Material: **Modoc (Lutuami) (Lutuami):** used root as warp and woof in baskets. **Mohave:** used bark as cradle mattress. **Owens Valley Paiute:** this material preferred for bows; bark used as covering of small winter houses. **Pomo:** roofing; used in basketry.

Medicine: **Washo:** the branches, at times the berries only, burned as a fumigant after illness; the fumes from burning twigs, when inhaled, were believed to clear up headaches and colds. **Modoc (Lutuami) (Lutuami):** smoke inhaled for mouth sores, throat irritations, coughs, colds; infusion drunk for coughs, colds.

Misc: **Kawaiisu (Tehachapi):** named but not used.

Citation(s): Barrett and Gifford 1933; Barrows 1900; Garth 1953; Merrill 1923; Steward 1933; Train et al1941; Zigmond 1981; Strike 1994; Barrett 1908.

Terminology: **Miwok (Central Sierra):** setekine. **Atsugewi:** mahuop. **Cahuilla:** is-wut. **Kwaiisu:** sina?awa?apɨ.

Citation(s): Barrett and Gifford 1933; Garth 1953; Barrows 1900; Zigmond 1981.

Note(s): Food values – berries:

	(wet)	(dry)
Moisture	25.3%	
Sugar		
Reducing	19.6%	26.2%
Non-reducing	0.0%	0.0%
Starch	0.0%	0.0%
Hemicellulose	12.2%	9.3%

Citation(s): Yanovsky 1938.

695 *Juniperus osteosperma*

Synonym(s): *Juniperus californica utahensis, Juniperus utahensis*

Common Name(s): Desert Juniper. Juniper Berry.

Usage: Food: **Tubatulabal:** used berries.

Material: **Panamint Shoshone (Koso):** made bows from trunk or large limbs of tree that has died and seasoned while standing.

Medicine: **Washo:** the branches, at times the berries only, burned as a fumigant after illness; the fumes from burning twigs, when inhaled, were believed to clear up headaches and colds.

Citation(s): Coville 1938; Train et al 1941; Voegelin 1938.

Terminology: **Washo:** puh-ahl.

Citation(s): Train et al 1941.

Note(s): fruit ripens during the autumn of the second season.

Note(s): Food values – berries:

	(wet)	(dry)
Moisture	25.1%	
Sugar		
Reducing	6.5%	8.5%
Starch	0.0%	0.0%
Hemicellulose	9.3%	12.1%

Citation(s): Yanovsky 1938; Sargent 1922.

696 *Justicia californica*

Family: *Acanthaceae*

Synonym(s): *Beloperone californica*

Common Name(s): Chuparosa.

Usage: Food: **Cahuilla:** may have sucked the flower

for its nectar. **Diegueño:** sucked nectar from flowers.

Citation(s): Bean and Saubel 1972; Strike 1994.

Terminology: **Cahuilla:** pisily, "sweet."

Citation(s): Bean and Saubel 1972.

697 *Keckiella breviflora*

Family: *Plantaginaceae*

Common Name(s): Small-Flowered Penstemon.

Usage: Misc: **Kawaiisu (Tehachapi):** named but not used.

Citation(s): Zigmond 1981.

Terminology: **Kawaiisu (Tehachapi):** ka?miČarabɨ.

Citation(s): Zigmond 1981.

698 *Keckiella breviflora var. breviflora*

Common Name(s): Small-Flowered Beard-Tongue.

Usage: Medicine: **Miwok, Maidu:** steeped and drunk for colds, to treat open sores..

Misc: **Tubatulabal:** considered weeds.

Citation(s): Barrett and Gifford; Voegelin 1938; Strike 1994.

Terminology: **Miwok (Central Sierra):** lulusu.

Citation(s): Barrett and Gifford 1933.

699 *Keckiella cordifolia*

Common Name(s): Climbing Penstemon.

Usage: Medicine: **Chumash:** leaves, boiled, decoction drunk for colds, runny nose, washed sores, wounds; leaves, fresh, poultice applied to sores, wounds.

Citation(s): Timbrook 2007.

Terminology: **Chumash (Barbareño):** tenech.

Citation(s): Timbrook 2007.

700 *Keckiella corymbosa*
Synonym(s): *Penstemon cordifolius*
Common Name(s): Scraggly Penstemon.

Usage: Medicine: **Kawaiisu (Tehachapi):** root, mashed, used as poultice, applied to swollen limbs.
Citation(s): Strike 1994.

701 *Kelloggia galioides*
Family: *Rubiaceae*

Usage: Misc: **Miwok:** named but not used.
Citation(s): Barrett and Gifford 1933.
Terminology: **Miwok (Central Sierra):** pusukulu.
Citation(s): Barrett and Gifford 1933.

702 *Kopsiopsis hookeri*
Family: *Orobanchaceae*
Synonym(s): *Orobanche tuberorsa, Boschnaikia hookeri*
Common Name(s): Cancer Plant.

Usage: Food: **Luiseño:** the roots were eaten.
Citation(s): Sparkman 1908.
Terminology: **Luiseño:** mashal.
Citation(s): Sparkman 1908; Strike 1994.

703 *Lactuca* spp.
Family: *Asteraceae*
Common Name(s): Lettuce.

Usage: Food: **Luiseño:** stems, leaves eaten raw, or cooked.

Citation(s): Strike 1994.

704 *Laminaria* spp.
Family: *Laminariceae*
Common Name(s): Kelp

Usage: Medicine: **Costanoan:** used in the preparation of enemas.
Citation(s): Rayburn 2012.

705 *Larrea tridentata*
Family: *Zygophyllaceae*
Synonym(s): *Larrea mexicana, Larrea divaricata*
Common Name(s): Creosote Bush.

Usage: Material: **Cahuilla:** a glue is obtained on the bark from an amber colored gum deposited by a small scale insect; used as firewood. **Panamint Shoshone (Koso):** used gum deposited on the bark by scale insect (*Carteria larreae*) mixed with pulverized rock and thoroughly pounded, heated before application. **Kawaiisu (Tehachapi):** made digging stick; good firewood, smokeless, clean burning, great heat.

Medicine: **Cahuilla:** leaves steeped and drunk for bowel complaints and for consumption, colds, chest infections, stomach cramps association with menstruation, decongestant for clearing lungs, cure for cancer; heavy doses used for inducing vomiting; poultice to heal open wounds, draw out poisons, prevent infections; crushed leaves applied to sores and wounds; liniment made from plant favorite cure for elderly afflicted with swollen limbs due to poor blood circulation; used to eradicate dandruff, as a hair wash, disinfectant, body deodorizer. **Kawaiisu (Tehachapi):** boil leaves to wash on sore and aching parts of the body. **Chemehuevi (Southern Paiute):** considered their major medicinal plant.

Citation(s): Barrows 1900; Coville 1892; Zigmond 1981; Bean and Saubel 1972; Strike 1994.

Terminology: **Cahuilla:** á-tu-kul, atukul. **Kawaiisu (Tehachapi):** yata(m)bɨ.

Citation(s): Barrows 1900; Zigmond 1981; Bean and Saubel 1972.

Note(s): decoction of plant has some effect to enhance immune responses.

Citation(s): Davicino et al 2007.

706 *Lasthenia* spp.
Family: *Asrteraceae*

Usage: Food: **Miwok (Coast), Luiseño, Maidu:** seeds eaten, often in pinole.

Citation(s): Strike 1994.

707 *Lasthenia californica* spp. *californica*
Family: *Asteraceae*
Synonym(s): *Baeria chrysostoma*
Common Name(s): Goldfields.

Usage: Food: **Cahuilla:** seeds collected in June, parched, ground into flour, made into mush.

Misc: **Tubatulabal:** considered as weeds.
Citation(s): Voegelin 1938; Bean and Saubel 1972.
Terminology: **Cahuilla:** aklakul.
Citation(s): Bean and Saubel 1972; Strike 1994.

708   *Lasthenia glabrata*
Common Name(s): Gold Fields.

Usage: Food: **Cahuilla:** seeds pounded into a very
fine flour and eaten dry or mixed with
water as mush.
Citation(s): Barrows 1900; Bean and Saubel 1972.
Terminology: **Cahuilla:** ák-lo-kal, aklukal.
Citation(s): Barrows 1900; Bean and Saubel 1972.

709   *Lathyrus* spp.
Family: *Fabaceae*
Common Name(s): Wild Pea.

Usage: Food: **Sinkyone:** parched, made into pinole;
stored for Winter. **Maidu, Yuki:** plants
cooked, eaten.
Medicine: **Wiyot:** good medicine for diarrhea.
Maidu, Yuki: older plants boiled, used
to poultice swollen joints.
Misc: **Pomo (Southwestern):** named but not
used.
Citation(s): Merriam 1967; Nomland 1935; Goodrich et al 1980;
Strike 1994.
Terminology: **Pomo (Southwestern):** ?ama·la mára, "rabbit
food."
Citation(s): Goodrich et al 1980.
Note(s): large quantities of the seeds, or an exclusive diet,
either raw or cooked, or in the form of flour, produce
a paralytic syndrome in man; a diet composed almost

exclusively will bring on the symptoms after 4 to 8 weeks; amounts greater than one-third or one-half the total dietary intake will result in symptoms after a proportionately longer period of time; smaller quantities appear innocuous; symptoms consist of partial or total paralysis, usually confined to legs, but in severe cases attacking the arms also; characteristic posture is with the feet turned in, toes down.

Citation(s): Kingsbury 1964; Liener 1980.

710 *Lathyrus jepsonii* ssp. *californicus*
Synonym(s): *Lathyrus watsoni*

Usage: Food: **Yuki:** cooked and eaten for greens when only about 3 inches tall.

Medicine: **Yuki:** older plants boiled and applied as a poultice to swollen joints.

Citation(s): Chesnut 1902.

711 *Lathyrus palustris*
Common Name(s): Pea.

Usage: Food: **Karok:** eaten as greens in the Spring when it is tender.

Citation(s): Schenck and Gifford 1952.
Terminology: **Karok:** kushteituk.
Citation(s): Schenck and Gifford 1952.

712 *Lathyrus vestitus*
Common Name(s): Wild Pea.

Usage: Food: **Miwok:** eaten as greens; seeds eaten raw.

Medicine: **Costanoan:** roots used as emetic, as a panacea, a general remedy; said to be

somewhat intoxicating; root, decoction, used.
Citation(s): Barrett and Gifford 1933; Strike 1994; Rayburn 2012.
Terminology: **Miwok (Central Sierra):** lulumati; **(Southern Sierra):** lu̱'lu̱met.
Citation(s): Barrett and Gifford 1933.

713   *Lathyrus vestitus* var. *vestitus*
Synonym(s): *Lathyrus laetiflorus*
Common Name(s): Wild Pea.

Usage: Food: **Cahuilla:** said that this plant was used by food although did not remember which parts were used.
Citation(s): Bean and Saubel 1972.
Terminology: **Cahuilla:** chawak se'ish (?).
Citation(s): Bean and Saubel 1972.
Note(s): seeds are poisonous and can cause paralysis; slow and weak pulse, shallow breathing, convulsions. [also see discussion in 709 above]. Experimental work with chicks and rats produced lameness in chick, lameness with bone deformities in rats.
Citation(s): Hardin and Arena 1974; Kingsbury 1964.

714   *Layia fremontii*
Family: *Asteraceae*

Usage: Food: **Maidu:** seeds eaten.
Citation(s): Strike 1994.

715   *Layia glandulosa*
Common Name(s): Tidy-Tips.

Usage: Food: **Luiseño:** the seeds were used for food.
Cahuilla: seeds gathered by women

June to August, ground into flour, cooked with other grounds seeds in mush.

Citation(s): Sparkman 1908; Bean and Saubel 1972.

Terminology: **Luiseño:** solisal.

Citation(s): Sparkman 1908.

716 *Layia platyglossa*

Synonym(s): *Blepharipappus platyglossus*

Usage: Food: **Yuki:** seeds used in pinole.

Citation(s): Chesnut 1902.

717 *Layia platyglossa*

Synonym(s): *Layia platyglossa* ssp. *campestris*

Usage: Food: **Cahuilla, Chumash, Miwok (Coast), Luiseño, Maidu:** seeds eaten.

Citation(s): Strike 1994.

718 *Lemna minor*

Family: *Araceae*

Common Name(s): Duckweed.

Usage: Medicine: **Maidu:** used as a diuretic, general tonic.

Citation(s): Strike 1994.

719 *Lepechinia calycina*

Family: *Lamiaceae*

Synonym(s): *Sphacele calycina*

Common Name(s): Pitcher Sage.

Usage: Medicine: **Miwok:** decoction of leaves drunk to abate fever, ague, headache; one cupful

was the dose and was often sufficient to cure. **Maidu:** decoction of leaves drunk for fever, chills, headache, to relieve colds.
Citation(s): Barrett and Gifford 1933; Strike 1994.
Terminology: **Miwok (Central Sierra):** watakka'iyu.
Citation(s): Barrett and Gifford 1933.

720 *Lepidium* spp.

Family: *Brassicaceae*

Usage:	Food:	**Maidu:** leaves eaten as greens.
	Material:	**Maidu:** leaves boiled, solution used as hair shampoo, to prevent hair loss.
	Medicine:	**Maidu:** used to prevent, or cure, a scurvy-like illness; as a counter-irritant, treatment for rashes from poison oak.
	Misc:	**Owens Valley Paiute:** named but not used.

Citation(s): Steward 1933; Strike 1994.
Terminology: **Owens Valley Paiute:** ătsăha'ma'ᵃ.
Citation(s): Steward 1933.

721 *Lepidium lasiocarpum*

Common Name(s): Hairy-Pod Peppergrass.

Usage:	Food:	**Kawaiisu (Tehachapi):** seeds, ripe in June, pounded, mixed with water, drunk as beverage.

Citation(s): Zigmond 1981.

722 *Lepidium nitidum*

Common Name(s): Peppergrass.

Usage:	Food:	**Luiseño:** the seeds are eaten; the leaves

are used for greens. **Chumash:** seeds, tossed in basket with hot coals, ground, eaten as pinole.

Material: **Cahuilla:** leaves boiled, allowed to set until a brownish-colored solution was present; used to wash hair, keep scalp clean, prevent baldness.

Medicine: **Chumash:** leaf, or two, decoction, treated diarrhea, dysentery, to counteract a strong laxative.

Citation(s): Sparkman 1908; Bean and Saubel 1972; Timbrook 2007.

Terminology: **Luiseño:** pakil. **Cahuilla:** pakil. **Chumash (Barbareño, Ynezeño):** khaksh; **(Ventureñ0):** 'iqma'y.

Citation(s): Sparkman 1908; Bean and Saubel 1972; Timbrook 2007.

723 *Lepidium virginicum*
Common Name(s): Tall Pepper-Grass.

Usage: Misc: **Karok:** named but not used.
Citation(s): Schenck and Gifford 1952.
Terminology: **Karok:** chantinihtunechash, "little tick."
Citation(s): Schenck and Gifford 1952.
Note(s): occasional introduction from East Coast United States.
Citation(s): Munz and Keck 1965.

724 *Lepidospartum squamatum*
Family: *Asteraceae*

Usage: Misc: **Tubatulabal:** considered weeds. **Kawaiisu (Tehachapi):** named but not used.
Citation(s): Voegelin 1938; Zigmond 1981.

Terminology: **Kawaiisu (Tehachapi):** pašivapɨ.
Citation(s): Zigmond 1981.

725 *Leptosiphon androsaceus*
Family: *Polemoniaceae*
Synonym(s): *Gilia androsacea montana, Linanthus
androsaceus*

Usage: Misc: **Miwok:** named but not used.
Citation(s): Barrett and Gifford 1933.
Terminology: **Miwok (Central Sierra):** suppatawa.
Citation(s): Barrett and Gifford 1933.

726 *Leptosiphon aureus* ssp. *aureus*
Synonym(s): *Linanthus aureus*

Usage: Misc: **Tubatulabal:** considered weeds.
Citation(s): Voegelin 1938.

727 *Leptosiphon ciliatus*
Synonym(s): *Gilia ciliata, Linanthus ciliatus*

Usage: Medicine: **Pomo, Maidu:** used in the form of an
infusion as a remedy for coughs and
colds in children; a cold decoction was
drunk instead of water to purify the
blood.
Citation(s): Chesnut 1902; Strike 1994.

728 *Leptosyne bigelovii*
Family: *Asteraceae*
Synonym(s): *Coreopsis bigelovii*

Usage: Food: **Tubatulabal:** used leaves. **Kawaiisu
(Tehachapi):** in the Spring it is an

important food source; cut off at the base before blooming; eaten fresh, or cooked and fried in grease and salt.
Citation(s): Voegelin 1938; Zigmond 1981.
Terminology: **Kawaiisu (Tehachapi):** tɨhɨvidɨbɨ.
Citation(s): Zigmond 1981.

729 *Leptosyne maritima*

Synonym(s): *Coreopsis maritima*
Common Name(s): Tickseed, Sea Dahlia.

Usage: Medicine: **Diegueño:** root, boiled, liquid drunk to relieve stomach aches.
Citation(s): Strike 1994.

730 *Lessingia germanorum*

Family: *Asteraceae*
Common Name(s): Valley Lessingia.

Usage: Medicine: **Kawaiisu (Tehachapi):** when dry in August-September, the outer skin is peeled off and pounded into small pellets, five of which are burned on body when there is pain.
Citation(s): Zigmond 1981.
Terminology: **Kawaiisu (Tehachapi):** narag^wanɨ(m)bɨ.
Citation(s): Zigmond 1981.

731 *Letharia vulpina*

Family: *Palmeliaceae*
Common Name(s): Wolfbane, Yellow Tree Lichen.

Usage: Material: **Maidu:** yellow dye used to paint faces.
          Medicine: **Maidu:** decoction used to poultice open sores, reduce inflammations.

Citation(s): Strike 1994.

732 *Lewisia nevadensis*
Family: *Montiaceae*

Usage: Food: **Maidu (Northern):** roots eaten.
Citation(s): Dixon 1905.

733 *Lewisia rediviva*
Common Name(s): Bitter Root.

Usage: Food: **Northern Paiute (Paviotso):** the root
was pulled and boiled "like macaroni;"
it could also be dried.
Citation(s): Kelly 1932.
Terminology: **Northern Paiute (Paviotso):** kañü'tc'.
Citation(s): Kelly 1932.
Note(s): Food values – berries:

	(wet)	(dry)
Moisture	8.7%	
Sugar		
Reducing	5.6%	6.1%
Non-reducing	3.2%	3.5%
Starch	34.2%	37.5%
Hemicellulose	3.6%	3.9%

Citation(s): Yanovsky 1938.

734 *Ligusticum* spp.
Family: *Apiaceae*

Usage: Food: **Maidu:** roots, stems, shoots eaten.
Material: **Yuki:** root used to ward off
rattlesnakes; roots, pulverized, used to
stun fish.

Medicine: **Pomo (Southwestern):** tea used to cure anemia. **Northern Paiute (Paviotso):** roots, crushed, used to poultice sprains, bruises.

Misc: **Owens Valley Paiute, Mono Lake Paiute:** named but not used; considered as poisonous.

Citation(s): Steward 1933; Strike 1994.

Terminology: **Owens Valley Paiute:** pa'wonova. **Mono Lake Paiute:** haukinup.

Citation(s): Steward 1933.

735 *Ligusticum apiifolium*

Synonym(s): *Ligusticum apiodorum*
Common Name(s): Lovage.

Usage: Medicine: **Karok:** the roots are soaked in water and the water drunk; used for a person who lacks appetite. **Pomo (Southwestern):** root was boiled and the decoction drunk to stop or check hemorrhage from the lungs.

Citation(s): Gifford 1967; Schenck and Gifford 1952; Goodrich et al 1980.

Terminology: **Pomo (Southwestern):** tulebachwa, "hummingbird" (tule), "angelica," (bachwa). **Karok:** sakusuhish, kusuiva.

Citation(s): Gifford 1967; Schenck and Gifford 1952; Goodrich et al 1980.

736 *Ligusticum grayii*

Common Name(s): Wild Plum, Wild Parsley.

Usage: Food: **Atsugewi:** leaves eaten in the Spring when tender; soaked in water and then cooked in an earth oven; might be stored in cooked form.

Material: **Atsugewi:** used for poisoning fish; root pulverized and powder poured into a quiet pool.

Medicine: **Pomo (Southwestern):** root was boiled and the decoction drunk to stop or check hemorrhage from the lungs.

Citation(s): Garth 1955; Strike 1994.

Terminology: **Atsugewi:** bóhom.

Citation(s): Garth 1953.

737 *Lilium occidentale*
Family: *Liliaceae*
Common Name(s): Eureka Lily.

Usage: Food: **Karok:** bulb baked in the earth oven and eaten.

Citation(s): Schenck and Gifford 1952.

738 *Lilium pardalinum*
Common Name(s): Tiger Lily.

Usage: Food: **Yana:** seed-like portions steamed in earth oven for about an hour and eaten. **Karok:** most highly regarded of the bulbs; dug in the Fall and cooked in earth oven. **Atsugewi:** cooked in earth oven and eaten immediately.

Material: **Miwok:** flowers worn as wreaths.

Misc: **Miwok:** considered bulbs as poisonous.

Citation(s): Barrett and Gifford 1933; Sapir and Spier 1943; Schenck and Gifford 1952; Strike 1994.

Terminology: **Miwok (Northern Sierra):** palauda; **(Central Sierra):** palauta. **Yana:** dā̄i'waki. **Atsugewi:** skapɳw.

Karok: matayish, "mountain tayish."
Citation(s): Barrett and Gifford 1933; Sapir and Spier 1943;
Garth 1953; Schenck and Gifford 1952.

739 *Lilium parvum*
Common Name(s): Small Tiger Lily.

Usage: Food: **Mono Lake Paiute:** used bulb.
Citation(s): Steward 1933.
Terminology: **Mono Lake Paiute:** masai'da.
Citation(s): Steward 1933.

740 *Lilium rubescens*
Common Name(s): Chaparral Lily.

Usage: Misc: **Karok:** named but not used.
Citation(s): Schenck and Gifford 1952.
Terminology: **Karok:** xiripipich, "throw away famine."
Citation(s): Schenck and Gifford 1952.

741 *Lilium washingtonianum*

Usage: Food: **Maidu (Northern):** roots used.
Citation(s): Dixon 1905.

742 *Limnanthes alba*
Family: *Limnanthaceae*
Common Name(s): Meadow Foam.

Usage: Food: **Maidu:** seeds eaten.
Citation(s): Strike 1994.

743 *Limonium* spp.

Family: *Plumbanginaceae*
Common Name(s): March Rosemary, Sea Lavender.

Usage: Medicine: **Costanoan:** decoction used to relieve internal injuries, venereal problems cure urinary ailments, as a blood tonic; leaves pulverized, placed in nostrils, sneezing relieved nasal congestion.
Citation(s): Strike 1994.

744 *Limonium californicum*

Common Name(s): Sea Lavender.

Usage: Food: **Diegueño:** leaves fresh, or dried, eaten.
        Medicine: **Costanoan:** decoction, used to clean the blood, as a remedy for venereal disease, for urinary problems; leaves, powdered, placed in nostrils to cause sneezing, to relieve congestion.
Citation(s): Strike 1994; Rayburn 2012.

745 *Linanthus dichotomus*

Family: *Polemoniaceae*
Common Name(s): Evening Snow.

Usage: Misc: **Kawaiisu (Tehachapi):** named but not used.
Citation(s): Zigmond 1981.
Terminology: **Kwaiisu:** tutuvɨnivɨ.
Citation(s): Zigmond 1981.

746 Linanthus pungens

Family: *Poleminiaceae*
Synonym(s): *Leptodactylon pungens* ssp. *pulchriflorum*
Common Name(s): Granite Gilia.

Usage: Medicine: **Maidu:** decoction used to bathe swellings, sore eyes, scorpion stings.
Citation(s): Strike 1994.

747 Linum lewisii var. lewisii

Family: *Linaceae*
Synonym(s): *Linum perenne* spp. *lewsii*

Usage: Medicine: **Maidu:** leaves used to poultice swellings; seed poultice used to treat inflammations, bruises; root infusion used as eye medicine; stem infusion relieved stomach aches, intestinal disorders.
Citation(s): Strike 1994.

748 Lithophragma affine

Family: *Saxifragaceae*
Synonym(s): *Tellima affinis, Lithophragma affinis*

Usage: Medicine: **Yuki, Maidu:** possibly chewed root to relieve colds or stomach ache.
Misc: **Miwok:** named but not used.
Citation(s): Barrett and Chesnut 1933; Chesnut 1902; Strike 1994.
Terminology: **Miwok (Central Sierra):** epesenu.
Citation(s): Barrett and Gifford 1933.

749   Lithospermum californicum
   Family: *Boraginaceae*
   Common Name(s): California Puccoon, Gromwell
   Puccoon.

   Usage: Medicine: **Maidu:** possibly used as contraceptive.
   Citation(s): Strike 1994.

750   Lobaria pulmonaria
   Family: *Lobariaceae*
   Common Name(s): Lung Lichen, Oak Lungs.

   Usage: Material: **Maidu:** obtained a brown or yellow
   dye.
   Citation(s): Strike 1994.

751   Loeseliastrum matthewsii
   Family: *Polemoniaceae*
   Synonym(s): *Langloisia matthewsii*

   Usage: Medicine: **Tubatulabal:** root boiled, one cup, three
   times a day, drunk to treat severe colds.
   Misc: **Tubatulabal:** considered weeds.
   Citation(s): Voegelin 1938; Strike 1994.
   Note(s): some difference as to data used (or references checked)
   between the two citations [author].

752   Logfia filanginoides
   Family: *Asteraceae*
   Synonym(s): *Festuca californica*
   Common Name(s): Bunchgrass.

   Usage: Misc: **Pomo (Southwestern):** named but not
   used.
   Citation(s): Goodrich et al 1980.

Terminology: **Pomo (Southwestern):** gahšim?.
Citation(s): Goodrich et al 1980.

753 *Lomatium* spp.

Family: *Apiaceae*
Synonym(s): *Peucedanum spp.*

Usage: Misc: **Miwok:** named but not used.
Citation(s): Barrett and Gifford 1933.
Terminology: **Miwok (Central Sierra):** sumsümi.
Citation(s): Barrett and Gifford 1933.

754 *Lomatium bicolor* var. *leptocarpum*

Synonym(s): *Lomatium leptocarpum*

Usage: Food: **Northern Paiute (Paviotso):** the root was eaten uncooked, either fresh, or dried.
Citation(s): Kelly 1932.
Terminology: **Northern Paiute (Paviotso):** tu·nu´·yu.
Citation(s): Kelly 1932.

755 *Lomatium californium*

Synonym(s): *Leptotaenia californica, Lomatium californica*
Common Name(s): Incense Root, Hog Fennel, Angelica.

Usage: Food: **Hupa:** the fresh shoots were eaten raw. **Karok:** the pounded roots are smoked; the roots are eaten raw. **Yuki:** steam eaten raw when young, and "it's good;" young shoots were also cooked and eaten. **Kawaiisu (Tehachapi):** provides Spring greens and is eaten raw.
       Material: **Karok:** tied around neck, repelled

rattlesnakes.

Medicine: **Hupa:** the doctor repeats prayers in ceremony while burning the root; during a burial ceremony an offering of tobacco and powdered root is blown from the hand as an offering to the divinity; a priest scatters the powdered root over the ground where the dance line is to stand prior to the start of the White Deer-Skin Dance. **Karok:** plant and roots soaked in water which is then drunk by person who does not feel like eating. **Yuki:** for a severe cold the dried root is ground and smoked in a pipe; root boiled, strain, and drunk as tea is also good for colds; for pains in hands and other parts of the body a sliver of root is lighted and pressed into the flesh until the spot is burned; it is tied around the neck to ward off sickness and rattlesnakes. **Chumash:** valued for its medicinal value and as a sort of talisman for the control of rattlesnakes; root chewed for headaches, chewed and rubbed on body for any sort of pain; scent is inhaled for headache; root, tea drunk for rheumatism, stomach ailments; applied as a poultice to sores, pains; root, chewed, as a tonic, treatment for flatulence, headache, neuralgia. **Kawaiisu (Tehachapi):** dried root boiled; liquid drunk hot as remedy for colds, but causes vomiting; chewed for sore throats; crumbled into water, rubbed on stomach for relief of stomach ache; believed that spirts (ghosts) were afraid of the root.

Misc: **Yurok:** considered a sacred plant, having a forked root from which the first woman was made.

Citation(s): Curtin 1957; Goddard 1903; Merriam 1967; Schenck and Gifford 1952; Timbrook 1987, 2007; Zigmond 1981; Strike 1994.

Terminology: **Yuki:** chi-en-chow'. **Hupa:** mûatcexōlen. **Karok:** muhish. **Kawaiisu (Tehachapi):** kayeezi. **Chumash (Barbareño, Ynezeño):** pa'; **(Ventureño):** chpa'.

Citation(s): Curtin 1957; Goddard 1903; Schenck and Gifford 1952; Zigmond 1981; Timbrook 2007.

756 *Lomatium canbyi*

Synonym(s): *Peucedanum canbyi*

Usage: Food: **Northern Paiute (Paviotso):** the root was cooked in an earth oven, either fresh or dried. **Maidu:** seeds, tuber eaten fresh, or cooked, or dried and stored.

Citation(s): Kelly 1932; Strike 1994.

Terminology: **Northern Paiute (Paviotso):** ha·pa'.

Citation(s): Kelly 1932.

Note(s): food values.

Starch	17.02%
Albuminoides	3.25%
Glucose	1.24%
Saccharose	10.66%
Mucilage	15.34%
Dextrin	0.40%
Resin	2.57%
Fat and wax	2.12%
Ash	2.12%
Moisture	7.90%

Cellulose and undetermined   35.30%
Citation(s): Trimble 1890.

757   *Lomatium dissectum* var. *multifidum*
Synonym(s): *Leptotaenia multifida*

Usage: Medicine: **Northern Paiute (Paviotso):** crushed root used for both sores and swellings; used as a poultice. **Washo:** the oily sap from the fresh sliced roots is carefully gathered and used on cuts and sores; if fresh roots not available, dried roots were boiled and the oil skimmed from the surface of the water; fresh root was ground to a pulp and applied to the severed umbilical cords of new-born babies. **Kawaiisu (Tehachapi):** root roasted, peeled, mashed, kneaded into water; cuts and open wounds dipped into infusion; root pounded into salve for sore limbs, cuts and wounds.

Citation(s): Kelly 1932; Train et al 1941; Zigmond 1981.
Terminology: **Northern Paiute (Paviotso):** do′·sa[abü]. **Washo:** dosa, doza. **Kawaiisu (Tehachapi):** tɨ(m)bɨgovɨ.
Citation(s): Kelly 1932; Train et al 1941; Zigmond 1981.

758   *Lomatium macrocarpum*
Common Name(s): Sheep Parsnip.

Usage: Food: **Northern Paiute (Paviotso):** the root was eaten fresh or dried. **Pomo (Southwestern):** young leaves eaten; sweet seed used to flavor tea and pinole. **Maidu:** seeds, root eaten cooked.

Citation(s): Kelly 1932; Goodrich et al 1980; Strike 1994.
Terminology: **Northern Paiute (Paviotso):** hu'ʼnibui. **Pomo (Southwestern):** duwi cʰáʔbu, "coyote carrot."
Citation(s): Kelly 1932; Goodrich et al 1980.
Note(s): seeds contain:

Ash	7.0%
Protein	25.9%
Oil	25.5%

Citation(s): Earle and Jones 1962.

759 *Lomatium mohavense*
Common Name(s): Desert Parsley.

Usage: Misc: **Kawaiisu (Tehachapi):** named but not used.
Citation(s): Zigmond 1981.
Terminology: **Kawaiisu (Tehachapi):** kɨɨvɨɨzi koroorɨ.
Citation(s): Zigmond 1981.

760 *Lomatium nudicaule*
Common Name(s): Pestle Parsnip.

Usage: Food: **Atsugewi:** leaves and tender stems eaten raw. **Maidu:** all parts eaten; leaves used as tea.

Medicine: **Maidu:** used as treatment for colds, sore throats, stomach aches, headaches, backaches, swollen joints, carbuncles; seed decoction facilitated childbirth.
Citation(s): Garth 1953; Strike 1994.
Terminology: **Atsugewi:** watini.
Citation(s): Garth 1953.

761 *Lomatium triternatum*

Usage: Food: **Atsugewi:** roots cooked in earth oven.
Citation(s): Garth 1953.
Terminology: **Atsugewi:** gusgi.
Citation(s): Garth 1953.

762 *Lomatium utriculatum*
Common Name(s): Bladder Parsnip.

Usage: Food: **Yuki:** young leaves eaten raw in May of June when they are still crisp. **Atsugewi:** plants above roots cooked, fried in grease and salt. **Maidu:** all parts eaten.

Medicine: **Kawaiisu (Tehachapi):** plant is boiled and liquid applied to swollen and broken limbs.

Citation(s): Chesnut 1902; Garth 1953; Zigmond 1981; Strike 1994.
Terminology: **Atsugewi:** kĪB.
Citation(s): Garth 1953.

763 *Lonicera* ssp.
Family: *Caprifiliaceae*

Usage: Medicine: **Costanoan:** decoction used for bathing swollen feet; infected sores; fruit, dried, decocted, used for cough syrup.
Citation(s): Rayburn 2012.

764 *Lonicera ciliosa*

Common Name(s): Orange Honeysuckle.

Usage: Food: **Maidu:** fruit eaten.
Citation(s): Strike 1994.

765 *Lonicera conjugialis*

Common Name(s): Bridal Honeysuckle.

Usage: Food: **Maidu, Modoc (Lutuami) (Lutuami):** fruit eaten.
Citation(s): Strike 1994.

766 *Lonicera hispidula*

Synonym(s): *Lonicera hispidula californica, Lonicera hispidula* var. *vacillans*
Common Name(s): California Honeysuckle, Hairy Honeysuckle.

Usage: Food: **Maidu:** fruit eaten.
      Material: **Miwok:** used as foundation in basketry; as wrapping material in coiled baskets. **Pomo (Southwestern):** hollow stems used for pipe stems; ashes used for tattooing. **Yuki:** used as basket foundation material.
      Misc: **Karok:** linked plant to myths dealing with Coyote.
Citation(s): Barrett and Gifford 1933; Schenck and Gifford 1952; Goodrich et al 1980; Strike 1994.
Terminology: **Miwok (Central Sierra):** pipenu. **Karok:** pinef tatapuwa, "coyote's trap."
Citation(s): Barrett and Gifford 1933; Schenck and Gifford 1952.

767 *Lonicera interrupta*

Common Name(s): Honeysuckle, Chaparral Honeysuckle.

Usage: Food: **Yuki:** children are fond of sucking the nectar from the flowers.

Material: **Yuki:** used stems as foundation in basketry. **Kawaiisu (Tehachapi):** hollow section of stem may serve as cigarette-type pipe.

Medicine: **Yuki:** a concentrated tea is made from the leaves and used as a wash for sore eyes. **Maidu:** leaf, stem, root decoction used as blood purifier. **Costanoan:** used as cough medicine, a lotion for skin irritations, sores, as a disinfectant.

Misc: **Miwok:** named but not used.

Citation(s): Barrett and Gifford 1933; Chesnut 1902; Merrill 1923; Zigmond 1981; Strike 1994.

Terminology: **Miwok (Central Sierra):** wa'kilwakīlu.

Citation(s): Barrett and Gifford 1933.

768 *Lonicera involucrata*

Common Name(s): Twinberry.

Usage: Food: **Maidu, Modoc (Lutuami):** fruit eaten.

Medicine: **Maidu:** leaves chewed as emetic; leaf decoction used to bathe swellings, as a liniment; bark decoction used to wash sore eyes, taken for colds; leaf or bark decoction used to relieve chest pain, other aches.

Misc: **Yurok:** named but not used; berries said to be poisonous.

Citation(s): Merriam 1967; Strike 1994.

769   *Lonicera subspicata* var. *denudata*
>   Common Name(s): Chaparral Honeysuckle.

>   Usage: Material: **Chumash:** sprigs, branches, made
>   brooms, brushes; coarsely twined utility
>   baskets, toy arrows.
>   Citation(s): Timbrook 2007.
>   Terminology: **Chumash (Barbareño, Ynezeño):** tu';
>   **(Ventureño):** chtu 'iqonon.
>   Citation(s): Timbrook 2007.

770   *Ludwigia palustris*
>   Family: *Onagraceae*
>   Common Name(s): Marsh Purslane.

>   Usage: Medicine: **Maidu:** used   to   treat   bronchial
>   ailments.
>   Citation(s): Strike 1994.

771   *Lupinus* spp.
>   Family: *Fabaceae*
>   Common Name(s): Lupine.

>   Usage: Food: **Luiseño:** used for greens. **Yuki:** the
>   young plants were roasted on stones
>   over a hot fire and eaten as greens.
>   **Maidu:** leaves, stems leached overnight
>   in running stream, eaten.
>   Medicine: **Owens Valley Paiute, Diegueño:** used
>   for bladder trouble and failure to
>   urinate.
>   Misc: **Coast Yuki:** named but not used.
>   Citation(s): Curtin 1957; Gayton 1948; Gifford 1939; Sparkman
>   1909; Steward 1933; Strike 1994.
>   Terminology: **Owens Valley Paiute:** kaivodup, kai'voidava.

Yuki: totse. **Luiseño:** mawut. **Chumash (Barbareño):** wala'laq'; **(Ventureño):** qlahaw'.
Citation(s): Steward 1933; Gifford 1939; Sparkman 1908; Timbrook 2007.

772 *Lupinus affinis*

Synonym(s): *Lupinus carnosulus*

Usage: Food: **Yuki:** young roasted leaves eaten as greens.
Citation(s): Chesnut 1902.

773 *Lupinus albifrons*

Common Name(s): Sliver Lupine, Bush Lupine.

Usage: Medicine: **Karok:** it is boiled and the patient steamed as a remedy for stomach trouble. **Karok, Maidu:** burned in sweathouses to relieve stomach ache.
Misc: **Tubatulabal:** named but not used.
Citation(s): Schenck and Gifford 1952; Voegelin 1938; Strike 1994.
Terminology: **Karok:** amtapara.
Citation(s): Schenck and Gifford 1952.

774 *Lupinus arboreus*

Common Name(s): Yellow-Flowered Lupine.

Usage: Material: **Pomo (Southwestern):** fibers from roots used for string; flowers used in wreaths in Flower Dance in Strawberry Festival in May.
Citation(s): Gifford 1967; Goodrich et al 1980.
Terminology: **Pomo (Southwestern):** kalkasa.
Citation(s): Gifford 1967; Goodrich et al 1980.

775 *Lupinus bicolor*

Synonym(s): *Lupinus micranthus, Lupinus polycarpus*

Usage: Food: **Miwok:** eaten for greens.

Misc: **Tubatulabal:** named but not used.

Citation(s): Voegelin 1938, Clark 1904.

776 *Lupinus chamissonis*

Common Name(s): Blue Dune Lupine.

Usage: Material: **Miwok (Coast):** roots made into cordage

Citation(s): Strike 1994.

777 *Lupinus latifolius*

Common Name(s): Lupine, Broad-Leaved Lupine.

Usage: Food: **Miwok:** leaves and flowers steamed in earth oven; preferred to *Lupinus densiflorus*; after steaming quantities were dried and stored for Winter use; dried leaves and flowers were usually boiled after soaking in water 3-4 hours to removed bitter taste; sometimes served as a relish with Manzanita (*Arctostaphylos* spp.) cider.

Material: **Miwok:** lined acorn leaching basket with leaves to keep meal from running through.

Misc: **Karok:** named but not used.

Citation(s): Barrett and Gifford 1933; Schenck and Gifford 1952.

Terminology: **Miwok (Central Sierra):** wataksa, tcī′ūtcīūwa; **(Southern Sierra):** wa′taksa. **Karok:** mahamtapparas, "mountain amtapparas."

Citation(s): Barrett and Gifford 1933; Schenck and Gifford 1952.

778 *Lupinus microcarpus* var. *densiflorus*
Synonym(s): *Lupinus densiflorus*
Common Name(s): Rose Lupine.

Usage: Food: **Miwok:** early in Spring the leaves and flowers were stripped from stalk, steamed in earth oven and eaten with acorn soup.
         Misc: **Tubatulabal:** named but not used.
Citation(s): Barrett and Gifford 1933; Voegelin 1938.
Terminology: **Miwok (Central Sierra):** tūlmī′ssa.
Citation(s): Barrett and Gifford 1933.

779 *Lupinus microcarpus* var. *microcarpus*
Synonym(s): *Lupinus subvexus*

Usage: Misc: **Tubatulabal:** named but not used.
Citation(s): Voegelin 1938.

780 *Lupinus odoratus*

Usage: Misc: **Tubatulabal:** named but not used.
Citation(s): Voegelin 1938.

781 *Lupinus sparsiflorus*

Usage: Misc: **Tubatulabal:** named but not used.
Citation(s): Voegelin 1938.

782 *Lycium andersonii*
Family: *Solanaceae*
Common Name(s): Box-Thorn.

Usage: Food: **Panamint Shoshone (Koso):** red berries dried and made into mush. **Kawaiisu (Tehachapi):** red fruit, edible in Spring, eaten fresh; juice squeezed out and drunk; berries dried and stored. **Owens Valley Paiute:** berries, ground, mixed with water. **Cahuilla:** berries eaten fresh or dried; dried berry boiled into mush, or ground into flour and mixed with water.

Citation(s): Coville 1892; Zigmond 1981; Bean and Saubel 1972; Strike 1994.

Terminology: **Kawaiisu (Tehachapi):** hu?upivɨ.

Citation(s): Zigmond 1981.

783 *Lycium cooperi*

Common Name(s): Peach Thorn, Desertthorn.

Usage: Misc: **Kawaiisu (Tehachapi):** named but not used.

Citation(s): Zigmond 1981.

Terminology: **Kawaiisu (Tehachapi):** canavɨ.

Citation(s): Zigmond 1981.

784 *Lycium fremontii*

Usage: Food: **Cahuilla:** berries eaten fresh or dried; dried berry boiled into mush, or ground into flour and mixed with water.

Citation(s): Bean and Saubel 1972.

785 *Lycium torreyi*

Common Name(s): Boxthorn Berries.

Usage: Food: **Tubatulabal:** used berries.
Citation(s): Voegelin 1938.

786 *Lycoperdon* spp.

Family: *Lycoperdaceae*
Common Name(s): Common Puffball, Devil's Snuff-
Box.

Usage: Medicine: **Yuki:** spores used to some extent to dry
up running sores; if powder got into
eyes, it would cause blindness.
Misc: **Pomo (Southwestern):** considered
poisonous.
Citation(s): Chesnut 1902; Curtin 1957; Goodrich et al 1980.
Terminology: **Yuki:** pó-hut. **Pomo (Southwestern):** ?ama
q'ála·ša.
Citation(s): Goodrich et al 1980; Curtin 1957.

787 *Lycoperdon perlatum*

Common Name(s): Gem-studded Puffball.

Usage: Food: **Miwok:** eaten.
Medicine: **Miwok:** sprinkled spores over wounds
to stop hemorrhaging.
Citation(s): Strike 1994.

788 *Lycoperdon sculptum*

Common Name(s): Sierra Puffball.

Usage: Food: **Miwok:** dried in sun for two or three
days; pulverized, boiled, and eaten with
acorn soup.

Citation(s): Barrett and Gifford 1933.
Terminology: **Miwok (Central Sierra):** potokele, patapsi.
Citation(s): Barrett and Gifford 1933.

789 *Lycium cooperi*
Family: *Solanaceae*
Common Name(s): Boxthorn.

Usage: Misc: **Tubatulabal:** considered weeds.
Citation(s): Voegelin 1938.

790 *Lycopus americanus*
Family: *Lamiaceae*
Common Name(s): Water Hoarhound.

Usage: Medicine: **Maidu:** used to simulate appetite, to relieve stomach ache.
Citation(s): Strike 1994.

791 *Lyonothamnus floribundus* ssp. *aspleniifolius*
Family: *Rosaceae*
Common Name(s): Santa Cruz Island Ironwood.

Usage: Material: **Chumash:** wood, principle wood for harpoons, shafts of canoe paddles, made pry bars for abalone, fighting knives, house posts; bark, boiled for black dye in basketry.
Citation(s): Timbrook 2007.
Terminology: **Chumash (Island):** wɨlɨ; **(Ventureño):** wɨ'lɨ.
Citation(s): Timbrook 2007.

792 *Lysichiton americanus*

Family: *Araceae*

Common Name(s): Skunk Cabbage.

Usage: Material: **Karok, Yurok:** leaves used as a temporary lining for open work baskets when used to hold berries.

Citation(s): Merriam 1967; Strike 1994.

793 *Lythrum californicum*

Family: *Lythraceae*

Common Name(s): California Loosestrife.

Usage: Material: **Kawaiisu (Tehachapi):** used for washing hair.

Medicine: **Kawaiisu (Tehachapi):** used in conjunction with *Solanum xanti.*

Citation(s): Zigmond 1981.

Terminology: **Kawaiisu (Tehachapi):** pɨɨsivɨ.

Citation(s): Zigmond 1981.

794 *Lythrum hyssopifolia*

Common Name(s): Grass Poly.

Usage: Medicine: **Maidu:** used to expedite healing, to reduce inflammation of mucus membranes.

Citation(s): Strike 1994.

795 *Macrocystis* spp.

Family: *Laminariaceae*

Common Name(s): Kelp.

Usage: Food: **Bear River:** salt made from dried sections, scraped off outside; stored in dried sections.

Material: **Miwok (Coast):** used kelp to lash points to arrows.

Medicine: **Bear River:** sores and boils treated with burned ashes to draw out pus and to heal the sore after being opened with a flint sliver. **Costanoan:** used in the preparation of enemas.

Citation(s): Nomland 1938; Strike 1994; Rayburn 2012.

796 *Macrocystis pyrifera*

Synonym(s): *Macrocystis Kutkeane, Macrocystis intergifolia*

Common Name(s): Giant Kelp, Whip Kelp.

Usage: Food: **Pomo:** eaten raw, mostly for salty flavor.

Citation(s): Barrett 1952.

Terminology: **Pomo (Eastern):** kácē, **(Southern):** pá′nēma.

Citation(s): Barrett 1952.

Note(s): research has suggested that the genus is monospecific with the authors proposing that the genus be collapsed back into a single species, *M. pyrifera*.

Citation(s): Demes et al 2009.

797 *Madia* spp.

Family: *Asteraceae*

Synonym(s): *Madaria* spp.

Common Name(s): Yellow-Blooming Tarry-Smelling Weed.

Usage: Food: **Maidu (Northern), Tolowa, Wintun, Central (Nomlaki):** seeds eaten. **Bear River:** seeds gathered by means of an elongated oval basketry fan. **Maidu,**

Southern (Nisenan): the seeds are eaten; the seeds are parched a little and then beaten into flour, eaten without further cooking, or made into bread or mush. **Yuki, Huchnom:** used seeds in pinole. **Yokuts (Northern Hill):** seeds pulverized and eaten dry, or added to Manzanita (*Arctostaphylos* spp.) cider for flavoring; never used in acorn mush. **Sinkyone:** seeds parched and ground into pinole. **Hupa:** burned areas where plant grew; seeds gathered from scorched plants crushed into flour.

Citation(s): Dixon 1905; Drucker 1937; Foster 1944; Gayton 1948; Goldschmidt 1951; Nomland 1935, 1938; Powers 1877; Strike 1994.

798 *Madia elegans*

Synonym(s): *Madia densifolia, Madia elegans* ssp. *densifolia*

Common Name(s): Common Madia, Tarweed, Gumweed.

Usage: Food: **Pomo, Pomo (Southwestern):** used seeds in pinole. **Miwok:** seeds gathered in midsummer, pulverized; the meal eaten dry.

Medicine: **Maidu:** used to treat urinary problems.

Citation(s): Barrett 1952; Barrett and Gifford 1933; Goodrich et al 1980; Chesnut 1902.

Terminology: **Pomo (Central):** dje′ca, **(Southwestern):** mucha kíli. **Miwok (Central Sierra, South Sierra):** yō′wa. **Karok:** maktunvechash, "little mank."

Citation(s): Barrett 1952; Barrett and Gifford 1933; Schenck and Gifford 1952; Strike 1994.

799 *Madia glomerata*
Common Name(s): Tar Weed.

Usage: Food: **Maidu (Northern):** seeds eaten.
Citation(s): Dixon 1905.

800 *Madia gracilis*
Synonym(s): *Madia dissitiflora*
Common Name(s): Gum-Weed.

Usage: Food: **Yuki:** seeds used in pinole. **Miwok:** seeds one of the most valued; gathered in August, parched, and pulverized.
Citation(s): Barrett and Gifford 1933; Chesnut 1902.
Terminology: **Miwok (Central Sierra):** etce'.
Citation(s): Barrett and Gifford 1933.

801 *Madia sativa*
Synonym(s): *Madia capitata*
Common Name(s): Tar Weed, Chile Tarweed.

Usage: Food: **Huchnom:** used to make bread. **Pomo:** used seeds in pinole. **Pomo (Southwestern):** seeds parched, pulverized ands eaten as pinole; the meal was moistened to keep people from choking on the dry meal; seeds stored raw and parched and pounded only when needed for food. **Yuki:** oil manufactured from seeds used for cooking. **Miwok:** seeds used for food.
Citation(s): Barrett 1952; Barrett and Gifford 1933; Chesnut 1902; Gifford 1967; Powers 1877; Goodrich et al 1980.
Terminology: **Pomo (Southwestern):** mushchakili; "grain" (muhca), "black" (kili); **Pomo (Central):** hō'mtca.

Citation(s): Gifford 1967, Barrett 1952.

Note(s): brought to California by the Spanish in the early 1800's and soon dominated the native species.

Citation(s): Goodrich et al 1980..

802 *Maianthemum* spp.

Family: *Ruppiaceae*

Synonym(s): *Smilacina* spp.

Common Name(s): False-Solomon's Seal.

Usage: Material: **Tubatulabal:** mashed slightly; thrown into quiet pools as fish poison.

Medicine: **Yana:** roots were pounded fine and applied as a cure for a swelling or boil.

Citation(s): Sapir and Spier 1943; Voegelin 1938.

803 *Maianthemum racemosa*

Synonym(s): *Smilacina amplexicaulis, Smilacina racemosa* var. *amplexicaulis*

Common Name(s): Fat Solomon.

Usage: Medicine: **Karok:** the root is put upon the navel of a child after the umbilical cord is out, to keep it well; or if the navel protrudes to make it grow right. **Costanoan:** leaf, decoction used as a contraceptive. **Maidu:** decoction used as a contraceptive, root used to regulate menstrual disorders, relieve kidney problems, heal wounds, as a heart tonic.

Citation(s): Schenck and Gifford 1952; Rayburn 2012.

Terminology: **Karok:** anupuhich, "imitation navel," pikawasahich, "imitation leather."

Citation(s): Schenck and Gifford 1952.

804 *Maianthemum stellatum*

Synonym(s): *Smilacina stellata, Smilacina sessilifolia,*
Smilacina stellata var. *sessilifolia*
Common Name(s): Slim Solomon.

Usage: Material: **Kawaiisu (Tehachapi):** root is mashed
and used as a fish stupefier. **Owens
Valley Paiute:** mashed with rock,
wrapped or sewn in worn-out baskets
and dipped into pools to form dams
and shaken to stupify fish.

Medicine: **Washo, Maidu:** a powder from
pulverized root used to stanch the
bleeding of wounds; the liquid from
mashed soaked roots said to have
antiseptic values in case of blood
poisoning; a concentration tea made
from boiled roots considered to be a
good tonic. **Maidu, Yana:** root used to
poultice swellings, boils. **Maidu:** root
infusion used to bathe sore, infected
eyes.

Citation(s): Train et al 1941; Zigmond 1981; Strike 1994,
Steward 1933.

Terminology: **Washo:** dama-<u>go</u>-go-yes, <u>she</u>-gimba. **Kawaiisu
(Tehachapi):** paaruurɨbɨ. **Owens Valley Paiute:**
tü`üꞟwava, tügʷüꞋva.

Citation(s): Train et al 1941; Zigmond 1981, Steward 1933.

805 *Malacothamnus* spp.

Family: *Malvaceae*
Synonym(s): *Malvastrum* spp.

Usage: Medicine: **Diegueño:** leaf decoction used as an
emetic; at onset of first menstruation,

young girls were put in a heated pit for several days with leaves and other herbs., could only touch skin or hair with special wooden or bone tool, no meat or salt could be eaten during this time. **Luiseño:** leaf decoction used as a contraceptive and as an emetic.
Citation(s): Strike 1994.

806 *Malacothamnus fasciculatus*
Common Name(s): Chaparral Mallow.

Usage: Material: **Chumash:** bark, fibers, made into cordage.
Citation(s): Timbrook 2007.
Terminology: **Chumash (Barbareño):** khman.
Citation(s): Timbrook 2007.

807 *Malacothrix californica*
Family: *Asteraceae*
Common Name(s): California Dandelion.

Usage: Food: **Luiseño, Maidu:** the seeds are eaten.
Misc: **Tubatulabal:** considered weeds.
Citation(s): Sparkman 1908; Voegelin 1938; Strike 1994.
Terminology: **Luiseño:** makiyal.
Citation(s): Sparkman 1908.

808 *Malosma laurina*
Family: *Amaranthaceae*
Common Name(s): Laurel Sumac.

Usage: Food: **Chumash:** berries, pounded, dried in sun, eaten raw.
Medicine: **Chumash:** bark, boiled, drank for

dysentery.

Citation(s): Timbrook 2007.

Terminology: **Chumash (Barbareño, Ynezeño, Ventureño):** walqaqsh.

Citation(s): Timbrook 2007.

809 *Malva* spp.

Family: *Malvaceae*
Common Name(s): Chinese-Weed.

Usage: Food: **Cahuilla:** seeds eaten fresh.

Citation(s): Bean and Saubel 1972.

Note(s): native of Eurasia; plants is a laxative, digestive.

Citation(s): Munz and Keck 1965; Stuhr 1933.

810 *Malva nicaeensis*

Common Name(s): Bull Mallow.

Usage: Medicine: **Costanoan:** root decoction used as hair rinse, used internally as an emetic, for migraine headache, to treat fever, especially in children; leaves, hot poultice placed on stomach, or head, to relieve pain.

Citation(s): Rayburn 2012.

Note(s): native from Eurasia.

Citation(s): Munz and Keck 1965.

811 *Malva parviflora*

Common Name(s): Chinese-Weed, Mallow, Cheeses.

Usage: Food: **Chumash:** seeds, fruit, eaten, especially popular with children.

          Medicine: **Miwok:** soft stems and flowers were steeped and used as a poultice on running sores, boils, and swellings.

Citation(s): Barrett and Gifford 1933; Timbrook 2007.

Terminology: **Miwok (Central Sierra): kasani. Chumash
(Barbareño, Ynezeño):** mal; **(Obispeño):** temala;
(Ventureño): malwash.

Citation(s): Barrett and Gifford 1933; Timbrook 2007.

812 *Malvella leprosa*

Family: *Malvaceae*

Synonym(s): *Sida hederacea*

Common Name(s): Alkali Mallow.

Usage: Medicine: **Maidu:** used as a laxative.

Citation(s): Strike 1994.

Note(s): introduced from South America; toxic to some
livestock.

Citation(s): Baldwin et al 2012.

813 *Mammillaria grahamii* var. *grahamii*

Family: *Cactaceae*

Synonym(s): *Mammillaria microcarpa*

Common Name(s): Pincushion Cactus, Fishhook
Cactus.

Usage: Misc: **Kawaiisu (Tehachapi):** named but not
used.

Citation(s): Zigmond 1981.

Terminology: **Kawaiisu (Tehachapi):** noog^{w}iyavɨ.

Citation(s): Zigmond 1981.

814 *Marah* spp.

Family: *Cucurbitaceae*

Synonym(s): *Echinocystis* spp.

Common Name(s): Manroot.

Usage: Food: **Shelter Cove Sinkyone:** seeds roasted and eaten. **Chumash:** leaves may have been eaten.

Citation(s): Baumhoff 1958; Strike 1994.

815   *Marah fabacea*

Synonym(s): *Echinocystis fabacea, Marah fabaceus* var. *agrestis*

Common Name(s): Man-Root.

Usage: Material: **Wintun:** used root as fish poison. **Pomo (Southwestern):** favorite fish drug because of its abundance, used in both fresh and salt water, in tidal pools various vertebrate fishes and octopus were taken.

   Medicine: **Pomo (Southwestern):** root pounded, mixed with pounded pepper nuts and skunk grease; applied to head to prevent baldness. **Costanoan:** seeds, roasted, eaten (probably) for kidney trouble. **Maidu:** infusion used to soothe irritated skin., treatment for rheumatic pains, venereal problems, boils, inflammations. **Kawaiisu (Tehachapi):** seeds, roasted, pounded, used a salve to cure baldness; roots, shell of fruit, used as a shampoo.

Citation(s): Gifford 1967; Heizer 1953; Goodrich et al 1980; Rayburn 2012; Strike 1994.

Terminology: **Pomo (Southwestern):** beheschata.

Citation(s): Gifford 1967; Goodrich et al 1980.

816 *Marah horridia*

Synonym(s): *Echinocystis horrida*
Common Name(s): Chilicothe.

Usage: Material: **Miwok (Southern Sierra):** used seeds and root as fish poison.

Medicine: **Kawaiisu (Tehachapi):** seeds, roasted until it turns black, mashed to make greasy salve, applied to skin irritations and sores. **Kawaiisu (Tehachapi), Tubatulabal:** rind juice rubbed on skin to cure rashes, sores; burned seeds rubbed on facial blemishes, on a newborn baby's navel.

Citation(s): Heizer 1953; Zigmond 1981; Strike 1994.

817 *Marah macrocarpa*

Synonym(s): *Echinocystis macrocarpa*
Common Name(s): Chilicothe.

Usage: Material: **Luiseño:** the seeds were used in the manufacture of red paint. **Chumash:** seeds, pierced, and strung into necklaces just for fun (children and adults); seeds, also used as gaming pieces (by boys).

Medicine: **Luiseño:** a purgative is made from the roots; seed oil used as lotion to reduce swelling. **Costanoan:** used for skin problems; seed paste used for pimples, skin sores. **Diegueño:** leaves boiled, used to relieve hemorrhoid pain, seeds, pulverized, mixed with water , used a facial paint. **Chumash:** seeds, made into paste, stored for later use; put into water, medicine for pregnant women for blood, to insure that children would be born all right, given to babies, young children as a purgative, sprinkled on tumor to make it go away; kernels, roasted until black and greasy, special paint used for curing someone who was seriously ill; seeds, or roots, toasted, thought to promote hair growth; plant, mixed with stone dust used to reduce inflamation, heal wounds, remove cataracts, induce menstruation, cure urinary disorders, boiled, drank to induce sweating.

Citation(s): Sparkman 1908; Strike 1994; Rayburn 2012; Timbrook 2007.

Terminology: **Luiseño:** enwish. **Chumash (Barbareño):** molo'wot'; **(Ynezeño):** cha'; **(Ventureño):** 'anmakhwaka'y.

Citation(s): Sparkman 1908; Timbrook 2007.

Note(s): native of Chile; *Marah* spp. root has purgative properties.

seeds contain:

Ash   5.0%
Protein 29.4%
Oil   54.4%

Citation(s): Munz and Keck 1965; Stuhr 1933; Earle and Jones 1962.

818 *Marah oregana*

Synonym(s): *Echinocystis oregana, Marah micrampelis*
Common Name(s): Hill Man-Root, Big Root.

Usage: Food: **Coast Yuki:** nut was eaten.
       Material: **Yuki:** root used to poison fish.

Medicine: **Yuki:** seeds and root highly prized as a specific against rheumatism and venereal diseases; fresh root rubbed over rheumatic joints or on boils and swellings; often roasted first, then mashed (either root or seeds); used for a certain complaint of the urinary organs, two seeds are eaten in the morning and two in the evenings before meals.

Misc: **Karok:** named but not used; said to be poisonous.

Citation(s): Chesnut 1902; Gifford 1939; Schenck and Gifford 1952.

Terminology: **Karok:** tuush. **Yuki:** bokumbal, bokum (nut).

Citation(s): Schenck and Gifford 1952; Gifford 1939.

Note(s): native of Chile; *Marah* spp. root has purgative properties.

Citation(s): Munz and Keck 1965; Stuhr 1933.

819 *Marah watsonii*

Common Name(s): Taw Manroot.

Usage: Medicine: **Maidu:** seeds eaten to cure kidney ailments.

Citation(s): Strike 1994.

820 *Marrubium vulgare*

Family: *Lamiaceae*

Common Name(s): Horehound.

Usage: Medicine: **Yuki:** used in the form of a tea for coughs. **Chumash:** used mostly for cough medicine and for sore throats, colds, coughs, and lung ailments.

Kawaiisu (Tehachapi): infusion of leaves and flowering tops drunk hot or cold for relief of coughing and colds. Cahuilla: plant boiled whole; infusion used to flush kidneys. Costanoan: decoction, used to treat coughs, including whooping cough.

Misc: Tubatulabal: considered weeds. Miwok: named but not used.

Citation(s): Barrett and Gifford 1933; Curtin 1957; Voegelin 1938; Timbrook 1987; Zigmond 1981; Bean and Saubel 1972; Rayburn 2012; Timbrook 2007.

Terminology: Miowk (Central Sierra): yotcitayu.

Citation(s): Barrett and Gifford 1933.

Note(s): common pestiferous weed in waste places and old fields; naturalized from Europe; dried leaves and flowering tops are carminative, expectorant, laxative, aperient, disphoretic, stimulant, tonic.

Citation(s): Munz and Keck 1965; Stuhr 1933.

821 *Matricaria discoidea*

Family: *Asteraceae*
Synonym(s): *Matricaria matricariodes, Chamomilla suaveolens*
Common Name(s): Pineapple Weed.

Usage: Medicine: Pomo: decoction of leaves and flowers used to check diarrhea. Maidu: decoction of leaves and flowers used to check diarrhea, reduce fever, to produce profuse sweating, taken as a tonic. Cahuilla: infusion made to settle upset stomach, cure diarrhea and colic. Costanoan: used to reduce fevers, as an anti-convulsive analgesic, disinfectant, gastrointestinal cure, pediatric medicine; seeds, salve, used for infected sores; decoction given for stomach pain, indigestion, fever; decoction mixed with crushed brick and urine to treat infant convulsions. **Diegueño:** tea used to reduce fevers, stomach ache, cure colds, relieve menstrual cramps; for babies flower tea, for adults whole plant tea. **Miwok (Coast):** seeds, flowers, used to reduce fevers. **Chumash:** tea, taken for stomach pains, nerves, induce perspiration, reduce pain, cramps, colic, to suppress menstruation, used in childbirth, to treat catarrh, dysentery; poultice applied to reduce inflammations.

Citation(s): Chesnut 1902; Bean and Saubel 1972; Strike 1994; Rayburn 2102; Timbrook 2007.

Note(s): plant is tonic, diaphoretic; apparently introduced in/from eastern states.

Citation(s) Stuhr 1933; Munz and Keck 1965.

822 *Medicago polymorpha*

Family: *Fabaceae*
Synonym(s): *Medicago hispida*
Common Name(s): Bur-Clover.

Usage: Food: **Cahuilla:** seeds harvested in Spring and early Summer; parched, ground, and made into mush.

Citation(s): Bean and Saubel 1972.

Note(s): native of southern Europe.

Citation(s): Munz and Keck 1965; Munz 1968.

823 *Medicago sativa*

Common Name(s): Alfalfa.

Usage: Medicine: **Costanoan:** leaves, heated, held to ear for earache.
Citation(s): Rayburn 2012.
Note(s): native of Old World.
Citation(s): Munz and Keck 1965.

824 *Melica bulbosa*

Family: *Poaceae*

Usage: Food: **Pomo (Southwestern):** corms washed, dried and eaten raw; either whole or pounded like pinole.
Citation(s): Gifford 1967.
Terminology: **Pomo (Southwestern):** achim.
Citation(s): Gifford 1967.

825 *Melica imperfecta*

Common Name(s): Small-Flowered Melic.

Usage: Food: **Kawaiisu (Tehachapi):** seeds pounded, cooked into mush.
Medicine: **Kawaiisu (Tehachapi):** discard child's milk teeth in a clump of the plant to ensure growth of new teeth.
Citation(s): Zigmond 1981; Strike 1994..

826 *Melilotus indicus*

Family: *Fabaceae*
Synonym(s): *Melilotus indica*

Usage: Medicine: **Pomo (Southwestern):** whole plant boiled; tea used as a purgative; a very strong laxative.

Misc: **Tubatulabal:** considered weeds.

Citation(s): Voegelin 1938; Goodrich et al 1980.

Terminology: **Pomo (Southwestern):** kawaꞏyúhso, "horse-clover."

Citation(s): Goodrich et al 1980.

Note(s): native of Eurasia.

Citation(s): Munz and Keck 1965.

827 *Melissa officinalis*

Family: *Lamiaceae*

Common Name(s): Lemon Balm.

Usage: Medicine: **Costanoan:** decoction used for infant colic, stomachache in general.

Citation(s): Rayburn 2012.

828 *Mentha* spp.

Family: *Lamiaceae*

Common Name(s): Horse Mint.

Usage: Medicine: **Maidu (Northern):** made into tea for cathartic properties. **Chumash:** boiled, let steam into eyes to treat sore eyes.

Citation(s): Dixon 1905; Timbrook 2007.

Terminology: **Owens Valley Paiute:** wai'yanuva. **Chumash (Ventureño):** 'alaqtaha.

Citation(s): Steward 1933; Timbrook 2007.

829 *Mentha arvensis*

Common Name(s): Tule Mint.

Usage: Food: **Northern Paiute (Paviotso):** a teas was produced by pouring hot water on dried leaves. **Owens Valley Paiute:**

leaves boiled into refreshing drink. **Kawaiisu (Tehachapi):** leaves brewed and drunk as non-medicinal beverage.

Medicine: **Owens Valley Paiute:** tea of entire plant, except roots, drunk to keep cool; also chewed. **Kawaiisu (Tehachapi):** leaves and stems placed on area of pain and swelling with heat provided by hot rock placed on hot. **Maidu:** leaves, crushed, used to poultice swellings, bruises; leaf decoction used to treat colds, sore throats, toothache, headache, fever, nausea, stomach ache, intestinal gas, reduce pain, provide a tranquilizing effect. **Modoc (Lutuami) (Lutuami):** leaf infusion used to relieve stomach ache, colds, coughs; root, crushed, used to poultice aches, pains; leaves, crushed, used to poultice sore eyes.

Citation(s): Kelly 1932; Steward 1933; Zigmond 1981; Strike 1994.

Terminology: **Owens Valley Paiute:** kwā'nāuva, wai'yanuva, kwa'nanuvüva.

Citation(s): Steward 1933.

830 *Mentha canadensis*

Synonym(s): *Mentha arvensis* var. *villosa*

Usage: Medicine: **Washo:** a half-cupful of strong tea made from the tops of the plant used to stop diarrhea.

Citation(s): Train et al 1941

Terminology: **Washo:** pah-do-lo-yi.

Citation(s): Train et al 1941.

Note(s): dried leaves and flowering plants are carminative,
stimulant.
Citation(s): Stuhr 1933.

831 *Mentha spicata*
Common Name(s): Spearmint.

Usage: Food: **Miwok:** hot tea of boiled leaves drunk
as a beverage. **Yuki:** used to make
beverage.

Medicine: **Miwok:** hot tea of boiled leaves was
drunk to relieve stomach trouble and
diarrhea.

Misc: **Yuki:** a handsome woman placed
leaves on her bosom as perfume; also
spread it among her garments for the
same reason.

Citation(s): Barrett and Gifford 1933; Curtin 1957.
Terminology: **Miwok (Central Sierra):** sisimela.
Citation(s): Barrett and Gifford 1933.
Note(s): naturalized from Europe; dried leaves and flowering
plants are carminative, stimulant.
Citation(s): Munz and Keck 1965; Stuhr 1933.

832 *Mentzelia* spp.
Family: *Loasaceae*
Common Name(s): Buena Mujer.

Usage: Food: **Tubatulabal:** seeds referred to as being
better than Chia (*Salvia columbariae*)
seeds "because they had more grease."
Miwok: seeds were pulverized and
eaten as pinole. **Cahuilla:** seeds
gathered from February to October,
parched, ground into flour for mush.

Medicine:　Miwok, Kawaiisu (Tehachapi): pulverized seeds mixed with water, or preferably Fox or Wild Cat grease and applied as a poultice.

Citation(s): Barrett and Gifford 1933; Voegelin 1938; Bean and Saubel 1972; Strike 1994.

Terminology: **Miwok (Central Sierra):** matcū̄'.

Citation(s): Barrett and Gifford 1933.

833　*Mentzelia affinis*

Usage:　Food:　**Kawaiisu (Tehachapi):** seed collected in June when flowers have fallen off, parched, ground, eaten (consistency like peanut butter).

Citation(s): Zigmond 1981.

Terminology: **Kawaiisu (Tehachapi):** ku?uvi.

Citation(s): Zigmond 1981.

834　*Mentzelia albicaulis*

Usage:　Food:　**Northern Paiute (Paviotso):** the seeds were eaten. **Tubatulabal:** used seeds. **Kawaiisu (Tehachapi):** seed collected in June when flowers have fallen off, parched, ground, eaten (consistency like peanut butter).

Citation(s): Kelly 1932; Voegelin 1938; Zigmond 1981.

Terminology: **Northern Paiute (Paviotso):** gu·ha'. **Kawaiisu (Tehachapi):** ku?uvi.

Citation(s): Kelly 1932; Zigmond 1981.

835 *Mentzelia congesta*

Usage: Food: **Kawaiisu (Tehachapi):** seed collected in June when flowers have fallen off, parched, ground, eaten (consistency like peanut butter).
Citation(s): Zigmond 1981.
Terminology: **Kawaiisu (Tehachapi):** ku?uvi.
Citation(s): Zigmond 1981.

836 *Mentzelia dispersa*

Usage: Food: **Kawaiisu (Tehachapi):** seed collected in June when flowers have fallen off, parched, ground, eaten (consistency like peanut butter).
Citation(s): Zigmond 1981.
Terminology: **Kawaiisu (Tehachapi):** ku?uvi.
Citation(s): Zigmond 1981.

837 *Mentzelia gracilenta*

Usage: Food: **Tubatulabal:** used seeds.
Citation(s): Voegelin 1938.

838 *Mentzelia involucrata*

Usage: Material: **Kawaiisu (Tehachapi):** children throw leaves at one another; they stick and are hard to remove.
Citation(s): Zigmond 1981.

839   *Mentzelia laevicaulis*
Synonym(s): *Mentzelia laevicaulia*

Usage: Medicine: **Wailaki, Maidu:** decoction of the leaves
used internally to relieve stomach ache;
used as a wash in some loathsome skin
diseases.
Citation(s): Chesnut 1902; Strike 1994.

840   *Mentzelia veatchiana*

Usage: Food: **Kawaiisu (Tehachapi):** seed collected
in June when flowers have fallen off,
parched, ground, eaten (consistency
like peanut butter).
Medicine: **Maidu, Wailaki:** leaf decoction drank
for stomach ache.
Citation(s): Zigmond 1981; Strike 1994.
Terminology: **Kawaiisu (Tehachapi):** ku?uvi.
Citation(s): Zigmond 1981.

841   *Microseris* spp.
Family: *Asteraceae*

Usage: Food: **Miwok:** eaten as greens.
Citation(s): Strike 1994.

842   *Microseris laciniata*

Usage: Food: **Miwok (Coast):** eaten as greens.
Citation(s): Strike 1994.

843   *Mimulus* spp.
Family: *Phrymaceae*

Usage: Medicine: **Miwok:** root used to make a tea with astringent properties; used as a cure for diarrhea.
Citation(s): Barrett and Gifford 1933.
Terminology: **Miwok (Central Sierra):** hosina, yasaññaña.
Citation(s): Barrett and Gifford 1933.

844   *Mimulus aurantiacus*
Synonym(s): *Mimulus puniceus*
Common Name(s): Sticky Monkey Flower, Bush Monkey Flower.

Usage: Medicine: **Pomo (Southwestern):** flower, stem, and leaves boiled, strained, used as eyewash for sore eyes. **Costanoan:** may have been crushed and placed on sores; decoction, used as kidney and bladder remedy. **Diegueño:** tea used to regulate menstrual periods.
Citation(s): Goodrich et al 1980; Rayburn 2012; Strike 1994.
Terminology: **Pomo (Southwestern):** hu?úˑwenu qʰale, "eye medicine plant."
Citation(s): Goodrich et al 1980.

845   *Mimulus cardinalis*

Usage: Medicine: **Karok:** soaked in water to make medicine to wash a new baby with.
         Misc: **Tubatulabal:** considered weeds.
Citation(s): Schenck and Gifford 1952; Voegelin 1938.
Terminology: **Karok:** patchimitkiin.
Citation(s): Schenck and Gifford 1952.

846 *Mimulus guttatus*

Synonym(s): *Mimulus luteus, Mimulus nasutus*
Common Name(s): Musk-Flower.

Usage: Food:
Wailaki: used as greens; ash from leaves used as a source of salt. **Miwok:** leaves were boiled. **Maidu, Southern (Nisenan):** eaten in the spring as greens.

Material:
Miwok: flowers used as wreaths,

Medicine:
Kawaiisu (Tehachapi): stems and leaves boiled, head covered and steam inhaled to relieve soreness in chest and back.

Misc:
Tubatulabal: considered weeds.

Citation(s): Barrett and Gifford 1933; Chesnut 1902; Voegelin 1938; Zigmond 1981; Powers 1877.
Terminology: **Miwok (Central Sierra):** puksa.
Citation(s): Barrett and Gifford 1933.

847 *Mimulus moschatus*

Common Name(s): Musk-Plant.

Usage: Food:
Miwok: when young, the plant was boiled; not stored for Winter use as quantity was always small. **Maidu:** young plants, boiled, eaten.

Citation(s): Barrett and Gifford 1933; Strike 1994.
Terminology: **Miwok (Central Sierra):** pokosa, yusunu.
Citation(s): Barrett and Gifford 1933.

848   *Mirabilis greenei*
        Family: *Nyctaginaceae*

Usage: Medicine: **Karok:** used with a formula to make a
                      new-born baby healthy.
Citation(s): Schenck and Gifford 1952.
Terminology: **Karok:** yupshitanachpirish, "knee and ankle."
Citation(s): Schenck and Gifford 1952.

849   *Mirabilis laevis*
        Synonym(s): *Mirabilis californica*

Usage: Medicine: **Luiseño:** a decoction of the leaves is
                      used as a purgative.
         Misc: **Tubatulabal:** considered weeds.
Citation(s): Sparkman 1908; Voegelin 1938.
Terminology: **Luiseño:** nanukvish, tisi.
Citation(s): Sparkman 1908.

850   *Monardella breweri* ssp. *glanduifera*
        Family: *Lamiaceae*
        Synonym(s): *Monardella lanceolata*

Usage: Food: **Luiseño:** a tea was used as a beverage.
         Medicine: **Luiseño:** a tea was used medicinally.
                      Miwok: a decoction of the leaves,
                      upper stems, and flowers was drunk for
                      colds and for headaches. **Maidu:** leaf
                      and flower decoction used to treat
                      colds, headaches, colic.
Citation(s): Barrett and Gifford 1933; Sparkman 1908; Strike
        1994.
Terminology: **Luiseño:** huvawut.
Citation(s): Sparkman 1908.

851 *Monardella candicans*

Usage: Food: **Tubatulabal:** used leaves and stalks.
Citation(s): Voegelin 1938.

852 *Monardella linoides*
Common Name(s): Norrowleaf Monardella.

Usage: Food: **Kawaiisu (Tehachapi):** leaves and flowers brewed for tea.
Citation(s): Zigmond 1981.
Terminology: **Kawaiisu (Tehachapi):** kaaruukɨtɨbɨ.
Citation(s): Zigmond 1981.

853 *Monardella odoratissima*

Usage: Food: **Miwok:** decoction of the leaves and flower heads used as a beverage. **Kawaiisu (Tehachapi):** leaves and flowers brewed for tea.

Material: **Karok:** picked and put inside hat for nice smell when one is going on a journey.

Medicine: **Karok:** sometimes used when a "sweat medicine" is prepared; with the right formula it can be made into a "love medicine" used by women. **Miwok:** decoction of the stems and flower heads drunk for colds, fevers, to relieve digestive upsets. **Washo:** decoction used to cure colds, relieve digestive upsets. **Maidu:** decoction used to cure colds, relieve digestive upsets, for sore eyes, drank to regulate menstrual cycles.

Citation(s): Barrett and Gifford 1933; Schenck and Gifford 1952; Zigmond 1981.
Terminology: **Karok:** smakat, "meadow hat." **Kawaiisu (Tehachapi):** kaaruukɨtɨbɨ, keevowadi.
Citation(s): Schenck and Gifford 1952; Zigmond 1981.

854 *Monardella villosa*

Common Name(s): Coyote-Mint, Pennyroyal.

Usage: Medicine: **Cahuilla:** tea made from leaves used for stomachache. **Costanoan:** decoction applied to deliberately made cuts on the back of a person having difficulty healing to help ease breathing, to draw of "bad blood;" salve, used for respiratory problems. **Maidu:** infusion used as blood purifier, cure for stomach aches. **Wintun:** tea used to treat colds.

Citation(s): Bean and Saubel 1972; Strike 1994; Rayburn 2012.

855 *Monardella sheltonii*

Synonym(s): *Monardella villosa* ssp. *sheltonii*

Usage: Food: **Maidu:** leaves collected when plant is in seed and used fresh and dry for tea.
Medicine: **Maidu:** tea used to cure colic and as a blood purifier.

Citation(s): Chesnut 1902.

856 *Monardella viridis*

Common Name(s): Common Monardella.

Usage: Food: **Kawaiisu (Tehachapi):** leaves and flowers brewed for tea.

Citation(s): Zigmond 1981.

Terminology: **Kawaiisu (Tehachapi)**: kaaruukɨtɨbɨ.
Citation(s): Zigmond 1981.

857 *Monotropa hypopitys*
Family: *Ericaceae*
Synonym(s): *Hypopithys lanuginosa*
Common Name(s): Pine-Sap.

Usage: Misc: **Karok:** named but not used.
Citation(s): Schenck and Gifford 1952.
Terminology: **Karok:** putin'nupara, "what-ought-to-be-
underground sticking out."
Citation(s): Schenck and Gifford 1952.

858 *Montia parvifolia*
Family: *Montiaceae*
Synonym(s): *Claytonia perfoliata, Montia perfoliata*
Common Name(s): Indian Lettuce, Wild Lettuce,
Miner's Lettuce.

Usage: Food: **Luiseño:** used for greens and also eaten
raw. **Miwok:** stems, leaves, and
blossoms eaten raw. **Wintun:** eaten raw
in the Spring when tender, sometimes
steamed. **Maidu, Southern (Nisenan),
Washo:** gathered in great quantities
and laid near nests of certain large red
ants, which circulate through it;
afterwards it is taken up, shaken out,
and eaten; the ants, running over it,
impart a sour taste to it and make it as
good as if vinegar had been put on it.
Yuki: whole plant eaten raw or cooked.
Cahuilla: plant eaten fresh, or boiled as
a green. **Chumash:** seeds, eaten.

Misc: **Tubatulabal:** considered weeds.

Citation(s): Barrett and Gifford 1933;Chesnut 1902; Du Bois 1935; Powers 1877; Sparkman 1908; Voegelin 1938; Bean and Saubel 1972; Strike 1994; Timbrool 2007.

Terminology: **Miwok (Central Sierra):** sestu. **Luiseño:** towish popa'kaw. **Cahuilla:** palsingat. **Chumash (Barbareño):** shilik'; **(Ynezeño):** shilik.

Citation(s): Barrett and Gifford 1933; Sparkman 1908; Bean and Saubel 1972; Timbrook 2007.

859   *Morella californica*

Family: *Myricaceae*
Synonym(s): *Myrtis communis, Myrica californica*
Common Name(s): California Wax Myrtle.

Usage: Material: **Karok:** used in basketry (sticks).

       Misc: **Coast Yuki, Pomo (Southwestern):** named but not used.

Citation(s): O.Neale 1932; Schenck and Gifford 1952; Gifford 1939; Goodrich et al 1980.

Terminology: **Karok:** kisrip. **Yuki:** kyuwü. **Pomo (Southwestern):** s'apʰa qʰale.

Citation(s): O'Neale 1932; Schenck and Gifford 1952; Gifford 1939; Goodrich et al 1980.

Note(s): *Myrtis communis* is native to southern Europe and north Africa; Lichvar and Kartesz (2009) list *Morella california* as the proper nomenclature for this plant on the west coast, including California.

860   *Muhlenbergia rigens*

Family: *Poaceae*
Synonym(s): *Epicampes rigens, Epicampes rigens californica, Cinna macroura*
Common Name(s): Deer Grass.

Usage: Material: **Miwok, Yokuts, Washo, Monache (Western Mono), Kitanemuk, (Tejon), Panamint Shoshone (Koso), Chemehuevi, Kawaiisu (Tehachapi), Tubatulabal, Luiseño, Cahuilla, Cupeño, Diegueño, Chumash:** the stalk was used as a coil foundation (of three reeds) for baskets. **Kawaiisu (Tehachapi):** stems gathered in November, form the multiple-rod foundation in coiled basketry; inserted stems in pierced ears to keep hole from closing. **Cahuilla:** stalk used as horizontal or foundation around which the coils were wrapped in basketry; the grass was used as the body of the coil in baskets..

Citation(s): Coville 1892; Curtis 1924; Dawson and Deetz 1965; Grant 1964; Merriam 1955; Merrill 1923; Zigmond 1981; Bean and Saubel 1972; Strike 1994; Timbrook 2007; Barrows 1900; Sparkman 1908.

Terminology: **Monache (Western Mono):** monop. **Luiseño:** yulalish. **Kawaiisu (Tehachapi):** sipu(m)bivɨ. **Cahuilla:** suul.

Citation(s): Gifford 1932; Sparkman 1908; Zigmond 1981; Bean and Saubel 1972.

861   *Nama demissum*

Family: *Boraginaceae*

Usage: Food: **Kawaiisu (Tehachapi):** seeds pounded, boiled into mush.

   Misc: **Tubatulabal:** considered weeds.

Citation(s): Voegelin 1938; Zigmond 1981.

Terminology: **Kawaiisu (Tehachapi):** tɨvimaasita.

Citation(s): Zigmond 1981.

862 *Navarretia* spp.
Family: *Polemoniaceae*

Usage: Food: **Miwok:** seeds gathered in August, parched and pulverized, eaten dry.
Citation(s): Barrett and Gifford 1933.
Terminology: **Miwok (Central Sierra):** hañu.
Citation(s): Barrett and Gifford 1933.

863 *Navarretia atractyloides*

Usage: Medicine: **Costanoan:** plants, toasted, powdered, or ashes, used as burn dressing (said to permit burns to heal without scarring).
Citation(s): Strike 1994; Rayburn 2012.

864 *Navarretia cotulifolioa*
Synonym(s): *Navarretia cotulaefolia*

Usage: Medicine: **Miwok:** boiled, the decoction applied to swellings.
Citation(s): Barrett and Gifford 1933.

865 *Navarretia leucocephala*
Common Name(s): White Navarretia.

Usage: Medicine: **Maidu:** decoction used to reduce swelling.
Citation(s): Strike 1994.

866 *Nemophila menziesii*

Family: *Boraginaceae*
Common Name(s): Baby Blue Eyes.

Usage: Medicine: **Pomo:** root decoction used to cure asthma.

           Misc: **Karok, Kawaiisu (Tehachapi):** named but not used (?).

Citation(s): Schenck and Gifford 1952; Zigmond 1981; Strike 1994.

Terminology: **Karok:** ˋatmahavnikaanich,"that which sees the first salmon coming." **Kawaiisu (Tehachapi):** kaawanavɨ, "rat-net."

Citation(s): Schenck and Gifford 1952; Zigmond 1981.

867 *Nereocystis luetkeama*

Family: *Lessoniaceae*
Common Name(s): Bull Kelp.

Usage: Food: **Pomo (Southwestern):** the thick part of the stalk cooked in oven, or in hot ashes, until puffy; strip dried for Winter.

           Medicine: **Pomo (Southwestern):** salty strip can be sucked on for colds to sooth throat and clear mucus.

           Material: **Pomo (Southwestern):** stem, shredded into slender pieces, used as cordage or fish line.

Citation(s): Goodrich et al.

Terminology: **Pomo (Southwestern):** cʰanama.

Citation(s): Goodrich et al.

Note(s): this species moved from Family *Laminariaceae* to Family *Lessoniaceae*.

Citation(s): Yoon et al 2001.

868 *Nicotiana* spp.

Family: *Solanaceae*
Common Name(s): Wild Tobacco.

Usage: Food:

Luiseño, Yuki, Monache (Western Mono): used for smoking. **Yokuts:** smoked or chewed a little. **Yokuts (Southern Valley):** gathered in abundance, ground with lime, made into paste, and formed into "cones or small loaves" wrapped in Tule (*Scirpus* spp.) leaves, and eaten; tobacco smoked alone. **Wappo:** not chewed; smoked usually in sweat house. **Wintun, Central (Nomlaki):** smoked; never used in any other way. **Cahuilla:** was chewed, smoked, or used in drinkable decoction depending upon purpose for which it was used; smoked in ritual settings; after fasting ritual by hunters, smoked to alleviate hunger.

Medicine:

Yokuts: used for cuts; tobacco and lime mixture frequently used as a salve or poultice applied to a painful spot; taken with water as an emetic. **Monache (Western Mono):** used to induce vomiting, which is thought to encourage dreams and insure good health. **Kawaiisu (Tehachapi):** dried, crushed into powder, mixed with lime; put on cuts, on bites, to stop bleeding or itching; as snuff it clears out a stuffy nose; wrapped in cloth, held between teeth, it deadens toothache; when swallowed causes vomiting and cleans out stomach. **Cahuilla:** water solution served an as emetic to induce vomiting; leaves employed as poultice to heal cuts, bruises, swellings, and other wounds; to alleviate earaches, smoke blown into ear, then covered with warm pad; leaves steamed, applied externally to lymph glands of neck to cure scrofula and other forms of neck inflammation; leaves placed on hot rock in sweat-house, steam inhaled to help rheumatism, or cure nasal congestion; weak tea drunk at time of menstruation to keep body free from unpleasant odors; before traveling, smoke was blown in the direction to be traveled to clear away all danger, to ensure the blessing of traveler's spiritual guides.

Misc: **Coast Yuki:** it did not grow in their territory. **Shasta:** was planted; seeds obtained from the *Gamutwa (?)*, who got them from the **Karok**, who got them from the **Yurok**. **Yokuts:** gathered on the west side of the Tulare River; not cultivated or pruned. **Monache (Western Mono):** most came from people to the south. **Wintun, Hill (Patwin), Wintun, River (Patwin):** gathered chiefly along streams. **Wintun, Central (Nomlaki):** gathered along creek banks where it grew wild; never cultivated. **Miwok, Wappo:** gathered wild, not planted. **Maidu, Southern (Nisenan):** burn ground clear in Winter, scattered seeds in ashes in Spring; picked leaves when yellow, dried.

Citation(s): Beals 1933; Curtin 1957; Driver 1936; Gayton 1948; Gifford 1939; Goldschmidt 1951; Holt 1946; Kroeber 1932; Powers 1877; Sparkman 1908; Zigmond 1981; Bean and Saubel 1972.

Terminology: **Yuki:** walmilbal. **Yokuts, Monache (Western Mono:** šógϴh, só'og. **Cahuilla:** pivat.

Citation(s): Gifford 1939; Gayton 1948; Bean and Saubel 1972.

869 *Nicotiana attenuata*

Common Name(s): Wild Tobacco.

Usage: Food: **Wiyot:** smoked in small pipes. **Tubatulabal:** gathered and used. **Cahuilla:** smoked after being pounded in small stone mortars. **Maidu (Northern):** used.

Medicine: **Chumash:** formerly eaten not only as a recreational drug but also to promote good health; dried, ground, and used plain or mixed with lime ashes from burned shells; tobacco mixed with water drunk to relieve pain in the stomach.

Misc: **Miwok:** cultivated.

Citation(s): Barrett and Gifford 1933; Barrows 1900; Dixon 1905; Loud 1918; Merriam 1955; Voegelin 1938; Timbrook 1987.

Terminology: **Northern Paiute (Paviotso):** puhi'-pa'mo', "green" (puhi); wi'si-pa'mo'. **Owens Valley Paiute:** pā'müpi. **Mono Lake Paiute:** pa'ʰmū. **Panamint Shoshone (Koso):** pāhū'mbi. **Luiseño:** pivat-isil, "coyote tobacco."

Citation(s): Kelly 1932; Steward 1933; Sparkman 1908.

Note(s): the alkaloid nicotine is extremely poisonous, causing

sever vomiting, diarrhea, slow pulse, dizziness, collapse, and respiratory failure.
Citation(s): Hardin and Arena 1974.

870 *Nicotiana clevelandii*

Usage: Medicine: **Chumash:** only recreational use, had some ritual and medicinal applications.
Citation(s): Timbrook 2007.

871 *Nicotiana quadrivalis*
Synonym(s): *Nicotiana plumbaginifolia, Nicotiana Bigelovii, Nicotiana bigelovii exalta*
Common Name(s): Tobacco, Wild Tobacco.

Usage: Food: **Hupa:** the plant is grown by scattering seeds in ashes of burned logs; the plant is smoked in a pipe. **Wiyot:** smoked in small wooden pipes. **Pomo:** gathered in early Summer and dried for use during other seasons of the year. **Yana:** used. **Tubatulabal:** gathered and used. **Miwok:** cultivated and used. **Sinkyone:** plants seeds in ashes of burned over places; used for smoking. **Maidu, Southern (Nisenan):** smoke alone or mixed with dried Manzanita (*Arctostaphylos* spp.) leaves. **Karok:** seed is planted in the ground after the vegetation has been burned off; the leaves are rubbed into fine power and stored for later use. **Chumash:** only recreational use, had some ritual and medicinal applications.

Medicine: **Kawaiisu (Tehachapi):** dried, crushed

into powder, mixed with lime; puts on cuts, on bites, to stop bleeding or itching; a snuff it clears out a stuffy nose; wrapped in cloth, held between teeth, it deadens toothache; when swallowed it causes vomiting and cleans out stomach. **Gabrielino:** infusion used to induce vomiting to cure fevers. **Costanoan:** used to cure toothache, earache, stomach ache, skin sores or irritations, a respiratory remedy, sedative, stimulant, emetic, hallucinogen, cathartic, a ceremonial medicine; smoke blown into ear for earache; leaves, fresh, chewed as an emetic; leaves, smoked, used as a general purgative in social/ritual contexts; commonly used by men.

Citation(s): Barrett 1952; Barrett and Gifford 1933; Baumhoff 1958; Chesnut 1902; Goddard 1903; Loud 1918; Merriam 1955, 1967; Powers 1877; Sapir and Spier 1943; Voegelin 1938; Zigmond 1981; Goodrich et al 1980; Strike 1994; Rayburn 2012; Harrington 1932; Schenck and Gifford 1952; Timbrook 2007.

Terminology: **Wintun:** lol. **Pomo (Southwestern):** ka·wa qʰale. **Karok:** avaraheira, "that which is smoked." **Chumash (Barbareño, Ynezeño, Ventureño):** show; **(Obispeño):** stuʼyiʼ.

Citation(s): Du Bois 1935; Goodrich et al 1980; Schenck and Gifford 1952; Timbrook 2007.

872 *Nolina bigelovii*

Family: *Ruscaceae*
Common Name(s): Nolina.

Usage: Food: **Cahuilla:** stalk baked in stone-lined roasting pit, covered with sand, left overnight.
Citation(s): Bean and Saubel 1972.
Terminology: **Cahuilla:** kuku'ul.
Citation(s): Bean and Saubel 1972.

873 *Nolina parryi*

Common Name(s): Mescal.

Usage: Food: **Tubatulabal:** used.
Citation(s): Voegelin 1938.

874 *Nothocalais* spp.

Family: *Asteraceae*
Synonym(s): *Microseris* spp., *Scorzonella* spp.

Usage: Food: **Northern Paiute (Paviotso):** the roots were cooked in an earth oven.
Citation(s): Kelly 1932.
Terminology: **Northern Paiute (Paviotso):** kwanü'pic.
Citation(s): Kelly 1932.

875 *Nothocalais alpestris*

Synonym(s): *Agoseris alpestris*

Usage: Misc: **Tubatulabal:** considered weeds.
Citation(s): Voegelin 1938

876 *Notholithocarpus densiflora*

Family: *Fagaceae*
Synonym(s): *Lithocarpus densiflora, Quercus densiflora*
Common Name(s): Tanbark Oak, Tan Oak.

Usage: Food: **Hupa:** most esteemed of the oaks. **Pomo:** used for making soup, gruel, and mush, rarely bread; **(Southwestern):** best like of the acorns. **Yurok:** one of the principle foods. **Shelter Cove Sinkyone:** form the principle food. **Shasta:** considered to be better than varieties than occur locally. **Yuki:** preferred to all other kinds. **Karok:** preferred.

Material: **Pomo (Southwestern):** strung acorns with shell, twirled to make music; acorn used in first fruits ceremony in October after first rainfall.

Medicine: **Pomo (Southwestern):** suck on acorn to sooth throat; infusion used asa disinfectant on sores. **Costanoan:** infusion used to sooth skin sores, especially on face; liquid held in mouth to tighten loose teeth. **Sinkyone:** infusion used to treat colds, stomach aches.

Citation(s): Barrett 1952; Baumhoff 1958; Chesnut 1902; Dixon 1907; Gifford 1967; Goddard 1903; Merriam 1967; Schenck and Gifford 1952; Goodrich et al 1980; Strike 1994; Rayburn 2012.

Terminology: **Pomo (Southwestern):** tish kale, "beautiful tree;" **(Southeastern):** tcetce'c būdū. **Karok:** xunyeip, xurish (shelled meat of acorn), xuntapan (acorn), xuran (shells of acorn). **Yuki:** sho-kish.

Citation(s): Gifford 1967; Barrett 1952; Schenck and Gifford 1952; Curtin 1957; Goodrich et al 1980.

877 *Notholithocarpus densiflora* var. echinoides

Synonym(s): *Lithocarpus densiflora* var. *echinoides*
Common Name(s): Scrub Tan Oak.

Usage: Food: **Karok:** if there is a shortage of *Notholithocarpus densiflorus* (#876) the people go to the mountains and collect these.
Citation(s): Schenck and Gifford 1952.
Terminology: **Karok:** xunyeiis (plants and acorn).
Citation(s): Schenck and Gifford 1952.

878 *Nuphar polysepala*

Family: *Nymphaeceae*
Synonym(s): *Nymphaea polysepala*
Common Name(s): Water Lily, Yellow Water-Lily.

Usage: Food: **Yuki:** seeds used. **Maidu:** leaves and buds eaten; roots boiled, or baked, eaten when other foods were scarce.
Material: **Modoc (Lutuami) (Lutuami):** used seed shell as black dye.
Medicine: **Pomo:** roots used as a remedy for stomach trouble. **Maidu:** roots, heated, or leaf poultice used to relieve pain.
Citation(s): Barrett 1952; Chesnut 1902; Merrill 1923; Strike 1994.
Terminology: **Pomo (Northern):** katīlaka'n.
Citation(s): Barrett 1952.

879   *Oemleria cerasiformis*

        Family: *Rosaceae*
        Synonym(s): *Osmaronia cerasiformis*
        Common Name(s): Oso Berry.

Usage: Food: **Shasta:** the berries were eaten.
        Misc: **Karok:** named but not used.
Citation(s): Barrett and Gifford 1933; Holt 1946; Schenck and
        Gifford 1952.
Terminology: **Shasta:** horaíhihú, "spider" (horai) "bush"
        (hihu). **Karok:** puraf. **Miwok (Central Sierra):**
        poloyina.
Citation(s): Barrett and Gifford 1933; Holt 1946; Schenck and
        Gifford 1952.

880   *Odontostomum hartwegii*

        Family: *Tecophilaceae*
        Common Name(s): Hartweg's Orchid.

Usage: Food: **Maidu:** bulbs cooked, eaten.
Citation(s): Strike 1994.

881   *Oenanthe sarmentosa*

        Family: *Apiaceae*
        Common Name(s): Twiggy Water Dropwort.

Usage: Food: **Miwok:** stem eaten raw. **Maidu:** stems
        and roots eaten.
        Medicine: **Maidu:** seeds, roots used as an emetic.
Citation(s): Barrett and Gifford 1933; Strike 1994.
Terminology: **Miwok (Central Sierra):** komani.
Citation(s): Barrett and Gifford 1933.

882   *Oenothera caespitosa* ssp. *marginata*
>   Synonym(s): *Oenothera caespitosa*, *Oenothera caespitosa*
>   var. *marginata*

Usage: Misc: **Tubatulabal:** considered weeds.
Citation(s): Voegelin 1938.

883   *Oenothera californica*
>   Synonym(s): *Oenothera pallida*

Usage: Misc: **Tubatulabal:** considered weeds.
Citation(s): Voegelin 1938.

884   *Oenothera villosa* ssp. *strigosa*
>   Synonym(s): *Oenothera hookeri*
>   Common Name(s): Evening Primrose.

Usage: Food: **Owens Valley Paiute:** used as food.
                      Pomo (Southwestern): flowers chewed
                      with gum to make gum yellow.
Usage: Misc: **Miwok:** named but not used.
Citation(s): Steward 1933; Goodrich et al 1980; Barrett and
          Gifford 1933.
Terminology: **Owens Valley Paiute:** ko'do'ova. **Pomo
          (Southwestern):** qosá?bu. **Miwok (Central Sierra):**
          sonta.
Citation(s): Steward 1933; Goodrich et al 1980; Barrett and
          Gifford 1933.

885   *Olneya tesota*
>   Family: *Fabaceae*
>   Common Name(s): Desert-Ironwood, Ironwood.

Usage: Food: **Cahuilla:** pod and seeds, gathered in
                      May to June, roasted, ground into flour.

Material: Cahuilla: because of extreme hardness,
              used for throwing sticks and clubs;
              excellent firewood.
Citation(s): Bean and Saubel 1972.

886   *Opuntia* spp.
      Family: *Cactaceae*

Usage: Food: **Luiseño:** the fruit is eaten both fresh
                    and dried; seeds ground into meal.
                    Cahuilla: fruit eaten. **Chumash:** fruit,
                    pulp, eaten fresh
        Material: **Cahuilla:** spines used as awls in
                    making baskets. **Chumash:** red juice
                    used a paint pigment; stems, exude
                    mucilaginous juice, used a binder in
                    paint pigments; spines, used in
                    tattooing.
        Medicine: **Costanoan:** fruit, cut in half, heated,
                    juice rubbed on areas affected by
                    rheumatism, fruit, warm, used as
                    poultices. **Chumash:** sap is believed to
                    heal wounds, beautify complexion;
                    flower, tea, taken for the heart.
Citation(s): Barrows 1900; Curtis 1924; Merrill 1923; Sparkman
        1908; Rayburn 2012; Timbrook 2007.
Terminology: **Luiseño:** navut. **Cahuilla:** navit, navityuluku (the
        fruit). **Chumash (Barbareño):** khɨkhɨ; **(Ynezeño):** khɨ;
        (Obispeño): tqɨ'ɨ; **(Ventureño):** khɨ'ɨl.
Citation(s): Sparkman 1908; Timbrook 2007.

887　*Opuntia basilaris*

Synonym(s): *Opuntia basilaras*
Common Name(S): Prickly Pear.

Usage:　Food:　**Tubatulabal:** receptacle eaten. **Cahuilla:** young fruit gathered in early Summer and cooked or steamed with hot stones in a pit for twelve hours; buds collected, cooked, eaten immediately, or dried for storage; joints cut into small pieces eaten as greens, or boiled in water, mixed with other food; seeds ground into edible mush; the plant considered one of the most desirable of foods. **Panamint Shoshone (Koso):** in early May and June the flat fleshy joints as well as the buds, blossoms, and immature fruit were collected and dried; prepared for eating by boiling and adding salt. **Kawaiisu (Tehachapi):** early in Spring, usually about March, the buds were removed and cooked.

Citation(s): Barrows 1900; Coville 1892; Voegelin 1938; Zigmond 1981; Bean and Saubel 1972.
Terminology: **Cahuilla:** manal.
Citation(s): Bean and Saubel 1972.

888　*Opuntia engelmannii* var. *engelmannii*

Synonym(s): *Opuntia Engelmannii* var. *littoralis,*
Opuntia Lindheimeri var. *littoralis,*
Opuntia occidentalis var. *littoralis,*
Opuntia occidentalis var. *megacarpa*
Common Name(s): Mexican Prickly Pear, Prickly Pear.

Usage: Food: **Tubatulabal:** receptacle eaten. **Cahuilla:** fruit obtained from other Indian groups; fruit gathered May and June; joints also diced and eaten.
Citation(s): Voegelin 1938; Barrows 1900, Bean and Saubel 1972.
Terminology: **Cahuilla:** qexe'yily.
Citation(s): Bean and Saubel 1972.

889 *Opuntia ficus-indica*

Synonym(s): *Opuntia megacantha*
Common Name(s): Tuna, Mission Cactus.

Usage: Food: **Cahuilla:** pads gathered when young, May until August, diced, boiled and eaten, or dried for storage; buds picked, eaten fresh, or dried for storage.

Medicine: **Cahuilla:** fruit said to be excellent purgative; plugs made from plant inserted into wounds as healing agents.
Citation(s): Bean and Saubel 1972.
Terminology: **Cahuilla:** navet.
Citation(s): Bean and Saubel 1972.
Note(s): native of tropical America.
Citation(s): Munz and Keck 1965.

890 *Orobanche californica*

Family: *Orobanchaceae*
Common Name(s): California Broomrape.

Usage: Food: **Maidu:** roots eaten.
Citation(s): Strike 1994.

891   *Orobanche cooperi*
> Synonym(s): *Aphyllon ludovicianum, Orobanche*
> *ludoviciana* var. *cooperi*
> Common Name(s): Cancer Root, Broom Rape.

Usage: Food: **Cahuilla:** roots gathered in the Spring before the plant blossoms and while the roots are young and tender; roasted in coals; root gathered, April to June, peeled and eaten.
Citation(s): Barrows 1900; Bean and Saubel 1972
Terminology: **Cahuilla:** meslam.
Citation(s): Bean and Saubel 1972.

892   *Orobanche fasciculata*
> Common Name(s): Cancer Root.

Usage: Food: **Maidu:** entire plant eaten.
Citation(s): Strike 1994.

893   *Orobanche uniflora*
> Common Name(s): Broomrape.

Usage: Medicine: **Maidu:** used to treat bronchial problems, intestinal upsets, toothache, rheumatic pain.
Citation(s): Strike 1994.

894   *Orthilia secunda*
> *Family: Ericaceae*
> *Synonym(s): Pyrola secunda*
> Common Name(s): Serrated Wintergreen.

Usage: Medicine: **Maidu:** decoction used as eyewash.
Citation(s): Strike 1994.

895 *Orthocarpus* spp.

Family: *Orobanchaceae*

Usage: Medicine: **Costanoan:** foliage, decoction, used for cough remedy.
Citation(S): Rayburn 2012.

896 *Osmorhiza* spp.

Family: *Apiaceae*
Common Name(s): Snake Root.

Usage: Medicine: **Miwok:** chewed and put on snake bite.
Citation(s): Barrett and Gifford 1933.
Terminology: **Miwok (Central Sierra):** kawibe.
Citation(s): Barrett and Gifford 1933.

897 *Osmorhiza berteroi*

Synonym(s): *Osmorhiza nuda, Osmorhiza nuda brevipes, Osmorhiza chilensis*
Common Name(s): Sweet Cicely.

Usage: Food: **Miwok:** leaves were boiled. **Maidu:** leaves, root, tender tops eaten. **Karok:** leaves, stems eaten in Spring.

Medicine: **Karok:** the roots are one of the most important medicines; can be dried and kept in the house; used for almost anything. **Hupa:** used by priest to purify grave-diggers so they could go among their fellows.
Citation(s): Barrett and Gifford 1933; Goddard 1903; Schenck and Gifford 1952; Strike 1994.
Terminology: **Miwok (Central Sierra):** tcuyuma. **Hupa:** kimaūLûkay, "medicine fat." **Karok:** kishwuf.
Citation(s): Barrett and Gifford 1933; Goddard 1903; Schenck

and Gifford 1952.

898 *Osmorhiza brachypoda*
Common Name(s): Sweet Cicely.

Usage: Food: **Kawaiisu (Tehachapi):** plant is mashed, put into water, strained; infusion used for washing hair.

Medicine: **Kawaiisu (Tehachapi):** root is boiled, drunk for relief of colds and coughs.

Citation(s): Zigmond 1981.

899 *Osmorhiza occidentalis*

Usage: Medicine: **Washo:** the root decoction is taken as a cure for colds and other pulmonary diseases as well as pneumonia and influenza.

Citation(s): Train et al 1941.

Terminology: **Washo:** oo-<u>chu</u>-lee mah-<u>too</u>.

Citation(s): Train et al 1941.

900 *Oxalis* spp.
Family: *Oxalidaceae*
Common Name(s): Sorrell.

Usage: Food: **Miwok:** roots eaten.

Medicine: **Shelter Cove Sinkyone:** used for poultices.

Citation(s): Baumhoff 1958; Clark 1904.

901 *Oxalis californica*

Synonym(s): *Oxalis albicans* spp. *californica*
Common Name(s): California Wood-Sorrel.

Usage: Food: **Chumash:** leaves, eaten, sour.
Citation(s): Timbrook 2007.
Terminology: **Chumash (Barbareño):** ʼaqnipshkáy, "sour."
Citation(s): Timbrook 2007.

902 *Oxalis oregana*

Common Name(s): Wood Sorrell, Redwood Sorrell.

Usage: Food: **Hupa:** the leaves used to line fire pits used to cook *Chlorogalum pomeridianum*; also eaten mixed with *C. pomeridianum*, said to improve flavor. **Pomo (Southwestern):** young leaves chewed in minute quantity for their sour taste, not swallowed.

Medicine: **Karok:** used with a formula for anyone who does not feel like eating. **Pomo (Southwestern), Sinkyone:** whole plant was boiled and the decoction used to wash parts of the body affected with rheumatism.

Misc: **Coast Yuki:** named but not used.
Citation(s): Gifford 1939; Goddard 1903; Schenck and Gifford 1952; Goodrich et al 1980.
Terminology: **Yuki:** hewushet. **Pomo (Southwestern):** moch kale, "sour plant." **Karok:** takannafich.
Citation(s): Gifford 1939, 1967; Schenck and Gifford 1952; Goodrich et al 1980.

903 *Paeonia brownii*
Family: *Paeoniaceae*

Usage: Medicine: **Northern Paiute (Paviotso), Owens Valley Paiute:** seeds soaked in water to make cough medicine. **Washo:** a decoction from the boiled roots taken for tuberculosis. **Cahuilla(?):** root reduced to powder used that way or in decoction; cure colds, sore throats, chest pains. **Costanoan:** roots, decocted, used to treat stomach ache; plants, boiled with burned meat, drunk, after adding oil, to cure indigestion and/or constipation used to treat pneumonia.

Citation(s): Kelly 1932; Train et al 1941; Bean and Saubel 1972; Rayburn 2012.

Terminology: **Northern Paiute (Paviotso):** bati'p'. **Washo:** <u>doo</u>-<u>yah</u>-gum-hoo, <u>tue</u>-ago-nomo.

Citation(s): Kelly 1932; Train et al 1941.

904 *Paeonia californica*
Common Name(s): Western Peony.

Usage: Medicine: **Diegueño:** tea used to treat gastrointestinal, lung, kidney ailments. **Salinan:** root chewed, juice swallowed to treat colds, to treat paralysis, cure rashes from poison oak; tea relieved headaches. **Chumash:** root, tea, or chewed, treatment of menstrual disorders, neuralgia.

Citation(s): Strike 1994; Timbrook 2007.

Terminology: **Chumash (Barbereño, Ynezeño, Ventureño):** mim.

Citation(s): Timbrook 2007.

905   *Palafoxia aripa*
        Family: *Asteraceae*
        Synonym(s): *Palafoxia linearis*
        Common Name(s): Spanish Needles.

Usage: Material: **Cahuilla:** used to make yellow dye.
Citation(s): Bean and Saubel 1972.
Terminology: **Cahuilla:** tesqal.
Citation(s): Bean and Saubel 1972.

906   *Panicum* spp.
        Family: *Poaceae*

Usage: Misc: **Mono Lake Paiute:** named but not used.
Citation(s): Steward 1933.
Terminology: **Mono Lake Paiute:** muhi'dap.
Citation(s): Steward 1933.

907   *Pancium urvilleanum*
        Common Name(s): Panic Grass.

Usage: Food: **Cahuilla:** seeds, singed to remove hair, boiled for several hours to make gruel.
Citation(s): Bean Saubel 1972.
Terminology: **Cahuilla:** sangval.
Citation(s): Bean and Saubel 1972.
Note(s): seeds contain:
        ash   4.0%
        protein 5.8%
        oil, some starch, no tannin or alkaloids.
Citation(s): Bean and Saubel 1972.

908   *Parkinsonia florida*

Family: *Fabaceae*
Synonym(s): *Parkinsonia terreyana, Cercidium floridum*
Common Name(s): Palo Verde.

Usage: Food: **Cahuilla:** slender bean ground and
                      cooked into atole; beans picked July to
                      August.
Citation(s): Barrows 1900; Bean and Saubel 1972.
Terminology: **Cahuilla:** o-o-wit, u'uwet.
Citation(s): Barrows 1900; Bean and Saubel 1972.
Note(s): fruit ripening and falling approximately in July.
Citation(s): Sargent 1922.

909   *Parmelia saxicola*

Family: *Parmeliaceae*
Common Name(s): Greenish-Grey Lichen.

Usage: Medicine: **Maidu, Southern (Nisenan):** colic is
                      treated with a tea made from this plant.
Citation(s): Powers 1877.
Note(s): abundant in Oak and Hickory, and Oak and Pine
         forests; found mostly on sandstones and serpentines,
         and to a lessor extent on quartzite and granites.
Citation(s): Schneider 1898.

910   *Pastinaca sativa*

Family: *Apiaceae*
Common Name(s): Common Parsnip.

Usage: Misc: **Karok:** named but not used (?).
Citation(s): Schenck and Gifford 1952.
Note(s): native of Europe; escaped from cultivated gardens.
Citation(s): Munz and Keck 1965; Jepson 1925.

911　*Paxistima myrsinites*

        Family: *Celastraceae*
        Synonym(s): *Pachystima myrsinites*
        Common Name(s): Oregon Boxwood.

Usage: Food: **Karok:** berries eaten when ripe, but not preserved.
          Medicine: **Maidu:** leaves, boiled, used to poultice pain, inflammation.

Citation(s): Schenck and Gifford 1952; Strike 1994.
Terminology: **Karok:** maxapuris, "mountain huckleberry."
Citation(s): Schenck and Gifford 1952.

912　*Pedicularis attollens*

        Family: *Orobachaceae*

Usage: Medicine: **Washo, Maidu:** a tea from boiled leaves could be taken as a tonic; used raw or boiled as a poultice for cuts, sores, and swellings (?).

Citation(s): Train et al 1941; Strike 1994.
Terminology: **Washo:** wem-she.
Citation(s): Train et al 1941.

913　*Pedicularis densiflora*

Usage: Food: **Yuki:** children are found of sucking the honey from flowers. **Maidu:** nectar eaten.

Citation(s): Chesnut 1902; Strike 1994.

914 *Pellaea andromedaefolia*
Family: *Pteridaceae*
Common Name(s): Coffee Fern.

Usage: Medicine: **Diegueño:** tea used to relieve menstrual cramps.
Citation(s): Strike 1994.

915 *Pellaea brewerii*
Common Name(s): Fern.

Usage: Food: **Yokuts:** used as a beverage, like tea.
Citation(s): Powers 1877.

916 *Pellaea mucronata*
Synonym(s): *Pellaea ornithopus*
Common Name(s): Tea Fern, Bird-Claw Fern, Bird's-Foot Fern.

Usage: Food: **Luiseño, Kawaiisu (Tehachapi):** a tea is drunk as a beverage.
            Material: **Maidu, Miwok:** rhizome fibers used to create brown patterns in basketry.
            Medicine: **Luiseño:** a decoction of the fronds is used medicinally. **Miwok:** steeped in hot water and drunk to stop nose-bleed; drunk as a Spring medicine and blood purifier. **Maidu, Costanoan:** decoction used to stop hemorrhages, as an emetic, a wash for skin problems, especially on face; leaves collected in early Winter, made tea to reduce fever. **Costanoan:** decoction, used to cough up "bad blood."
Citation(s): Barrett and Gifford 1933; Sparkman 1908; Zigmond

1981; Strike 1994; Rayburn 2012.
Terminology: **Luiseño:** wikunmal. **Miwok (Northern Sierra):** pē′sippēsa.
Citation(s): Sparkman 1908; Barrett and Giffford 1933.

917   *Pellaea mucronata var. californica*
Synonym(s): *Pellaea compacta*

Usage: Food: **Tubatulabal:** used leaves and stalks.
Citation(s): Voegelin 1938.

918   *Peltigera canina*
Family: *Peltigeraceae*
Common Name(s): Dog Lichen.

Usage: Material: **Maidu:** yielded yellow dye.
         Medicine: **Maidu:** used as purgative, to eliminate intestinal obstructions.
Citation(s): Strike 1994.

919   *Penstemon centranthifolius*
Common Name(s): Scarlet Bugler.

Usage: Material: **Cahuilla:** gathered for use as decoration at funerals or church affairs.
         Medicine: **Costanoan:** poultice used for skin disorders, as a disinfectant; treatment for deep, infected sores.
Citation(s): Bean and Saubel 1972; Strike 1994; Rayburn 2012.
Terminology: **Cahuilla:** tuchilychungva.
Citation(s): Bean and Saubel 1972.

920 *Penstemon deustus*

Usage: Medicine: **Northern Paiute (Paviotso):** mashed leaves used as poultice for sores. **Miwok, Maidu:** used to treat colds, open sores.
Citation(s): Kelly 1932; Strike 1994.
Terminology: **Northern Paiute (Paviotso):** na·mogu·d.
Citation(s): Kelly 1932.

921 *Penstemon gracilentus*
Common Name(s): Graceful Penstemon.

Usage: Medicine: **Miwok, Maidu:** used to treat colds, open sores.
Citation(s): Strike 1994.

922 *Penstemon heterophyllus*

Usage: Misc: **Miwok:** named but not used.
Citation(s): Barrett and Gifford 1933.
Terminology: **Miwok (Central Sierra):** yutumpula.
Citation(s): Barrett and Gifford 1933.

923 *Penstemon laetus*

Usage: Medicine: **Karok, Maidu:** one of the ingredients for making medicine for a person who is grieving and can not get over it; used both as a "steam" medicine and as a drink.
Misc: **Tubatulabal:** considered weeds. **Kawaiisu (Tehachapi):** named but not used.
Citation(s): Schenck and Gifford 1952; Voegelin 1938; Zigmond

1981; Strike 1994.
Terminology: **Karok:** ichiwohannahuich. **Kawaiisu**
(**Tehachapi**): caaruuwagadɨbɨ.
Citation(s): Schenck and Gifford 1952; Zigmond 1981.

924 *Penstemon palmeri* var. *palmeri*

Usage: Misc: **Tubatulabal:** considered weeds.
Citation(s): Voegelin 1938.

925 *Penstemon rostriflorus*
Synonym(s): *Penstemon bridgesii*
Common Name(s): Scarlet Penstemon.

Usage: Medicine: **Kawaiisu (Tehachapi):** root is mashed, mixed with a little water, applied as poultice to swollen limbs.
Citation(s): Zigmond 1981.
Terminology: **Kawaiisu (Tehachapi):** agakidɨ (agakidɨ, "red").
Citation(s): Zigmond 1981.

926 *Pentachaeta aurea*
Family: *Asteraceae*
Synonym(s): *Chaetopappa aurea*

Usage: Material: **Cahuilla:** plant pollen used as a cosmetic for women.
Citation(s): Bean and Saudel 1972.

927 *Pentagramma pallida*
Family: *Pteridaceae*
Synonym(s): *Ceropteris triangularis, Gymnogramma triangularis, Pityrogramma triangularis*
Common Name(s): Goldenback Fern, Gold Fern.

Usage: Material: **Pomo, Hupa:** stem used as black
                    pattern in baskets.
        Medicine: **Karok:** "talked to" in order to mitigate
                    the afterpains in childbirth. **Maidu:**
                    stalks chewed to relieve toothache; to
                    reduce pain following childbirth.
                    Miwok: chewed for toothache; the quid
                    being kept near the troublesome tooth.
Citation(s): Barrett and Gifford 1933; Merrill 1923; Schenck and
        Gifford 1952; Voegelin 1938; Strike 1994.
Terminology: **Miwok (Central Sierra):** pilpilka. **Karok:**
        apshikkhamnakuuishich, "black legs."

928 *Peraphyllum ramosissium*
Family: *Rosaceae*

Usage: Misc: **Owens Valley Paiute:** named but not
                    used.
Citation(s): Steward 1933.
Terminology: **Owens Valley Paiute:** tüsa'va, tüsa'pᵃ.
Citation(s): Steward 1933.

929 *Perideridia* spp.
Family: *Apiaceae*
Synonym(s): *Carum* spp.
Common Name(s): Wild Potato.

Usage: Food: **Miwok:** roots eaten.
Citation(s): Clark 1904.

930 *Perideridia bolanderi*
Synonym(s): *Eulophus bolanderi*

Usage: Food: **Miwok:** bulb cooked in basket by stone
                    boiling, then peeled and eaten; served

as a substitute for acorns when they are used up in June; made mush at such times; sun dried and pounded with skin on, at times eaten raw. **Atugewi:** placed roots in shallow baskets with damp sand and worked with feet until skins came off; dried on flat rocks; stored for Winter; when used, pounded in mortar and made into bread or soup.

Citation(s): Barrett and Gifford 1933; Garth 1953.

Terminology: **Miwok (Central Sierra):** olasi.

Citation(s): Barrett and Gifford 1933.

931 *Perideridia gairdneri*

Synonym(s): *Carum Gairdneri*

Common Name(s): Anise, Squaw-Root.

Usage: Food: **Pomo:** when fresh and new in Spring the tops are eaten as greens; the seeds are used for pinole; the roots are eaten raw, cooked, or used for pinole. **Yana:** roots roasted in earth oven all day and eaten. **Karok:** roots dried until skin gets loose, then cooked in earth oven, covered with Maple (*Acer*) leaves, and eaten. **Miwok:** boiled and eaten like potato. **Cahuilla:** tuberous roots eaten raw or cooked. **Modoc (Lutuami) (Lutuami):** root eaten.

Citation(s): Barrett 1952; Barrett and Gifford 1933; Sapir and Spier 1943; Schenck and Gifford 1952; Bean and Saubel 1972; Strike 1994.

Terminology: **Karok:** upva'amayav, "good tasting thing which is dug." **Yana:** ku's·iki (roots). **Miwok (Central Sierra):** tuñi. **Pomo (Central):** cbū'tū; cabū'tū,

(Southeastern): xūbū'xai.

Citation(s): Schenck and Gifford 1952; Sapir and Spier 1943; Barrett and Gifford 1933; Barrett 1952.

932 *Perideridia kelloggii*

Synonym(s): *Carum Kelloggii*

Common Name(s): Sweet Anise, Anise.

Usage: Food: **Wiyot:** the stalks were eaten after the skin was removed; it make the lips black. **Pomo:** used seeds in pinole; tubers and roots eaten raw; cooked like bread or used for pinole, **(Southwestern):** foliage eaten raw as greens. **Miwok:** bulb eaten. **Yuki:** tubers eaten; young plants eaten raw as "clover."

Material: **Pomo:** outer root fibers made into compact cylindrical brushes used "for combs," **(Southwestern):** roots were tied in a bundle to form a brush for the hair or for cleaning mush baskets. **Yuki:** brush made from fibers.

Medicine: **Pomo (Southwestern):** white flowers were boiled and the decoction drunk to stop vomiting.

Citation(s): Barrett 1952; Barrett and Gifford 1933; Chesnut 1902; Curtin 1957; Gifford 1967; Loud 1918; Goodrich et al 1980.

Terminology: **Yuki:** mö-sin'. **Pomo (Central):** ca'bun, cbū'tū, **(Southeastern):** xūbū'xai, **(Southwestern):** chibuta'. **Wiyot:** siswileatkok. **Miwok (Central Sierra):** sakasu, **(Southern Sierra):** sa'kkasu.

Citation(s): Curtin 1957; Barrett 1952; Gifford 1967; Loud 1918; Barrett and Gifford 1933; Goodrich et al 1980.

933 *Perideridia oregana*

Synonym(s): *Carum oreganum*

Common Name(s): Eops.

Usage: Food: **Northern Paiute (Paviotso):** the roots were peeled, eaten raw, or boiled, or dried in the sun and stored. **Modoc (Lutuami):** root eaten.

Citation(s): Kelly 1932; Strike 1994.

Terminology: **Northern Paiute (Paviotso):** ya'pa'.

Citation(s): Kelly 1932.

934 *Perideridia pringlei*

Synonym(s): *Eulophus Pringlei*

Usage: Food: **Yana:** roots were roasted in earth oven all day and eaten. **Kawaiisu (Tehachapi):** tuberous roots washed, rubbed to remove skin, boiled.

Citation(s): Sapir and Spier 1943; Zigmond 1981.

Terminology: **Yana:** dju'pp'a. **Kawaiisu (Tehachapi):** ya(m)barabɨ.

Citation(s): Sapir and Spier 1943; Zigmond 1981.

935 *Peritoma arborea*

Family: *Cleomaceae*

Synonym(s): *Isomeris arborea*

Common Name(s): Bladder-Pod.

Usage: Food: **Cahuilla:** little pods gathered and cooked in a small hole in the ground with hot stones; usage forgotten by modern Cahuilla as a food source. **Kawaiisu (Tehachapi):** flowers are eaten.

Citation(s): Barrows 1900; Zigmond 1981; Bean and Saubel
1972.
Terminology: **Kawaiisu (Tehachapi): sɨpitɨbɨ.**
Citation(s): Zigmond 1981.

936 *Persicaria amphibia*

Family: *Polygonaceae*
Synonym(s): *Polygonum amphibium* var. *stipulaceum,*
Polygonum coccineum
Common Name(s): Water Smartweed, Swampweed,
Swamp Knotweed.

Usage: Medicine: **Maidu:** young leaves eaten; decoction
used as a tonic, blood purifier,
aphrodisiac. **Miwok:** decoction used as
a tonic, blood purifier, aphrodisiac.
Citation(s): Strike 1994.

937 *Persicaria lapathifolium*

Synonym(s): *Polygonum lapathifolium*

Usage: Material: **Yokuts:** used plant as fish poison.
Misc: **Miwok:** named but not used.
Citation(s): Barrett and Gifford 1933; Heizer 1953.
Terminology: **Miwok (Central Sierra):** tcostowina.
Citation(s): Barrett and Gifford 1933.

938 *Persicaria maculosa*

Synonym(s): *Polygonum persicaria*

Usage: Misc: **Tubatulabal:** considered weeds.
Citation(s): Voegelin 1938.
Note(s): native of Europe.
Citation(s): Munz and Keck 1965.

939 *Petasites frigidus* var. *palmatus*

Family: *Asteraceae*
Synonym(s): *Petasites palmata, Petasites palmatus*
Common Name(s): Sweet Coltsfoot.

Usage: Food: **Yuki, Pomo, Wailaki:** used ashes for source of salt. **Maidu:** steams and leaves eaten.

Medicine: **Karok:** used for sickly baby; a "steaming" medicine used with a charm. **Maidu:** root, when dry and coarsely grated, is applied to boils and running sores to dry them up; valued root in the first stages of consumption and grippe.

Citation(s): Chesnut 1902; Schenck and Gifford 1952.
Terminology: **Karok:** kafichkamshash, "big kaf."
Citation(s): Schenck and Gifford 1952.

940 *Phacelia* spp.

Family: *Boraginaceae*

Usage: Medicine: **Kawaiisu (Tehachapi):** roots boiled; infusion drunk warm for general health; relieves coughs and colds.

Misc: **Mono Lake Paiute:** named but not used.

Citation(s): Steward 1933; Zigmond 1981.
Terminology: **Mono Lake Paiute:** tüma'nava.
Citation(s): Steward 1933.
Note(s): leaves suspected of causing dermatitis.
Citation(s): Hardin and Arena 1974.

941 *Phacelia californica*
Common Name(s): Purple Fiddleneck.

Usage: Food: **Miwok (Coast):** made a beverage.

Medicine: **Pomo (Southwestern):** fresh leaves crushed; juice rubbed on cold sores and impetigo. **Costanoan:** root, decoction, used to treat fever.

Citation(s): Goodrich et al 1980; Strike 1994; Rayburn 2012.

Terminology: **Pomo (Southwestern):** tʰúnkuhtu.

Citation(s): Goodrich et al 1980.

942 *Phacelia cicutaria* var. *hispica*
Synonym(s): *Phacelia hispica*

Usage: Misc: **Tabatulabal:** considered weeds.

Citation(s): Voegelin 1938.

943 *Phacelia distans*
Common Name(s): Fern Phacelia, Wild Heliotrope.

Usage: Food: **Kawaiisu (Tehachapi):** greens eaten in Spring.

Citation(s): Zigmond 1981.

Terminology: **Kawaiisu (Tehachapi):** yuhʷitɨbɨ.

Citation(s): Zigmond 1981.

944 *Phacelia heterophylla*
Common Name(s): Various-Leaved Bluebell.

Usage: Medicine: **Miwok, Maidu:** the plant, dried, was pulverized and put on fresh wounds.

Citation(s): Barrett and Gifford 1933.

Terminology: **Miwok (Central Sierra):** tawimuyu.

Citation(s): Barrett and Gifford 1933.

945 Phacelia ramosissima

Usage: Food: **Luiseño:** used for greens. **Kawaiisu (Tehachapi):** stems and leaves gathered in Spring, steam-cooked.

Medicine: **Kawaiisu (Tehachapi):** root boiled, drunk for gonorrhea; it causes vomiting; used to cure colds, coughs, alleviate lethargy. **Costanoan:** used to reduce fevers. **Pomo (Southwestern):** leaves, crushed, rubbed on sores, impetigo.

Citation(s): Sparkman 19808; Zigmond 1981; Strike 1994.
Terminology: **Luiseño:** sikimona.
Citation(s): Sparkman 1908.

946 Phacelia tanecetifolia

Synonym(s): *Phacelia tenecilifolia*

Usage: Misc: **Tubatulabal:** considered weeds.
Citation(s): Voegelin 1938.

947 Philadelphus spp.

Family: *Hydrangeaceae*

Usage: Material: **Achomawi:** used for some spear tips.
Citation(s): Merriam 1967.

948 Philadelphus Lewisii

Synonym(s): *Philadelphus lewisii* ssp. *californicus, Philadelphus Gordonianus, Philadelphus lewisii* ssp. *gordonianus*
Common Name(s): Syringa, Mock Orange.

Usage: Material: **Hupa:** the main shaft of arrows was

made from the straight shoots; sticks used as rattle in ceremonies, such as puberty rites for young girls. **Chimariko:** made arrows and fish spears, with bone points. **Miwok:** made arrows; used rods in coiled basketry. **Wintun:** made arrows of the straight shoots. **Yurok:** used for arrows. **Maidu:** roots used in basketry; leaves, roots used as soap. **Maidu (Northern):** made arrow shafts of the wood. **Pomo:** made baby-carrying baskets from stems. **Karok:** considered best source for arrow shafts; made straight tobacco pipe by removing pith; made gambling staves; young shoots are the most important source of arrow shafts, which have *Amerlanchier pallida* tips; tobacco pipes made from twigs. **Yuki:** straight branches were made into arrows; older, less pithy, wood used for bows.. **Yuki, Wailaki:** stem used as warp in baskets; younger, very pithy shoots used for arrows.

Citation(s): Clark 1904; Dixon 1905, 1910; Goddard 1903; Merriam 1955, 1967; Strike 1994; Chesnut 1902; Curtin 1957; Merrill 1923; Schenck and Gifford 1952.

Terminology: **Yuki:** kun'-le. **Karok:** xawish, "arrow wood."

Citation(s): Curtin 1957; Schenck and Gifford 1952.

949 *Pholisma arenarium*

Family: *Boraginaceae*

Synonym(s): *Pholistoma arenarium*

Usage: Food: **Kawaiisu (Tehachapi):** gathered in

February-March, baked, sometimes eaten raw.
Citation(s): Zigmond 1981.
Terminology: **Kawaiisu (Tehachapi):** tu?uvɨ.
Citation(s): Zigmond 1981.

950 *Pholistoma auritum*
Family: *Boraginaceae*
Common Name(s): Fiesta Flower.

Usage: Misc: **Kawaiisu (Tehachapi):** named but not used.
Citation(s): Zigmond 1981.
Terminology: **Kawaiisu (Tehachapi):** kaawanavɨ, "rat's net or web."
Citation(s): Zigmond 1981.

951 *Pholistoma membranaceum*
Synonym(s): *Ellisia membranacea*
Common Name(s): Fiddle Neck.

Usage: Food: **Tabatulabal:** used leaves. **Yokuts:** eaten with Salt Grass (*Distichlis spicata*) while in tender stage.
Citation(s): Gayton 1948; Voegelin 1938.

952 *Phoradendron* spp.
Family: *Viscaceae*
Common Name(s): Oak Mistletoe.

Usage: Food: **Chimariko:** smoked as a substitute for tobacco.
Citation(s): Powers 1877.

953   *Phoradendron bolleanum*

Synonym(s): *Phoradendron bolleanum* var. *densum,*
Phoradendron bolleanum var. *pauciflorum*

Usage: Material: **Kawaiisu (Tehachapi):** boiled together
with yata(m)bi (*Larrea tridentata*);
infusion used as a wash on limbs for
relief of rheumatism.

Misc: **Kawaiisu (Tehachapi):** named but not
used.

Citation(s): Zigmond 1981.

Terminology: **Kawaiisu (Tehachapi):** wohodɨbɨ sanapɨceeka,
"*Pinus sabiniana,* its mistletoe;" wa?adabɨa
sanapɨceeka, "*Juniprus californica,* its mistletoe;"
puugusivɨa sanapɨceeka, "*Abies concolor,* its
mistletoe."

Citation(s): Zigmond 1981.

Note(s): some problem with various identities of the plant and
usage [author's comment].

954   *Phoradendron californicum*

Usage: Food: **Cahuilla:** pink-white waxy berries
washed and mixed with small quantity
of ashes to counteract the viscidity,
boiled in earth pot for a few minutes.

Material: **Cahuilla:** used as a black dye; made by
boiling.

Citation(s): Curtis 1924; Bean and Saubel 1963, 1972.

Terminology: **Cahuilla:** chayal.

Citation(s): Bean and Saubel 1972.

955   *Phoradendron juniperinum*

Synonym(s): *Phoradendron ligatum*
Common Name(s): Mistletoe.

Usage: Food: **Cahuilla:** berries sometimes eaten
                      fresh.
         Medicine: **Maidu (Northern):** used now and then.
                      Cahuilla: pounded into flour, sprinkled
                      on wounds to aid healing; mixed with
                      water, used to bathe sore or infected
                      eyes.
Citation(s): Dixon 1905; Bean and Saubel 1972.
Terminology: **Cahuilla:** chayal.
Citation(s): Bean and Saubel 1972.
Note(s): a few berries are usually associated with only
             moderate abdominal pain, diarrhea; large amount
             have been fatal.
Citation(s): Fuller and McClintock 1986.

956   *Phoradendron serotinum*

Synonym(s): *Phoradendron flavescens*

Usage: Misc: **Kawaiisu (Tehachapi):** named but not
                      used.
Citation(s): Zigmond 1981.
Terminology: **Kawaiisu (Tehachapi):** havatɨbɨa sanapɨceeka,
             "*Plananus racemosa*, its mistletoe."
Citation(S): Zigmond 1981.

957   *Phoradendron serotinum* spp. *macrophyllum*

Synonym(s): *Phoradendron macrophyllum, Phoradendron
                 leucarpum* spp. *macrophyllum*
Common Name(s): Bigleaf Mistletoe.

Usage: Medicine: **Chumash:** tea,   "drank   to   make

themselves (women) sterile," drunk to
bring on menstruation; perhaps to
induce abortion.

Citatiuon(s): Timbrook 2007.

958 *Phoradendron serotinum* ssp. *tomentosum*

Synonym(s): *Phoradendron coloradense, Phoradendron
tomentosum* ssp. *macrophyllum*

Usage: Material: **Cahuilla:** used as a block dye; made by
boiling.

Medicine: **Maidu, Pomo:** used to induce
abortions. **Diegueño:** decoction us as
treatment for dandruff, to destroy
vermin, prevent hair loss. **Kawaiisu
(Tehachapi):** used externally to reduce
rheumatic pain.

Citation(s): Curtis 1924; Strike 1994.

959 *Phoradendron villosum*

Synonym(s): *Phoradendron flavescens (?), Phoradendron
flavescens* var. *villosum*
Common Name(s): Common Mistletoe, Douglas Oak
Mistletoe.

Usage: Medicine: **Pomo (Southwestern):** leaves boiled
and the decoction drunk to bring on
delayed menstruation. **Pomo:** chewed
leaves "all day long" to relieve
toothache; tea made of leaves used to
produce abortion. **Kawaiisu
(Tehachapi):** if brewed and drunk
during pregnancy, apparently during
the first two months, will cause an
abortion. **Sinkyone, Wailaki:** used as

medicinal tea, as a head and body wash. **Diegueño:** used to bathe sore muscles. **Maidu:** tea drunk to ease delivery of baby. **Wintun:** used to ease childbirth, to stimulate delayed menstruation; considered more effective for curing toothache than other mistletoe. **Chumash:** tea, "drank to make themselves (women) sterile," drunk to bring on menstruation; perhaps to induce abortion.

Misc: **Karok:** named but not used (?).

Citation(s): Chesnut 1902; Gifford 1967; Schenck and Gifford 1952; Zigmond 1981; Goodrich et al 1980; Strike 1994; Timbrook 2007.

Terminology: **Pomo (Southwestern):** kopina'. **Karok:** 'anach'uhish, "crow seed." **Kawaiisu (Tehachapi):** ma?ahnid+b+a sanap+ceeka, "*Quercus douglasii,* its mistletoe." **Chumash (Barbareño):** shlamaulasha'w; **(Ynezeño):** stumuku'n.

Citation(s): Gifford 1967; Schenck and Gifford 1952; Zigmond 1981; Goodrich et al 1980; Timbrook 2007.

Note(s): plant contains beta-phenylethylamine and tyramine; produces symptoms of acute gastroenteritis and cardiovascular collapse.

Citation(s): Kingsbury 1964; Hardin and Arena 1974.

960 *Phragmites australis*

Family: *Poaceae*

Synonym(s): *Phragmites communis, Phragmites vulgaris, Phragmites communis* var. *berlanderi*

Common Name(s): Cane, Reed, Common Reed, Carizzo Grass.

Usage: Food: **Owens Valley Paiute:** most important

source for sugar; green cane gathered in Summer when leaves are thick. **Panamint Shoshone (Koso):** in early Summer, commonly in June, plants are cut and dried in the sun; plant is ground and shifted, the moist sticky flour is molded into a thick gum-like mass and eaten. **Kawaiisu (Tehachapi):** sugar crystals that form on plant are gathered and eaten. **Chumash:** plant, ripened in June, cut, dried, cleaned, pounded, made into cakes.

Material: **Cahuilla:** used for string/cordage; used for house thatching; shafts of arrows made from stems. **Modoc (Lutuami) (Lutuami):** used the stem for warp and woof and for white pattern in baskets. **Achomawi:** used the stem for white pattern in baskets. **Panamint Shoshone (Koso):** made arrows from stems. **Tubatulabal:** used for arrows. **Kawaiisu (Tehachapi):** made two-piece arrow with wood foreshaft inserted into reed; straight, rigid, hollow stems indispensable in marking arrows, fire drill, pipes, and other items. **Chumash:** house thatching, tubes for tobacco, cigarettes, knives, paintbrush handles, game counter sticks, arrows, splints, whistles for dancers in Devil Dance and Barracuda Dance.

Citation(s): Barrows 1900; Coville 1892; Merrill 1923; Sparkman 1908; Steward 1933; Bean and Saubel 1972; Strike 1994.

Terminology: **Cahuilla:** wish, pakhal. **Owens Valley Paiute:** hau've.

Citation(s): Sparkman 1908; Steward 1933; Bean and Saubel
1972; Zigmond 1981; Timbrook 2007; Strike 1994.
Common Name(s): Carizzo Grass, Common Reed.
Terminology: **Chumash (Barbareño):** 'ekhpe'w; **(Ynezeño):**
'ekhpew; **(Ventureño):** topo. **Cahuilla:** wish, pakhal.
Owens Valley Paiute:
hau've.
Citation(s): Timbrook 2007; Sparkman 1908; Steward 1933;
Bean and Saubel 1972.

961 *Phyllospadix scouleri*
Family: *Zosteraceae*

Usage: Misc: **Yuki:** named but not used.
Citation(s): Gifford 1939.
Terminology: **Yuki:** ultam.
Citation(s): Gifford 1939.

962 *Phyllospadix torreyi*
Common Name(s): Surf-Grass, Seagrass.

Usage: Food: **Chumash:** gather, when available, to
eat in sweathouse.

Material: **Chumash:** knee pads for canoes,
sometimes as mattresses, thatch,
cordage, nets, twined mats.
Citation(s): Timbrook 2007.
Terminology: **Chumash (Barbareño):** shkash; **(Ventureño):**
chkapsh.
Citation(s): Timbrook 2007.

963   *Physocarpus capitatus*

Family: *Rosaceae*
Common Name(s): Nine-Bark.

Usage: Food: **Miwok:** berries eaten raw.
         Material: **Karok:** make arrows from shoots.
                      Miwok (Coast): stems pounded for
                      fiber to used for cordage.
Citation(s): Barrett and Gifford 1933; Schenck and Gifford 1952;
         Strike 1994.
Terminology: **Miwok (Central Sierra):** hemekine. **Karok:**
         tapashxavich, "real xavish."
Citation(s): Barrett and Gifford 1933; Schenck and Gifford 1952.

964   *Picea* spp.

Family: *Pinaceae*
Common Name(s): Spruce.

Usage: Material: **Bear River:** made tight baskets for such
                      as eating, water, and cooking of smooth
                      roots; made wefts in loose baskets from
                      root fibers; weft in cradles; made canoe
                      paddles.
Citation(s): Nomland 1938.

965   *Picea sitchensis*

Common Name(s): Lowland Spruce.

Usage: Material: **Hupa:** a fiber from roots was imported
                      from the coast; used as a woof in the
                      making of baskets. **Tolowa:** used root as
                      woof and rim binding in baskets.
                      **Whilkut (Chilula), Nongatl, Lassik,
                      Wailaki:** used root as woof in baskets.
                      Yurok: roots used as body material in

baskets. **Wiyot:** used as twining material.
Citation(s): Goddrad 1903; Merriam 1967; Merrill 1923; O'Neale 1932; Strike 1994.
Terminology: **Yurok:** teiwolite'po (roots).
Citation(s): O'Neale 1932.

966 *Pinus* spp.
Family: *Pinaceae*

Usage: Food:

Chimariko; Wintun, Central (Nomlaki): used nuts. **Tolowa:** used nuts traded from inland. **Cahuilla:** nuts, after roasting, had shells removed by rolling a mano over them, winnowed in a basket prior to eating or cooking; eaten whole, or ground and made into mush

Material:

Yurok, Wiyot, Pomo, Yokuts: root used as woof in baskets. **Shasta:** made house posts. **Cahuilla:** wood used as firewood; needles used as basketry material; pitch used as adhesive for baskets, pottery attaching arrowpoints to shafts; bark used as roofing material.

Medicine:

Atsugewi: sugar obtained from sap, boiled and given to person unable to urinate; good for stomach trouble, especially for young children; gum was pounded, chewed, and then used as a poultice, or to draw boils.

Citation(s): Dixon 1910; Drucker 1937; Garth 1953; Gayton 1948; Goldschmidt 1951; Merrill 1923; Bean and Saubel 1972.
Terminology: **Yokuts (Northern Hill):** pa'an (nut). **Cahuilla:**

wexet. **Chumash** (except for *Pinus monophylla)*
(Barbareño, Ynezeño): tak, tomol; **(Ventureño):**
tsɨkɨnɨn.
Citation(s): Gayton 1948; Bean and Saubel 1972; Timbrook 2007.

967 *Pinus attenuata*

Synonym(s): *Pinus tuberculata*
Common Name(s): Knob-Cone Pine.

Usage: Material: **Karok:** the black seeds from the cones
are gathered and used to make beads
and ornaments for dresses. **Hupa:** the
pine nut shells were strung on twine,
along with twined *Xerophyllum tenax*
leaves.
Citation(s): Goddard 1903; Schenck and Gifford 1952.
Terminology: **Karok:** ishshwikipis, oskihich (nut), "imitation
sugar pine nuts."
Citation(s): Schenck and Gifford 1952.

968 *Pinus contorta* ssp. *murrayana*

Synonym(s): *Pinus Murrayana*
Common Name(s): Tamarack Pine.

Usage: Food: **Maidu (Northern):** in times of want or
in early Spring ate the inner bark and
sap. **Modoc (Lutuami):** in times of want
or in early Spring ate the inner bark .
    Medicine: **Maidu (Northern):** bark used for
cathartic properties.
Citation(s): Dixon 1905; Strike 1994.

969 *Pinus edulis*

Common Name(s): Nut-Pine, Silver Pine.

Usage: Food: **Maidu, Southern (Nisenan):** the nuts are a choice article of food; in the Spring, if food is scarce, they eat the buds on the ends of the limbs, the inner bark, and the core of the cone, which is something like a cabbage-stalk when green.

Medicine: **Maidu, Southern (Nisenan):** burned and beaten to powder, or crush raw, and spread on in a plaster, the nuts form a specific for a burn or a scald; the pitch, and Mistletoe (*Arceuthobium*) which grows on this pine, are very valuable in the estimation of the people for coughs, colds, and rheumatism; set on fire it makes a dense smudge which the patient allows to come in contact with all parts of his body; the pitch is used to treat arrow-wounds, or wounds and sores of any kind; the cone-core and Bunch Grass are boiled for a hair dye; a tar, made from hot pitch and burned acorns, to be worn by widows in mourning. **Bear River:** resin used to poultice wounds, sores.

Citation(s): Powers 1877; Strike 1994.

970 *Pinus jeffreyi*

Synonym(s): *Pinus ponderosa Jeffreyi*
Common Names(s): Jeffrey Pine.

Usage: Material: **Karok:** roots are used for weaving tobacco baskets.
Citation(s): Schenck and Gifford 1952.
Terminology: **Karok:** isvirip.
Citation(s): Schenck and Gifford 1952.

971 *Pinus lambertiana*

Common Name(s): Sugar Pine.

Usage: Food: **Wintun:** only ripe nuts, gathered in Fall, boiled to remove pitch; resin gathered Summer mornings before it melted, eaten as candy. **Wintun, Hill (Patwin):** used nuts, eaten raw or baked. **Wappo:** nuts eaten. **Yuki:** sweet exudation of the tree was chewed as gum; the nuts were eaten. **Huchnom, Atsugewi:** used nuts. **Yokuts (Northern Hill):** used nuts, gathered in August. **Pomo (Southwestern):** nuts were eaten; pitch chewed as gum. **Maidu, Northern:** nuts and sugar eaten. **Maidu, Southern (Nisenan):** cones gathered in July and August before they began to harden, and eaten. **Kawaiisu (Tehachapi):** seeds gathered and eaten; sap, drained through a hole cut in the tree, dries "like powdered sugar," and eaten. **Karok:** unhulled seeds steamed with leaves of Wild Grape (*Vitis californica*), or California Bay

(*Umbellularia californica*), in earth oven, dried, stored. **Shasta:** seeds pulverized, made into small cakes, often adding dried, crushed Salmon. **Miwok (Central Sierra):** seeds pulverized into creamy consistency, eaten on special occasions, or presented to visitors. **Monache (Western Mono):** seeds eaten with tobacco (*Nicotiana*).

Material: **Hupa, Lassik, Yurok, Wiyot, Karok, Chimariko:** used root as woof in baskets. **Karok:** used planks in sweathouses; pitch as adhesive. **Pomo (Southwestern):** pitch used in whistles.

Medicine: **Kawaiisu (Tehachapi):** sugar, ground fine, mixed with mother's milk, serves as eye medicine; eyes will clear if applied to sore eyes and allowed to remain for half-hour.

Citation(s): Barrett 1952; Barrett and Gifford 1933;Beals 1933; Chesnut 1902; Curtin 1957; Clark 1904; Dixon 1905, 1907; Driver 1936; Du Bois 1935; Barrows 1900; Foster 1944; Gifford 1967; Goddard 1903; Kroeber 1932; Merrill 1923; O'Neale 1932; Schenck and Gifford 1952; Zigmond 1981; Strike 1994.

Terminology: **Miwok (Northern Sierra):** caña′ku, sa′ñagu; **(Central Sierra):** hi′ñatcī. **Yuki:** shūk-ōl, shūk-tui-ōl, pōl′-cum (nuts). **Pomo (Southwestern):** chuye kale; **(Southestern):** maka′bats xale, "sweet or sugar tree," ha′kui (nuts). **Yurok:** waxpe′u (roots), karamametsaa (roots). **Kaork:** oskiip, ous (nuts). **Kawaiisu (Tehachapi):** wayahakatɨbɨ.

Citation(s): Barrett and Gifford 1933; Curtin 1957; Gifford 1967; Barrett 1952; O'Neale 1932; Schenck and Gifford 1952; Zigmond 1981.

972 *Pinus monophylla*
Common Name(s): One-Leaf Pine, Piñon.

Usage: Food: **Owens Valley Paiute:** gathered pine nuts in the Fall, usually enough to last through the Winter and into the Summer. **Tubatulabal:** gathered and used. **Miwok:** grows principally on the eastern side of the Sierras; considered best seed food on the eastern side. **Washo:** gathered nuts over a period of a month to six weeks is early Autumn; ground into meal and cooked as mush; considered most important seed food. **Cahuilla:** gathered seeds in late Fall; most used variety. **Panamint Shoshone (Koso):** gathered nuts. **Yokuts (Northern Hill):** eaten roasted and mashed into rich mush. **Monache (Western Mono):** gathered in late Summer (August-September). **Northern Paiute (Paviotso):** nuts of minor importance; few trees available. **Kawaiisu (Tehachapi):** nuts eaten; dried, eaten in gruel. **Chumash:** nuts eaten, best source, cooked in baskets with hot stones, cracked, eaten whole, or pounded into dry flour.

Material: **Kawaiisu (Tehachapi), Panamint Shosone (Koso):** used pitch as waterproofing in baskets. **Cahuilla:** a gum for adhesive purposes obtained from sap. **Chumash:** pitch, used as adhesive, sealant; caulking compound for canoes, sometimes binder in paint.

Medicine: **Washo:** resin swallowed whole as pills in treating venereal disease. **Chumash:** sap, applied externally for colds and pains.

Citation(s): Barrett 1917; Barrows 1900; Clark 1904; Coville 1892; Curtis 1924; Dutcher 1893; Gayton 1948; Kelly 1932; Merrill 1923; Steward 1933; Train et al 1941; Zigmond 1981; Landberg 1965; Timbrook 2007.

Terminology: **Owens Valley Paiute:** tūvā. **Washo:** a̲h̲-gum, w̲a̲h̲-pee. **Kawaiisu (Tehachapi):** tɨvapi. **Chumash (Barbareño, Ynezeño, Ventureño):** posh.

Citation(s): Steward 1933; Train et al 1941; Zigmond 1981; Timbrook 2007.

973 *Pinus monticola*

Common Name(s): Silver Pine.

Usage: Medicine: **Maidu:** resin, or bark, infusion used to treat stomach ache, cough, sores. As a blood purifier; women chewed resin to enhance fertility.

Citation(s): Strike 1994.

974 *Pinus muricata*

Common Name(s): Bishop Pine.

Usage: Food: **Pomo (Southwestern):** nuts eaten fresh, dried for Winter use.

Material: **Pomo (Southwestern):** roots used in basketry; wood made poor firewood; roots used in fish traps; pitch used like glue.

Citation(s): Gifford 1967; Goodrich et al 9180.

Terminology: **Pomo (Southwestern):** kunum.

Citation(s): Gifford 1967.

975 *Pinus ponderosa*
Common Name(s): Yellow Pine.

Usage: Food: **Shasta:** nuts were in demand. **Pomo:** gummy exudation is used for chewing. **Miwok:** gathered, but infrequently, for nuts. **Maidu (Northern):** nuts eaten. **Kawaiisu (Tehachapi):** seed kernels eaten raw. **Modoc (Lutuami):** in times of want or in early Spring ate the inner bark.

Material: **Hupa:** house floors covered with planks; a fiber was obtained from the roots for the woof in making baskets. **Yurok:** made sea-lion clubs. **Miwok:** covered floor of ceremonial house with fresh needles; used roots as woof, foundation, and wrapping in baskets. **Achomawi, Wintun, Maidu:** used root as woof in baskets. **Maidu (Northern):** used roots in making large burden baskets. **Karok:** used root fibers in basketry. **Shasta:** used root for twining in basketry. **Pomo:** pitch used for adhesive.

Medicine: **Hupa:** a medicine is made from the bark with which to bathe a sick person by the shaman. **Pomo:** pitch considered valuable for medicinal properties.

Citation(s): Barrett and Gifford 1933; Chesnut 1902; Dixon 1905, 1907; Goddard 1903; Merriam 1966; Merrill 1923; O'Neale 1932; Schenck and Gifford 1952; Waterman 1920; Zigmond 1981.

Terminology: **Karok:** sarum, ishividip (roots). **Miwok (Central Sierra):** wa'ssa. **Hupa:** diltcwag. **Kawaiisu**

(Tehachapi): yɨvi(m)bi.
Citation(s): Schenck and Gifford 1933; O'Neale 1932; Barrett and Gifford 1933; Goddard 1903; Zigmond 1981.

976 *Pinus quadrifolia*

Synonym(s): *Pinus Parryana*

Usage: Food: **Cahuilla:** gathered seeds in late Fall
Citation(s): Barrows 1900.

977 *Pinus sabiniana*

Common Name(s): Digger Pine, Grey Pine.

Usage: Food: **Huchnom, Atsugewi, Pomo, Yuki:** used nuts. **Pomo (Southwestern):** nuts eaten fresh, dried for Winter use, eaten whole, or pounded into flour for pinole. **Wintun:** unripe nuts eaten late in May, unshelled; ripe gathered in Autumn; stored, shelled or unshelled; eaten with acorn soup. **Wintun, Hill (Patwin):** used nuts; eaten raw or baked. **Wailaki:** pitch gum highly prized for chewing. **Miwok:** seeds mostly used on the western side of the Sierra. **Maidu (Northern):** nuts eaten; **Maidu, Southern (Nisenan):** nuts used, 1st preference. **Hupa:** the seeds are eaten; the cones are eaten in June while still green, the cone split and the central portion with the soft unripe seeds is eaten raw. **Wiyot:** the nuts were eaten. **Tubatulabal:** gathered seeds. **Shasta:** nuts were always in demand. **Kawaiisu (Tehachapi):** seeds eaten fresh;

	pounded, mixed with water, or boiled.
Material:	**Miwok:** needles used for thatching, bedding, floor covering; bark used for house covering; sprouts used in basketry. **Hupa:** a fiber was obtained from the roots for the woof in making of baskets. **Wiyot:** the nut were used a beads for decorating women's skirts. **Yuki:** root used as woof and wrapping in baskets; pitch used to stick feathers over body to appear more formidable in battle; made pitch soot for tattooing. **Wailaki, Wappo:** used root as woof in baskets. **Karok:** the nuts are used as beads to decorate dance dresses. **Pomo:** used in roofing, in basketry.
Medicine:	**Pomo:** pitch used to cover sores; drunk infusion of the bark for consumption; burn small twigs and leaves into ashes and a patient lies wrapped in blankets on ashes and steams as a cure for rheumatism and bruises (takes 8 hours for cure). **Shasta:** sugar was eaten as a cathartic. **Wailaki:** older persons chew pitchy material as a cure for rheumatism. **Miwok:** charcoal from the nut meats is crushed and applied to sores, burns, and abrasions. **Costanoan:** pitch, chewed, as a treatment for rheumatism.

Citation(s): Barrett 1908, 1952; Barrett and Gifford 1933; Beals 1933; Chesnut 1902; Clark 1904; Dixon 1905, 1907; Du Bois 1935; Foster 1944; Goddard 1903; Kroeber 1932; Loud 1918; Merrill 1923; Schenck and Gifford 1952; Voegelin 1938; Zigmond 1981; Goodrich et al 1980; Rayburn 2012.

Terminology: **Miwok (Plains):** sa′kkü; **(Northern Sierra):** ca′kü, ka′wil, sa′kü; **(Southern Sierra):** sa′kū. **Pomo (Southeastern):** xo′cil xale, "nut tree;" **(Southwestern):** cuye qʰale. **Karok:** axhiyushiip. **Kawaiisu (Tehachapi):** wohodɨbɨ.

Citation(s): Barrett and Gifford 1933; Barrett 1952; Schenck and Gifford 1952; Zigmond 1981; Goodrich et al 1980.

Note(s): nuts - food values

      Rich, fatty oil   51.05%
      Crude Protein 28.05%
    (Analyzed by Dr. Walter C. Blasdale - University of California).

Citation(s): Chesnut 1902.

978   *Pinus torreyana*

Common Name(s): Torrey Pine.

Usage: Food: **Diegueño:** seeds used to add flavor to flour made from less palatable seeds.

Citation(s): Strike 1994.

979   *Plagiobothrys* spp.

Family: *Boraginaceae*
Synonym(s): *Allocarpa* spp.

Usage: Misc: **Owens Valley Paiute:** named but not used.

Citation(s): Steward 1933.
Terminology: **Owens Valley Paiute:** tuwanava.
Citation(s): Steward 1933.

980 *Plagiobothrys fulvus* var. *campestris*
Synonym(s): *Plagiobothrys campoestris*

Usage: Food: **Yuki:** the crisp, tender shoots and flowers furnish a rather pleasant, sweet, and aromatic food; seeds are gathered for pinole; the coloring matter at the base of the young leaves is used by women and children to stain their cheeks crimson.

Citation(s): Chesnut 1902.

981 *Plagiobothrys jonesii*

Usage: Misc: **Kawaiisu (Tehachapi):** named but not used.

Citation(s): Zigmond 1981.

Terminology: **Kawaiisu (Tehachapi):** na?aganɨ(m)bɨ.

Citation(s): Barrett and Gifford 1933; Schenck and Gifford 1952.

982 *Plagiobothrys nothofulvus*
Common Name(s): Popcorn Flower.

Usage: Food: **Yuki:** the young leaves eaten as greens.

Material: **Tubatulabal:** young girls sometimes rubbed mashed roots over cheeks to color face red; this was done merely "to make themselves pretty." **Kawaiisu (Tehachapi); Chumash:** juice, obtained at the junction of stem and root, is used as "paint" to make up face "just for fun."

Citation(s): Curtin 1957; Voegelin 1930; Zigmond 1981; Timbrook 2007.

Terminology: **Yuki:** kō-kes. **Kawaiisu (Tehachapi):**

na?agani(m)bɨ. **Chumash (Barbareño):** kʼáʼnay.
Citation(s): Curtin 1957; Zigmond 1981; Timbrook 2007.

983 *Plantago ssp.*

Family: *Plantiaginaceae*
Common Name(s): Plantain.

Usage: Medicine: **Chumash:** leaves, applied to sores to draw out poison; leaves, dipped in olive oil, placed on neck to treat toothache; poultice, used for boils, sore throat, abscesses, swellings.
Citation(s): Timbrook 2007.

984 *Plantago erecta*

Synonym(s): *Plantago hookeriana* var. *californica*
Common Name(s): California Plantain.

Usage: Medicine: **Maidu:** used as a diuretic, emetic, purgative. **Panamint Shoshone (Koso):** used to regulate mensuration, relieve nasal congestion, sooth skin irritations, cure fever, ease stomach ache; seeds used to produce laxative; leaves, wet, used to poultice bites, itches, wounds, sores.
Citation(s): Strike 1994.

985 *Plantago lanceolata*

Usage: Medicine: **Kawaiisu (Tehachapi):** leaves brewed; infusion put in ear for relief of earache.
Misc: **Miwok:** named but not used.
Citation(s): Barrett and Gifford 1933; Zigmond 1981.
Terminology: **Miwok (Central Sierra):** supa.

Citation(s): Barrett and Gifford 1933.
Note(s): introduced from Europe.
Citation(s): Munz and Keck 1965.

986 *Plantago major*

Usage: Medicine: **Kawaiisu (Tehachapi):** leaves brewed; infusion put in ear for relief of earache. **Costanoan:** root, decoction, taken internally for constipation, to reduce fever.
Citation(s): Zigmond 1981; Rayburn 2012.
Note(s): native to Europe.
Citation(s): Baldwin et al 2012.

987 *Platanus racemosa*
Family: *Platanaceae*
Common Name(s): Sycamore, Western Sycamore.

Usage: Food: **Kawaiisu (Tehachapi):** bark boiled in water to make tea.; drunk with sugar.
Material: **Cahuilla:** limbs and branches employed in house construction; wooden bowls, seasoned in water, greased with meat or oil to prevent splitting. **Chumash:** used burls to make bowls, trays, containers; leaves used to wrap food for cooking, to line acorn leaching pits; poles in house construction. **Maidu:** made a foot drum, 5-6 feet long, about 2 feet diameter, split lengthwise.
Medicine: **Costanoan:** infusion used as a general remedy.
Citation(s): Zigmond 1981; Bean and Saubel 1972; Strike 1994; Rayburn 2012; Timbrook 2007.

Terminology: **Cahuilla:** sivily. **Chumash (Barbareño, Ventureño):** khsho'; **(Island):** qsho'; **(Ynezeño):** shonush; **(Obispeño):** teksu; **(Purisimeño):** 'aqsho'.
Citation(s): Bean and Saubel 1972; Timbrook 2007.

988   *Platystemon californicus*
Family: *Papaveraceae*
Synonym(s): *Platystemon californicum*

Usage: Food: **Yuki:** leaves eaten for greens.
         Misc: **Tubatulabal:** considered weeds.
Citation(s): Chesnut 1902; Voegelin 1938.

989   *Pleiacanthus spinosus*
Family: *Asteraceae*
Synonym(s): *Lygodesmia spinosa*

Usage: Medicine: **Cahuilla:** poultice used on hemorrhaging wounds to stop bleeding.
Citation(s): Strike 1994.

990   *Pleurotus ostreatus*
Family: *Pleurotaceae*
Common Name(s): Oyster Mushroom, Tree Mushroom.

Usage: Food: **Pomo (Southwestern):** baked.
Citation(s): Goodrich et al 1980.
Terminology: **Pomo (Southwestern):** šici.
Citation(s): Goodrich et al 1980.

991 *Pluchea sericea*

Family: *Asteraceae*
Synonym(s): *Pluchea borealis*
Common Name(s): Arrow Weed, Marsh Fleabane.

Usage: Food: **Cahuilla:** roots of young plants roasted and eaten.

Material: **Luiseño:** Arrows were sometimes made from the plant as well as furnishing material to roof houses. **Cahuilla:** made arrows; long, slender leaves used as roofing material, in walls as part of wattle and daub construction, ramadas, windbreaks, fences, granaries.

Citation(s): Curtis 1924; Sparkman 1908; Bean and Saubel 1972..
Terminology: **Luiseño:** hangla. **Cahuilla:** hangal.
Citation(s): Sparkman 1908; Bean and Saubel 1972..

992 *Poa* spp.

Family: *Poaceae*

Usage: Food: **Diegueño, Maidu, Luiseño:** seeds eaten.
Citation(s): Strike 1994.

993 *Poa annua*

Common Name(s): Blue Grass.

Usage: Food: **Karok:** used seeds (?).
Citation(s): Schenck and Gifford 1952.
Terminology: **Karok:** ʼachichtunveechas, "little lice."
Citation(s): Schenck and Gifford 1952.
Note(s): introduced from Europe.
Citation(s): Munz and Keck 1965.

994 *Poa pratensis*

Usage: Material: **Miwok:** stems used as a brush to dip in Manzanita (*Arctostaphylos* spp.) Cider which was then sucked from the stems.
Citation(s): Schenck and Gifford 1933.
Terminology: **Miwok (Central Sierra):** tepute.
Citation(s): Barrett and Gifford 1933.
Note(s): often considered to be wholly a European, Mediterranean Eurasia, introduction, but many montane races, at least, are native.
Note(s): seeds contain:

Ash	2.7%
Protein	9.0%
Oil	2.0%

Citation(s): Munz and Keck 1965; Earle and Jones 1962; Baldwin et al 2012.

995 *Poa secunda*

Synonym(s): *Poa nevadensis*

Usage: Food: **Northern Paiute (Paviotso):** the seeds were eaten.
  Misc: **Kawaiisu (Tehachapi):** named but not used.
Citation(s): Kelly 1932; Zigmond 1981.
Terminology: **Northern Paiute (Paviotso):** sopi′bü (seed).
  Kawaiisu (Tehachapi): hugʷišiivɨ.
Citation(s): Kelly 1932; Zigmond 1981.

996 *Pogogyne douglasii*

Family: *Lamiaceae*
Synonym(s): *Pogogyne parviflora, Pogogyne douglasii*
ssp. *parviflora*

Usage: Food: **Yuki:** seeds used as sweet aromatic ingredient for pinole. **Maidu:** used leaves for tea.

Medicine: **Maidu:** fresh leaves placed in bruised condition over abdomen as a counter-irritant for pains in the stomach and bowels.

Citation(s): Chesnut 1902.

997 *Polygala cornuta*

Family: *Polygalaceae*
Common Name(s): Horned Milkweed.

Usage: Medicine: **Miwok, Maidu:** a strong, very bitter decoction served as an emetic; in dilute form used for coughs, colds, and pains.

Citation(s): Barrett and Gifford 1933; Strike 1994.

Terminology: **Miwok (Central Sierra):** kitma. **Karok:** ikutunvaxarharas.

Citation(s): Barrett and Gifford 1933; Schenck and Gifford 1952.

998 *Polygonum* spp.

Family: *Polygonaceae*
Common Name(s): Knotweed.

Usage: Food: **Owens Valley Paiute:** used as food.

Citation(s): Steward 1933.

Terminology: **Owens Valley Paiute:** tīa'kāraη ā'va.

Citation(s): Steward 1933.

999 *Polypogon monspeliensis*

Family: *Poaceae*

Common Name(s): Rabbit's Foot, Beard Grass.

Usage: Food: **Tabatulabal:** seeds used.

Misc: **Kawaiisu (Tehachapi):** named but not used.

Citation(s): Voegelin 1938; Zigmond 1981.

Terminology: **Kawaiisu (Tehachapi):** pa?yɨgʷasivɨ, "Kangaroo-rat tail."

Citation(s): Zigmond 1981.

Note(s): introduced from Europe, native to south and west Europe.

Citation(s): Munz and Keck 1965; Baldwin et al 2012.

1000 *Polypodium californicum*

Family: *Polypodiaceae*

Common Name(s): California Polypody.

Usage: Medicine: **Wailaki:** rub juice from the root on sores for its healing effect and on the body as a cure for rheumatism. **Salinan:** used to treat internal injuries, kidney ailments. **Diegueño:** used to stop internal bleeding. **Maidu:** root used to produce a diuretic, expectorant, to stop diarrhea, coughs; leaves, bruised, rubbed on sores, aching joints. **Chumash:** roots, tea, drunk, or as an external wash, for injuries.

Citation(s): Chesnut 1902; Strike 1994; Timbrook 2007.

Terminology: **Chumash (Barbareño, Ventureño):** kepeye; **(Island):** kepey; **(Ynezeño):** peye.

Citation(s): Timbrook 2007.

1001 *Polyporus* spp.

Family: *Polyporaceae*
Common Name(s): Mushroom.

Usage: Food: **Pomo:** highly esteemed; boiled.
                   Tubatulabal: cut into strips, dried, pounded, stored; used to flavor food during cooking.
Citation(s): Chesnut 1902; Strike 1994.

1002 *Polystichum dudleyi*

Family: *Dryopteridaceae*
Synonym(s): *Polystichum aculeatum*
Common Name(s): Sword Fern.

Usage: Misc: **Coast Yuki:** named but not used.
Citation(s): Gifford 1939.
Terminology: **Yuki:** cheplas.
Citation(s): Gifford 1939.

1003 *Polystichum munitum*

Common Name(s): Sword Fern.

Usage: Food: **Maidu:** rhizomes peeled, baked, eaten.
       Material: **Pomo (Southwestern):** fronds used for lining the acorn-leaching basin; also to line earth oven for acorn bread.
       Medicine: **Maidu:** young frond tips chewed to soothe sore throats, to ease childbirth; rhizome decoction used to wash sores; spores poultice used to heal sores, burns.
       Misc: **Karok:** fonds used in a game to see who has the longest wind; played by adults with no gambling.

Citation(s): Gifford 1967; Schenck and Gifford 1952; Goodrich
et al 1980; Strike 1994.
Terminology: **Karok:** tiptip hich, "imitation tiptip,"
tiptip'unuhyaachas, "round tiptip." **Pomo
(Southwestern):** chamaoda.
Citation(s): Schenck and Gifford 1952; Gifford 1967; Goodrich
et al 1980.

1004 *Polytrichum juniperinum*
Family: *Polytrichaceae*
Common Name(s): Hair-Cap Moss.

Usage: Medicine: **Maidu:** infusion used as an diuretic.
Citation(s): Strike 1994.

1005 *Populus* spp.
Family: *Salicaceae*
Common Name(s): Cottonwood Tree.

Usage: Material: **Luiseño:** apron made from inner bark.
Medicine: **Costanoan:** outer bark, made into
syrup, used in setting broken bones.
Misc: **Shasta:** named but not used.
Citation(s): Holt 1946; Sparkman 1908; Rayburn 2012.

1006 *Populus fremonti*
Common Name(s): Cottonwood.

Usage: Material: **Luiseño:** inner bark used to make
apron-like garment worn behind
women. **Yuki:** wood used to a slight
extent for fuel. **Cahuilla:** used as
firewood; made wooden mortars
Medicine: **Yuki:** a decoction of the bark is used as
a wash for bruises and cuts; tea made of

bark or leaves used for sore throat, colds, cuts, and sores. **Kawaiisu (Tehachapi):** bark boiled; decoction washed broken limbs. **Cahuilla:** leaves and bark boiled, poultice used to relieve swelling caused by muscle strain, or limb soaked in solution of hot water, barks and leaves; treating cuts; soaked handkerchief in solution tied around head to cure headache; solution used to treat saddle sore and swollen legs of horses.

Citation(s): Chesnut 1902; Curtin 1957; Sparkman 1908; Zigmond 1981; Bean and Saubel 1972.

Terminology: **Luiseño:** avahut. **Yuki:** pät'-mil. **Cahuilla:** lavalvanat.

Citation(s): Sparkman 1908; Curtin 1957; Bean and Saubel 1972.

1007 *Populus tremuloides*

Common Name(s): Aspen.

Usage: Medicine: **Shasta:** bark boiled to make treatment for chills and fever. **Maidu:** sap, root, bark, buds made into general tonic, used to treat ailments.

Misc: **Miwok:** named but not used.

Citation(s): Barrett and Gifford 1933; Holt 1946; Strike 1994.

Terminology: **Miwok (Central Sierra):** taktakkalu.

Citation(s): Barrett and Gifford 1933.

1008 *Populus trichocarpa*

Synonym(s): *Populus balsamifera* spp. *trichocarpa*

Common Name(s): Cottonwood, Black Cottonwood.

Usage: Material: **Hupa:** the root used for fire sticks; a

fiber was obtained from the root for the woof in making baskets. **Karok:** roots used for baskets; leaf buds in Spring used as glue to stick feathers on arrows. **Chumash:** house poles, bowls, trays, containers, sometimes dugout canoes; fiber, women's skirts.

Medicine: **Karok:** a love medicine is made of the leaves. **Maidu:** sap, root, bark, buds made into general tonic, used to treat ailments.

Misc: **Yuki:** named but not used. **Pomo (Southwestern):** named but not used.

Citation(s): Chesnut 1902; Goddard 1903; Merriam 1967; Merrill 1923; O'Neale 1932; Schenck and Gifford 1952; Goodrich et al 1980; Strike 1994; Timbrook 2007.

Terminology: **Karok:** ashappip. **Pomo (Southwestern):** puhláp^{h}law. **Chumash (Barbareño):** qweleqwe'l; **(Ynezeño):** qweleqwel; **(Ventureño):** khwelekhwel.

Citation(s): Schenck and Gifford 1952; Goodrich et al 1980; Timbrook 2007.

1009 *Portulaca oleracea*

Family: *Portulacaceae*
Common Name(s): Common Purslane.

Usage: Food: **Luiseño:** used for greens.
Citation(s): Sparkman 1908.
Terminology: **Luiseño:** pokut.
Citation(s): Sparkman 1908.
Note(s): naturalized from Europe.
Citation(s): Munz and Keck 1965.

1010 *Postelsia palmaeformis*

Family: *Laminariaceae*
Common Name(s): Sea Palm.

Usage: Food: **Pomo:** eaten raw, mostly for its salty
                      flavor. **Pomo (Southwestern):** eaten at
                      any time of the year; stalks cooked on a
                      flat rock or in hot ashes until color
                      changed from brown to blue; also
                      chewed raw. **Coast Yuki:** stalks were
                      cooked and eaten.
Citation(s): Barrett 1952; Gifford 1939, 1967; Goodrich et al
        1980.
Terminology: **Pomo (Southern):** kaiyē; **(Central):** kaiye spa;
        (Southwestern):   kaijê.   **Yuki:** ukhenchembal,
        "cormorant leaves."
Citation(s): Barrett 1952; Gifford 1939, 1967; Goodrich et al
        1980.

1011 *Potamogeton diversifolius*

Family: *Potamogetonaceae*
Common Name(s): Pondweed.

Usage: Material: **Kawaiisu (Tehachapi):** stems dried to
                      remove fibers to make cordage.
Citation(s):Zigmond 1981.
Terminology: **Kawaiisu (Tehachapi):** tɨviwɨʔvɨbɨ.
Citation(s): Zigmond 1981.

1012 *Potentilla anserina* ssp. *pacifica*

Family: *Rosaceae*
Synonym(s): *Potentilla egedii* var. *grandis*
Common Name(s): Silver-Backed Cinquefoil.

Usage: Food: **Miwok (Coast):** seeds used in pinole.

Citation(s): Strike 1994.

1013 *Potentilla gracilis*
Synonym(s): *Potentilla gracilis rigida*
Common Name(s): Five Finger.

Usage: Misc: **Karok, Kawaiisu (Tehachapi):** named
but not used.
Citation(s): Schenck and Gifford 1952; Zigmond 1981.
Terminology: **Karok:** mahuxnahich, "mountain imitation
strawberry." **Kawaiisu (Tehachapi):** motoobi
po?ova?adɨ, "motoobi near water."
Citation(s): Schenck and Gifford; Zigmond 1981.

1014 *Potentilla gracilis* ssp. *fastigiata*
Synonym(s): *Potentilla gracilis* ssp. *nuttallii*
Common Name(s): Slender Cinquefoil.

Usage: Food: **Miwok (Coast):** made beverage.
Citation(s): Strike 1994.

1015 *Potentilla norvegica*
Synonym(s): *Potentilla norvegica* spp. *monspeliensis*
Common Name(s): Rough Cinquefoil.

Usage: Medicine: **Maidu:** used as physic.
Citation(s): Strike 1994.

1016 *Proboscidea altheaefolia*
Family: *Martyniaceae*
Common Name(s): Unicorn Plant.

Usage: Food: **Cahuilla:** seeds eaten.
Material: **Cahuilla:** hook-like thorns used as a
tool in mending baskets and broken

pottery.
Citation(s): Bean and Saubel 1972.
Terminology: **Cahuilla:** akawat.
Citation(s): Bean and Saubel 1972.

1017 *Proboscidea louisianica*

Synonym(s): *Proboscidea parviflora*
Common Name(s): Unicorn Plant, Black Devil's Claw.

Usage: Material: **Kawaiisu (Tehachapi):** used as black pattern in coiled basketry. **Chemehuevi (Southern Paiute), Kawaiisu (Tehachapi), Monache (Western Mono), Owens Valley Paiute, Tubatulabal:** used in coiled baskets.

Citation(s): Zigmond 1981; Strike 1994.
Note(s): native of southern states.
Citation(s): Munz and Keck 1965.

1018 *Proboscideae lutea*

Family: *Martyniaceae*
Synonym(s): *Martynia probosciea, Ibicella lutea*
Common Name(s): Unicorn Plant, Devil's Horns.

Usage: Material: **Monache (Western Mono), Kitanemuk (Tejon), Chemehuevi, Kawaiisu (Tehachapi), Tubatulabal:** used seed pods as black pattern in baskets. **Panamint Shoshone (Koso):** used seed pods as black pattern in baskets; used flexible horns of seed pods in basketry.
        Misc: **Tubatulabal:** considered weeds (?).

Citation(s): Coville 1892; Merrill 1923; Voegelin 1938.
Note(s): some difference in what various ethnographers "learned" from the Tubatulabal [author].

1019 *Prosartes parvifolia*

Family: *Liliaceae*
Synonym(s): *Disporum Hookeri, Disporum smithii*
Common Name(s): Fairy Lantern, Fairy Bells.

Usage: Medicine: **Costanoan:** fruit used to relieve kidney ailments.
Misc: **Karok:** named but not used.
Citation(s): Schenck and Gifford 1952; Strike 1994; Rayburn 2012
Terminology: **Karok:** pottat'tui.
Citation(s): Schenck and Gifford 1952.

1020 *Prosopis glandulosa*

Family: *Fabaceae*
Synonym(s): *Prosopis juliflora*
Common Name(s): Mesquite, Western Honey Mesquite.

Usage: Food: **Cahuilla:** blossoms picked, roasted in pit of heated stones, "squeezed into balls ready for eating; bean pods pounded or crushed in wooden mortars; pulpy extract drunk as thirst impels; mature dried pods eaten as picked; ground into meal, dampened with water, left for a day to harden, sometimes made into balls, mostly into cakes; eaten dry, made into mush, mixed with water to form a beverage; dried bean pods and cakes stored.
Material: **Cahuilla:** trunk made into wooden mortars; superior firewood, kindling for cooking sweat-houses; limbs used for bow making, arrow foreshaft

attached with mesquite gum; bark pounded, rubbed, pulled to form soft fiber for weaving women's shirt, diapers for babies, carrying net for pottery; large limbs used as corner posts for houses, rafters, granary posts; leaves as thatching; thorns for tattooing skin.

Medicine: **Cahuilla:** gum, diluted with water, used as wash for open wounds, sores, and for treating sore eyes.

Citation(s): Saubel and Bean 9172.

Terminology: **Cahuilla:** ily, selkulat (the blossom food balls), "blossoms made of." kakhat (food cakes).

Citation(s): Bean and Saubel 1972.

Note(s): beans per 100 pounds contain:

crude protein	8.34 pounds
carbohydrates	52.02 pounds
fats	2.4 pounds.

Citation(s): Bean and Saubel 1972.

1021 *Prosopis glandulosa* var. *torreyana*
Common Name(s): Mesquite.

Usage: Food: **Luiseño:** the beans are ground into meal and used for food to a limited extent in some localities. **Kamia:** considered most important food plant. **Cahuilla:** gathered beans in July and August; pod and all are pounded up into meal and then soaked and eaten; whole dried seed pod stored in granaries made of *Artemesia ludovianca.* **Panamint Shoshone (Koso):** gathered in Fall and cached until Spring;

pounded dry pods into flour and made small cakes. **Kawaiisu (Tehachapi):** seeds, pounded, molded into cakes.

Material: **Mohave:** used bark as woof in baskets. **Cahuilla:** used as house framework; make mortars from the trunk; used plant sap as gum (glue).

Citation(s): Barrows 1900; Coville 1892; Curtis 1924; Gifford 1931; Merrill 1923; Sparkman 1908; Bean and Saubel 1963; Zigmond 1981.

Terminology: **Luiseño:** ela. **Cahuilla:** pechita menyikish (harden cakes of ground meal). **Kawaiisu (Tehachapi):** opi(m)bɨ.

Citation(s): Sparkman 1908; Bean and Saubel 1963; Zigmond 1981.

1022 *Prosopis pubescens*

Common Name(s): Screwbean, Tornillo.

Usage: Food: **Cahuilla:** pods picked when ripe, dried (prepared same as #1021 above). **Kamia:** spirally coiled pod collected in July from the ground around the plant; considered second to *Prosopis juliflora* var. *Torreyana*.

Material: **Cahuilla:** best bows made from seasoned limbs; also used for construction materials, corner posts, rafters, foundations; smaller limbs sometimes for bows.

Medicine: **Cahuilla:** roots and bark said to have medicinal value.

Misc: **Kawaiisu (Tehachapi):** named but not used.

Citation(s): Barrows 1900; Curtis 1924; Gifford 1931; Bean and

Saubel 1963; Zigmond 1981.
Terminology: **Cahuilla:** qwinyal, kwinyal. **Kawaiisu (Tehachapi):** kʷiyara.
Citation(s): Bean and Saubel 1972; Sparkman 1908; Zigmond 1981.

1023 *Prunella vulgaris* ssp. *lanceolata*
Family: *Lamiaceae*
Common Name(s): All-Heal, Selfheal.

Usage: Medicine: **Maidu:** treated burns, bruises, diarrhea, hemorrhages, sore throats, colic.
Citation(s): Strike 1994.

1024 *Prunus andersonii*
Family: *Rosaceae*
Common Name(s): Desert Peach.

Usage: Food: **Cahuilla:** fruit picked in June, boiled, sweetened with sugar, put in jars as jelly; remembered as a highly prized food.
        Misc: **Mono Lake Paiute:** named but not used.
Citation(s): Bean and Saubel 1972; Barrows 1900; Steward 1933.
Terminology: **Cahuilla:** chawkal. **Mono Lake Paiute:** sana'vi.
Citation(s): Bean and Saubel 1972; Steward 1933.

1025 *Prunus emarginata*
Common Name(s): Bitter Cherry.

Usage: Medicine: **Diegueño:** leaf decoction used on bruises, scars, swellings, sprains, treat facial rashes or abrasions. **Maidu:** made a tonic and general cure-all.

Citation(s): Strike 1994.

1026 *Prunus fasciculata*
Common Name(s): Desert Almond.

Usage: Material: **Kawaiisu (Tehachapi):** straight twigs used as arrow foreshafts; served as drill in fire-drill.
Citation(s): Zigmond 1981.
Terminology: **Kawaiisu (Tehachapi):** tivo?onɨbɨ.
Citation(s): Zigmond 1981.

1027 *Prunus fremontii*
Common Name(s): Desert Apricot.

Usage: Food: **Modoc (Lutuami) (Lutuami);** fruit eaten.
Citation(s): Strike 1994.

1028 *Prunus ilicifolia*
Synonym(s): *Cerasus ilicifolia*
Common Name(s): Wild Plum, Wild Cherry, Islaya.

Usage: Food: **Luiseño:** the fruit is eaten; the kernels are ground into flour and used for food. **Cahuilla:** plums picked in August, dried; pits broken open and extracted, crush in mortar, leached in sand basket, boiled into atole; fruit pressed to make drink. **Salinan, Kitanemuk (Tejon), Cupeño:** fruit and meat from pit eaten. **Esselen:** fruit, pulp, eaten; kernels made into powder used to flavor other foods and to make a soup. **Chumash:**

fruit, picked late Summer, pulp, eaten;
kernel, highly valued food source,
leached to remove hydrocyanic acid,
boil until done, mashed, made round
balls (could keep as long as a week).

Medicine: **Cahuilla:** bark made infusion to cure colds.

Misc: **Pomo (Southwestern):** named but not used.

Citation(s): Barrows 1900; Sparkman 1908; Goodrich et al 1980; Timbrook 1982; Landberg 1965; Bean and Saubel 1972; Breschini and Haversat 2004; Timbrook 2007.

Terminology: **Luiseño, Cahuilla:** chamish. **Pomo (Southwestern):** pʰagʰále. **Chumash (Barbareño, Ynezeño):** ʼakhtayukhash; **(Island):** wam; **(Obispeño):** chto; **(Ventureño):** ʼakhtatapɨsh.

Citation(s): Sparkman 1908; Goodrich et al 1980; Bean and Saubel 1972; Timbrook 2007.

1029 *Prunus subcordata*

Common Name(s): Sierra Plum, Wild Plum.

Usage: Food: **Maidu:** sometimes gathered for food; dried for long trips. **Maidu (Northern), Achomawi, Modoc (Lutuami) (Lutuami):** fruit eaten. **Miwok:** fruit eaten raw. **Atsugewi:** fruit prepared by removing seeds and drying pulp which might be pounded up and stored for Winter in small cakes.

Citation(s): Barrett and Gifford 1933; Chesnut 1902; Dixon 1905; Clark 1904; Garth 1953; Merriam 1967; Schenck and Gifford 1952; Strike 1994.

Terminology: **Miwok (Central Sierra):** yotoña. **Karok:** puumpurip, "plum purip." **Atsugewi:** batĭku.

Citation(s): Barrett and Gifford 1933; Schenck and Gifford 1952; Garth 1953.
Note(s): fruit ripens in August and September.
Citation(s): Sargent 1922.

1030 *Prunus virginiana* var. *demissa*

Synonym(s): *Cerasus demissa*, *Prunus demissa*
Common Name(s): Chokecherries, Western Choke-Berry, Chockberry, Western Choke Cherry, Chuckberrie.

Usage: Food: **Luiseño:** the fruit is eaten. **Northern Paiute (Paviotso):** the fruit was eaten fresh or dried. **Karok:** the berries are eaten ripe; they are not preserved. **Shasta:** berries eaten either fresh or dried. **Miwok:** berries eaten raw. **Cahuilla:** gathered the small red berries. **Atsugewi:** berries put in a tule (*Scirpus* spp.) basket when ripe and mashed, water added to form a paste which was eaten without cooking. **Maidu (Northern), Achomawi:** berries eaten. **Yuki:** berries eaten raw or cooked when ripe; eaten either fresh or dried. **Cahuilla:** fruit was eaten fresh, but was astringent and could cause stomach trouble; sun dried for future use; pit ground up, used as meal. **Kawaiisu (Tehachapi):** berries, ripe in August-September, eaten fresh.

Material: **Karok:** gum is used in fastening *Amelanchier pallida* foreshafts on end of arrows; also applied to surfaces of bows and arrows when design of *Delphinium* *decorum* paint is to be applied. **Miwok:** made ring for the hoop and pole game; also made gaming stick for the hand game. **Kawaiisu (Tehachapi):** stems used as arrows.

Medicine: **Yuki:** inner bark used as a tonic to check diarrhea and to relieve nervous excitability. **Karok:** the bark is scraped off twigs and placed beside the nose of a little baby when it has a cold. **Cahuilla, Karok:** tea used to cure colds. **Kawaiisu (Tehachapi):** tea used as laxative. **Maidu:** tea used to cure colds, decoction used as sedative, to arrest diarrhea, or a tonic. **Atsugewi:** leaves used to poultice cuts, sores, wounds, bruises; decoction to bathe wounds to promote rapid healing.

Citation(s): Barrett and Gifford 1933; Barrows 1900; Chesnut 1902; Curtin 1957; Dixon 1905, 1907; Garth 1953; Kelly 1932; Merriam 1967; Schenck and Gifford 1952; Sparkman 1908; Strike 1994; Bean and Saubel 1972; Zigmond 1981.

Terminology: **Miwok (Central Sierra):** pisakene; **(Southern Sierra):** pīha′kene. **Yuki:** chul-mäm. **Luiseño:** atut. **Cahuilla:** atut (the fruit). **Atsugewi:** cuiwap. **Northern Paiute (Paviotso):** do′·icabui. **Karok:** purip, puun (the berry). **Cahuilla:** atut.

Citation(s): Barrett and Gifford 1933; Curtin 1957; Garth 1953; Sparkman 1908; Kelly 1932; Schenck and Gifford 1952; Bean and Saubel 1972.

Note(s): the bark is tonic through its bitter principle, and more or less sedative through the hydrocyanic acid which it generates with water; for the latter effect, however, very large doses are required.

Citation(s): Stille and Maisch 1880

1031 *Pseudognaphalium biolettii*

Family: *Asteraceae*
Synonym(s): *Gnaphalium bicolor*
Common Name(s): Pearly Live-forever.

Usage: Medicine: **Diegueño:** boiled, used to poultice sores.
Citation(s): Strike 1994.

1032 *Pseudognaphalium californicum*

Synonym(s): *Gnaphalium decurrens* var. *californicum*
Common Name(s): California Everlasting.

Usage: Medicine: **Miwok:** leaves were bound on any swelling as a poultice; flowers and leaves were used as a poultice after heating them in a fire to make them sticky; decoction of the leaves drunk for colds and stomach trouble. **Costanoan:** used as an analgesic, treatment for colds, remedy for gastrointestinal problems; tea given for stomach pain, colds.
        Misc: **Karok:** named but not used.
Citation(s): Barrett and Gifford 1933; Schenck and Gifford 1933; Strike 1994; Rayburn 2012.
Terminology: **Karok:** mukiita.
Citation(s): Schenck and Gifford 1952.

1033 *Pseudognaphalium canescens*

Synonym(s): *Gnaphalium canescens*

Usage: Medicine: **Chumash:** reported to have been used for pulmonary disease, or to expel gas from the stomach.

Citation(s): Timbrook 2007.

1034 *Pseudognaphalium microcephalum*
Synonym(s): *Gnaphalium microcephalum*
Common Name(s): White Everlasting.

Usage: Material: **Pomo (Southwestern):** cottony flower tops used as lining in deer antler head disguises where it rested on forehead.

Medicine: **Karok, Maidu:** the plants is soaked in cold water and the eyes are washed with the infusion.

Citation(s): Schenck and Gifford 1952; Goodrich et al 1980; Strike 1994.

Terminology: **Karok:** ishkamakyannarav, "feather-to-put-in."
Pomo (Southwestern): naċolólda?bo· qʰale.

Citation(s): Schenck and Gifford 1952; Goodrich et al 1980.

1035 *Pseudognaphalium stramineum*
Synonym(s): *Gnaphalium chilense*
Common Name(s): Cotton-Batting Plant.

Usage: Material: **Pomo (Southwestern):** hunters used this plant as disguise in hunting deer.

Medicine: **Pomo (Southwestern):** plant boiled and applied as a poultice to a swollen face. **Kawaiisu (Tehachapi):** applied hot leaves or stems to relieve pain.

Misc: **Karok:** named but not used (?).

Citation(s): Gifford 1967; Schenck and Gifford 1952; Zigmond 1981.

Terminology: **Pomo (Southwestern):** nachalon dabo kale, nachalon (deer-antle disguise) dabo (lining of disguise) kale "bush." **Karok:** ˋamtapparas, "dusty."
Kawaiisu (Tehachapi): naragʷani(m)bɨ.

Citation(s): Gifford 1967; Schenck and Gifford 1952; Zigmond 1981.

1036 *Pseudostellaria jamesiana*

Family: *Caryophyllaceae*
Synonym(s): *Stellaria jamesiana*
Common Name(s): Sticky Starwort.

Usage: Foofd: **Maidu:** tubers eaten.
Citation(s): Strike 1994.

1037 *Pseudotrillium rivale*

Synonym(s): *Trillium rivale*

Usage: Misc: **Karok:** named but not used.
Citation(s): Schenck and Gifford 1952.
Terminology: **Karok:** pikvasahiic, "imitation feathers which stand up in headdress."
Citation(s): Schenck and Gifford 1952.

1038 *Pseudotsuga macrocarpa*

Family: *Pinaceae*

Usage: Material: **Chumash:** made digging sticks and planks.
         Medicine: **Tubatulabal:** sap heated, smoke inhaled to relieve sinus and asthma problems.
Citation(s): Grant 1964; Strike 1994.

1039 *Pseudotsuga menziesii* var. *menziesii*

Synonym(s): *Pseudotsuga mucronata, Pseudotsuga taxifolia*

Common Name(s): Red Fire, Douglas Fir, Douglas Fir.

Usage: Material: **Pomo:** used root as woof in baskets. **Hupa:** the pitch is heated over a fire ceremonially built during the Brush Dance; bark peeled from the tree by a priest in the afternoon before the "winter dance" could begin. **Hupa, Whilkut (Chilula):** used branches to smoke their bodies; gave good luck for salmon, deer, and wealth. **Whilkut (Chilula), Nongatl:** root used as woof in baskets. **Karok:** boughs used as fuel in sweat-house; boughs are considered both "good luck" and antiseptic; boughs used as "seasoning" when elk or deer meat is barbecued in the earth oven, being placed directly over the meat with a Maple (*Acer* spp.) leaf mat on top of them. **Pomo (Southwestern):** used for firewood; pitch used as glue. **Yuki:** when camping out, shelters made from branches.

Medicine: **Karok:** branches are mixed with branches of *Umbellularia californica* and *Pseudotsuga menziesii* laid in a pit of hot rocks, the patients in placed over this on a blanket and allowed to steam, good for colds and about any kind of sickness; a drink is administered to a woman to relieve the pains of childbirth, with a special charm. **Pomo:**

used leaves in sweat bath as cure for
rheumatism.

Misc: **Yuki:** named but not used(?).

Citation(s): Chesnut 1902; Curtin 1957; Gifford 1939, 1967;
Goddard 1903, 1914; Merrill 1923; Schenck and
Gifford 1952; Goodrich et al 1980.

Terminology: **Karok:** tapush. **Hupa:** neskin. **Pomo
(Southwestern):** kauwam. **Yuki:** nu, nex.

Citation(s): Schenck and Gifford 1952; Goddard 1903; Gifford
1967; Curtin 1957; Gifford 1939; Goodrich et al 1980.

Note(s): note the difference in reporting for **Yuki** – difference
reporting to the ethnographers.

1040 *Psorothamnus emoryi*

Family: *Fabaceae*
Synonym(s): *Dalea Emoryi polyadenia, Parosela
Emoryi, Dalea emoryi*
Common Name(s): Dye Bush, Indigo Bush.

Usage: Food: **Owens Valley Paiute:** eaten.

Material: **Chumash, Diegueño, Luiseño,
Cahuilla, Cupeño:** used as a yellow dye
(the stem). **Cahuilla:** branches steeped
in water form a light-yellowish-brown
dye; stalk used as foundation in
basketry.

Citation(s): Barrows 1900; Merrill 1923; Bean and Saubel 1972.

1041 *Psorothamnus fremontii* var. *fremontii*

Synonym(s): *Dalea fremontii*

Usage: Misc: **Kawaiisu (Tehachapi):** named but not
used.

Citation(s): Zigmond 1981.

Terminology: **Kawaiisu (Tehachapi):** pokoi.

Citation(s): Zigmond 1981.

1042 *Psorothamnus polyadenius*
Synonym(s): *Dalea polyadenia*

Usage: Material: **Cahuilla:** used to dye basketry materials yellow-brown.
Citation(s): Strike 1994.

1043 *Pteridium aquilinum* var. *pubescens*
Family: *Dennstaedtiaceae*
Synonym(s): *Pteridium aquilinum, Pteris aquilina, Pteridium languinosa, Pteridium Aquilinum* var. *languinosum*
Common Name(s): Brake Fern, Bracken.

Usage: Food: **Atsugewi:** leaves and tender stems eaten raw. **Tolowa:** root masses cooked in earth oven; said to be like milk and to have a fine flavor. **Cahuilla:** shoots scraped, boiled, eaten. **Costanoan:** decoction used as hair wash, root paste rubbed into scalp to encourage hair growth.

Material: **Pomo, Yokuts:** root used as black pattern in baskets. **Pomo (Southwestern):** root used in baskets. **Washo:** root used to provide black element in basketry design. **Karok:** used as bed upon which to place salmon just after they are caught and allowed to bleed before gutting; also used to wrap tobacco leaves when they are first gathered. **Pomo, Washo, Miwok, Monache (Western Mono), Kawaiisu (Tehachapi), Tubatulabal:**

root used as brown pattern in baskets. **Miwok, Maidu, Monache (Western Mono), Kitanemuk (Tejon), Kawaiisu (Tehachapi), Tubatulabal, Panamint Shoshone (Koso):** boiled root used for black pattern in baskets. **Chumash:** house thatching.

Medicine: **Yana:** roots were pounded fine and heated in a basket bucket; the poultice was applied over a burn. **Pomo (Southwestern):** when young fronds are curled, the juices are used as body deodorant.

Citation(s): Barrett 1917; Baumhoff 1958; Dixon 1905; Garth 1953; Gifford 1939, 1967; Merriam 1955, 1967; Merrill 1923; Sapir and Spier 1943; Schenck and Gifford 1952; Goodrich et al 1980; Bean and Saubel 1972; Rayburn 2012; Timbrook 2007.

Terminology: **Yuki:** cheplas. **Atsugewi:** watini. **Pomo (Southwestern):** me?oda, "fern." **Karok:** kataship. **Cahuilla:** welmat(?). **Chumash (Barbareño):** qɨch; **(Ventureño):** kich.

Citation(s): Gifford 1939, 1967; Garth 1953; Schenck and Gifford 1952; Goodrich et al 1980; Bean and Saubel 1972; Timbrook 2007.

Note(s): the plant contains potent carcinogenic substances; often symptoms of thiamine deficiency; cooking in hot water decreases but does not eliminate the substances.

Citation(s): Liener 1980.

1044 *Pterospora andromedea*

Family: *Pyrolaceae*
Common Name(s): Pinedrops.

Usage: Food: **Kawaiisu (Tehachapi):** it "tastes a little

sour;" baked, sometimes eaten raw.

Medicine: **Maidu:** used to stop bronchial hemorrhages.

Citation(s): Zigmond 1981; Strike 1994.

Terminology: **Kawaiisu (Tehachapi):** puhidu?uvɨ keevowadɨ, "mountain puhidu?uvɨ."

Citation(s): Zigmond 1981.

1045 *Pterygophora californica*

Family: *Alariaceae*

Common Name(s): Seaweed, Stalked Kelp.

Usage: Food: **Coast Yuki:** the streamers are cooked in coals and eaten.

Citation(s): Gifford 1939; Lane et al 2006.

Terminology: **Yuki:** balhetin, "streamers" (bal), "stem" (hetin).

Citation(s): Gifford 1939.

1046 *Purshia tridentata*

Usage: Medicine: **Washo:** prepared a physic by boiling ripe whole seed.

Citation(s): Train et al 1941.

Terminology: **Washo:** <u>bal</u>-mat-san.

Citation(s): train et al 1941.

1047 *Purshia tridentata var. glandulosa*

Family: *Rosaceae*

Synonym(s): *Purshia glandulosa*

Common Name(s): Antelope Brush, Dessert Bitterbrush.

Usage: Medicine: **Kawaiisu (Tehachapi):** leaves and inner bark boiled to produce yellowish "tea;" decoction serves as an emetic, or strong

laxative, an analgesic, to relieve menstrual cramps; remedy for venereal disease.
Citation(s): Zigmond 1981; Strike 1994.

1048 *Pycanthemum californicum*
Family: *Lamiaceae*
Common Name(s): Mountain Mint.

Usage: Medicine: **Miwok:** boiled as tea for colds.
Citation(s): Barrett and Gifford 1933.
Terminology: **Miwok:** holoyu.
Citation(s): Barrett and Gifford 1933.

1049 *Pyrola asarifolia* ssp. *asarifolia*
Family: *Ericaceae*
Synonym(s): *Pyrola californica* var. *purpurea, Pirola asarifolia incarnata, Pyrola asarifolia* var. *purpurea*

Usage: Medicine: **Karok:** used to cure a child who is unmanageable, too lively.
Citation(s): Schenck and Gifford 1952.
Terminology: **Karok:** achpush.
Citation(s): Schenck and Gifford 1952.
Note(s): Flavonol/Flavonoid chemicals are found in the plant. They possess and act as sedative action, tranquilizers.
Citation(s): Sandar et al 2011; Haber 1983.

1050 *Pyrola picta*
Synonym(s): *Pirola picta*
Common Name(s): White-Veined Shin-Leaf.

Usage: Medicine: **Karok:** used for child who is sick.
Maidu: leaf poultice used to treat sores,

bruises.

Citation(s): Schenck and Gifford 1952; Strike 1994.

Terminology: **Karok:** yumarepeisera.

Citation(s): Schenck and Gifford 1952.

1051 *Pyropia* spp.

Family: *Bangiaceae*

Synonym(s): *Porphyra* spp.

Common Name(s): Seaweed.

Usage: Food: **Hupa:** purchased from **Coast Yuki;**
boiled before eating. **Tolowa:** boiled and
eaten. **Miwok (Coast):** baked in hot
ashes, eaten.

Citation(s): Curtis 1924; Miller 2012; Strike 1994..

1052 *Pyropia lanceolata*

Synonym(s): *Porphyra lanceolata, Porphyra laciniata*

Common Name(s): Seaweed.

Usage: Food: **Pomo, Pomo (Southwestern):** eaten raw
or baked; valued chiefly for its salt.

Citation(s): Chesnut 1902; Goodrich et al 1980; Miller 2012.

1053 *Pyropia perforata*

Synonym(s): *Porphyra perforata*

Common Name(s): Seaweed.

Usage: Food: **Pomo:** eaten in dried form as cakes,
about 1 ft. in diameter and 3/8 in. Thick;
used as condiment with acorn soup and
similar foods. **Pomo (Southwestern):**
made into cake, cooked in earth oven;
stored in openwork basket for Winter
consumption. **Yuki:** dried in the sun;

before eating, dried leaves were heated before a fire to make brittle and easy to chew; eaten chiefly in the Summer. **Chumash:** eaten.

Citation(s): Barrett and Gifford 1952; Gifford 1939, 1967; Goodrich et al 1980; Miller 2012; Landberg 19654.

Terminology: **Pomo (Southwestern):** ŌtŌne, ôtônô; **(Southern):** ŌtŌ′nŌ; **(Central):** tŌ′nŌ; **(Southeastern):** xa′kīkûta, "water leaf." **Hupa:** la. **Yuki:** lilbal, "rock leaves."

Citation(s): Gifford 1967; Barrett 1952; Goddard 1903; Gifford 1939; Goodrich et al 1980; Miller 2012.

1054 *Quercus* spp.
Family: *Fragaceae*
Common Name(s): Oak.

Usage: Food: **Bear River:** boiled nuts in shells; shelled them, then began the leaching process. **Yuki, Atsugewi, Monache (Western Mono), Salinan, Wintun - Hill (Patwin), - River (Patwin), - Central (Nomlaki), Sinkyone, Wappo, Tolowa:** used acorns. **Chimariko:** used acorns.

Material: **Miwok:** made acorn mush stirrer. **Bear River:** made paddles for canoes. **Shasta:** made house posts. **Yokuts (Southern Valley):** oak balls contained a shreddy substance which is used as tinder in making fires and lighting pipes. **Cahuilla:** bark used to produce non-fading dyes; fallen leaves made excellent and warm mattress bedding

Medicine: **Shasta:** for a bruise, cut, sore, the bark was burned on a rock until the ashes were fine and white; the sore was washed with clean water and the ashes, when cold, were sprinkled on it. **Atsugewi:** galls were boiled and the resulting decoction was drunk to prevent blood poisoning or catching cold during childbirth. **Costanoan:** bark decoction, or insect galls, used for toothache, to tighten loose teeth; water from leaching process used as diarrhea remedy.

Misc: **Yokuts:** acorns were gathered from near Kingston.

Citation(s): Beals 1933; Barrett and Gifford 1933; Dixon 1910; Driver 1936; Drucker 1937; Du Bois 1935; Foster 1944; Garth 1953, Gayton 1948; Gifford 1932; Goldschmidt 1951; Holt 1946; Kroeber 1932; Mason 1912; Nomland 1935; 1938; Bean and Saubel 1972; Rayburn 2012.

Terminology: **Miwok (Central Sierra):** sakwuba, hakine.

Yokuts: pu"utuz (acorns of all species except *Quercus Kelloggii*), pai'ytn.

Citation(s): Barrett and Gifford 1933; Gayton 1948.

1055 *Quercus agrifolia*

Common Name(s): Live Oak, Red Oak, Field Oak, Encina, Pin Oak, Coast Live Oak.

Usage: Food: **Luiseño:** acorns esteemed for food; mush eaten either hot or cold. **Pomo (Southwestern):** used acorns for food. **Cahuilla:** used acorns. **Esselen, Chumash:** acorns, important staple food, required 2-3 leachings to remove tannic acid to make edible.

      Material: **Chumash:** stirring paddles, bows (from new shoots), hoop of hoop-and-pole game, parts of baby cradles, bowls; bark, favored firewood.

Medicine: **Costanoan, Yuki:** infusion used to stop diarrhea. **Chumash:** bark charcoal, mixed with water, drunk for indigestion, bowel trouble; bark, soaked in water, drunk to treat pustules, boils; oak gall juice, applied to any pustule, wound; charred oak galls, used as remedy for hemorrhoids.

Citation(s): Curtis 1924; Gifford 1967; Sparkman 1908; Bean and Saubel 1963; Goodrich et al 1980; Strike 1994; Breschini and Haversat 2004; Timbrook 2007.

Terminology: **Pomo (Southwestern):** yudji kale. **Luiseño:** wiashal. **Chumash (Barbareño, Ynezeño):** ku'w; **(Island):** kuwu; **(Obispeño):** tsuwu'; **(Purisimeño):** 'aku'w; **(Ventureño):** kuw.

Citation(s): Gifford 1967; Sparkman 1908; Goodrich et al 1980; Timbrook 2007.

1056 *Quercus chrysolpsis*

Common Name(s): Valparaiso Oak, Drooping Oak, Live Oak, Golden Cup Oak, Canyon Live Oak, Maul Oak, Pin Oak.

Usage: Food: **Luiseño:** acorns esteemed for food. **Hupa:** used in case of a short crop of *Lithocarpus densifolius*. **Pomo:** little used. **Tubatulabal, Karok, Cahuilla:** used acorns. **Shasta:** 3rd oak in order of preference. **Yuki:** little used due to inaccessibility of the trees. **Maidu (Northern):** one of the favorites. **Kawaiisu (Tehachapi):** acorns eaten, after leaching. **Esselen:** acorns, dried, stored, pounded, leached, made into mush

Material: **Luiseño:** a gambling toy is made from acorn (large) cups.

Citation(s): Barrett 1952; Chesnut 1902; Dixon 1905, 1907; Goddard 1903; Schenck and Gifford 1952; Sparkman 1908; Voegelin 1938; Holt 1946; Bean and Saubel 1963; Zigmond 1981; Breschini and Haversat 2004.

Terminology: **Karok:** xanputip, xanput (acorns). **Pomo (Southwestern):** lik!e'būdū; **(Eastern):** catca'm, "black-bodied live oak;" catca'm, "black-bodied live oak," catsa'k, "black-bodied live oak," kūca'; **(Northern):** kūca'; **(Central):** dūtci'. **Luiseño:** wiat. **Kawaiisu (Tehachapi):** pawi?a(m)bivɨ, wi?abi "acorn."

Citation(s): Schenck and Gifford 1952; Barrett 1952; Sparkman 1908; Zigmond 1981.

1057 *Quercus douglasii*

Common Name(s): Oak, Blue Oak, Winter Oak, Swamp Oak, Post Oak, Blue Oak.

Usage: Food: **Maidu, Southern (Nisenan):** used acorns [this entry by the ethnographer may have been confused with *Quercus lobata*]. **Pomo:** largely used for soup and bread. **Yana:** collected when ripe in the early Fall (generally sometime in September). **Tubatulabal:** collected and used. **Miwok:** used acorns; considered lowest grade; made watery soup. **Yuki:** largely used for soup and bread; eaten "in hard time." **Kawaiisu (Tehachapi):** acorns eaten after being leached.

Material: **Kawaiisu (Tehachapi):** used for firewood, ladles, and bowls. **Chumash:**

burl wood, made bowls.

Medicine: **Maidu:** leaves, chewed to relieve sore throats.

Citation(s): Barrett 1952; Barrett and Gifford 1933; Chesnut 1902; Curtin 1957; Powers 1877; Sapir and Spier 1943; Voegelin 1938; Zigmond 1981; Strike 1994; Timbrook 2007.

Terminology: **Miwok (Plains):** ala′wa, otca′pa; **(Northern Sierra):** ala′wa; **(Central Sierra):** wilisu. **Yuki:** mē′-lē. **Yana:** mauwā′yu'i, ma′uwāyu, "brush acorn." **Pomo (Central):** ūyū′ (larger than) kakū′l. **Kawaiisu (Tehachapi):** ma?ahnidɨ̵ɨ̵. **Chumash (Ynezeño):** tushqun; **(Purisimeño):** mish'kata.

Citation(s): Barrett and Gifford 1933; Curtin 1957; Sapir and Spier 1943; Barrett 1952; Zigmond 1981; Timbrook 2007.

1058 *Quercus dumosa*

Synonym(s): *Quercus dumosa revoluta*

Common Name(s): Curl-Leaf Oak, California Scrub Oak.

Usage: Food: **Luiseño:** acorn little esteemed for food. **Pomo:** rarely used for food. **Tubatulabal:** used. **Cahuilla:** used acorns. **Kawaiisu (Tehachapi):** acorns eaten after leaching. **Chumash:** acorns eaten, but considered poor food.

Material: **Chumash:** wood, made bows, arrow foreshafts, thatching needles.

Medicine: **Luiseño:** the gall nuts are used to doctor sores and wounds; they are said to possess powerful astringent properties.

Misc: **Pomo (Southwestern):** named but not used.

Citation(s): Barrett 1952; Barrows 1900; Chesnut 1902;
Sparkman 1908; Voegelin 1938; Bean and Saubel 1963;
Zigmond 1981; Goodrich et al 1980; Timbrook 2007.
Terminology: **Luiseño:** pawish. **Pomo (Northern):** batsō′m;
(Southwestern): xa′I būdū, ka?ba qʰale. **Kawaiisu
(Tehachapi):** mucitabɨ. **Chumash (Barbareño,
Ynezeño, Ventureño):** mɨs.
Citation(s): Sparkman 1908; Barrett 1952; Zigmond 1981;
Goodrich et al 1980; Timbrook 2007.

1059 *Quercus engelmanni*
Common Name(s): White Oak.

Usage: Food: **Luiseño:** acorns little esteemed for food;
from a deposit mode on this oak by a
scale insect a chewing gum is obtained.

Material: **Luiseño:** a fungus growing in decayed
wood is used for tinder, when fire was
kindled with flint and steel.

Citation(s): Sparkman 1908.
Terminology: **Luiseño:** tovashal.
Citation(s): Sparkman 1908.

1060 *Quercus garryana*
Common Name(s): Pacific Post Oak, Mountain White
Oak, Oregon Oak.

Usage: Food: **Maidu, Southern (Nisenan):** harvested
about middle of October; cooked in a
kind of mush in baskets by means of hot
rocks, or baked as bread in an
underground oven. **Yokuts:** favorite
acorn. **Hupa, Karok:** used in case of a
short crop of *Lithocarpus densiflorus*.
Shasta: considered 2nd in order of

preference. **Yuki:** used for soup only. **Pomo (Southwestern):** least liked of the acorns. **Kawaiisu (Tehachapi):** acorns eaten after leaching.

Medicine: **Karok:** bark on either side of a knot is pounded and rubbed on the abdomen and sides of a young mother before her first baby comes; she also drinks a little of the warm infusion of the bark in water.

Citation(s): Powers 1877; Chesnut 1902; Dixon 1907; Foster 1944; Gifford 1967; Goddard 1903; Schenck and Gifford 1952; Holt 1946; Zigmond 1981; Goodrich et al 1980.

Terminology: **Pomo (Southeastern):** tsafa′ būdū; **(Southwestern):** tsapa′ būdū, kaba′ kale, wiyi kale "white oak;" **(Eastern):** tsipa′ būdū; **(Central):** ūcī′ būdū. **Karok:** axaweiip, axawham (acorns). **Kawaiisu (Tehachapi):** tukʷavɨ.

Citation(s): Barrett 1952; Gifford 1967; Schenck and Gifford 1952; Zigmond 1981.

Note(s): erroneously identified as *Quercus gambelii* by Powers (1877) which closely resembles *Quercus Garryana* but does not grow in California but does in the neighboring states just to the east [author].

1061 *Quercus kelloggii*

Synonym(s): *Quercus californica, Quercus sonomensis*
Common Name(s): Oak, Black Oak, Kellog's Oak, California Black Oak.

Usage: Food: **Maidu, Southern (Nisenan):** eaten only if no others available; **(Northern):** one of the favorite acorns. **Luiseño:** the acorns of this tree are esteemed over those of other species; mush eaten either hot or cold. **Pomo:** considered 2nd best kind for bread and soup. **Tubatulabal, Esselen:** considered the best. **Shasta:** preferred of all varieties available to them. **Yuki:** considered 2nd best kind for bread and soup because they are especially rich in oil. **Miwok:** considered the best and the most nutritious. **Wintun:** used acorns. **Cahuilla:** used acorns; outstanding flavor, most gelatin-like consistency when cooked, good mush. **Achomawi:** favorite of the acorns. **Owens Valley Paiute:** used acorns but considered a minor food. **Karok:** used acorns but considered as less important then either *Quercus Garryana* or *Lithocarpus densifolius.* **Pomo (Southwestern):** used as food. **Yokuts:** most prized variety. **Kawaiisu (Tehachapi):** acorns eaten after leaching.

Material: **Pomo (Southwestern):** round fleshy insect galls are made into dark hair dye.

Citation(s): Foster 1944; Barrett and Gifford 1932; Voegelin 1938; Gifford 1939, 1967; Holt 1946; Dixon 1905, 1907; Bean and Saubel 1963, 1972; Zigmond 1981; Goodrich et al 1980; Breschini and Haversat 2004.

Terminology: **Luiseño:** kwila. **Yokuts:** ö$^{\prime\prime}$iün. **Miwok (Plains, Northern Sierra):** telē$^{\prime}$lī; **(Central Sierra):** tele$^{\prime}$lī. **Yuki:** nun, mom (acorn). **Karok:** xansipi, xansiip (acorn), houm (what acorns are made into). **Kawaiisu (Tehachapi):** k^{w}iiyavi (tree), k^{w}iiyava (acorn). **Pomo (Southwestern):** yuhši q^{h}ale. **Cahuilla:** qwinyily, wiwish (acorn mush).

Citation(s): Sparkman 1908; Gayton 1948; Barrett and Gifford 1933; Curtin 1957; Schenck and Gifford 1952; Zigmond

1981; Goodrich et al 1980; Bean and Saubel 1972.

Note(s): acorns average 2,265 calories per pound; wheat 1,497 calories.

Hulled acorns contain:

water	9.0%
protein	4.6%
fats	18.0%
fiber	11.4%
carbohydrates	55.5%
ash	1.6%

Citation(s): Bean and Saubel 1972.

1062 *Quercus lobata*

Common Name(s): White Oak.

Usage: Food: **Cahuilla:** eaten. **Pomo:** the main supply of acorns; eaten as soup or mush or bread; **(Southwestern):** not eaten. **Kawaiisu (Tehachapi):** acorns eaten after leaching. **Chumash:** acorns large and not very bitter, various comments as to being desirable or not. **Esselen:** important food source.

Medicine: **Costanoan, Yuki:** infusion used to stop diarrhea.

Citation(s): Barrows 1900; Barrett 1952; Gifford 1967; Zigmond 1981; Goodrich et al 1980; Strike 1994; Timbrook 2007.

Terminology: **Miwok (Plains):** sī′wek; **(Northern Sierra, Central Sierra):** mo′lla, lē′ka. **Yuki:** ku-yum-ōl, ki-yäm (acorn). **Pomo (Eastern):** batsŌ′m; **(Central, Eastern):** kakū′l; **(Southeastern):** kaba′ndū, mata′ būdū (shelled acorns); **(Southwestern):** s′apaʰa qʰale. **Kawaiisu (Tehachapi):** šiviidɨbɨ. **Chumash (Barbareño, Island, Ynezeño, Obispeño):** ta′; **(Ventureño):** ta.

Citation(s): Barrett and Gifford 1933; Curtin 1957; Barrett 1952;

Zigmond 1981; Goodrich et al 1980; Breschini and
Haversat 2004; Timbrook 2007.

1063 *Quercus palmeri*

Synonym(s): *Quercus dunnii*

Usage: Food: **Owens Valley Paiute:** used acorns, but
considered a minor food.
Citation(s): Steward 1933.

1064 *Quercus sadleriana*

Common Name(s): Oak, Deer Oak.

Terminology: **Karok:** yawish (the tree and the acorn).
Citation(s): Schenck and Gifford 1952.

1065 *Quercus wislizenia*

Common Name(s): Interior Live Oak, Large Shrub
Oak.

Usage: Food: **Maidu, Northern:** acorns eaten.
Kawaiisu (Tehachapi): acorns eaten
after leaching.
Citation(s): Dixon 1905; Zigmond 1981.
Terminology: **Miwok (Plains, Northern Sierra): s̲a-s̲a; (Central
Sierra):** sako'sa, sakasa, sasa. **Luiseño:** I'mushla.
Kawaiisu (Tehachapi): saasivɨ.
Citation(s): Barrett and Gifford 1933; Sparkman 1908; Zigmond
1981.

1066 *Ramalina menziersii*

Family: *Ramalinaceae*
Common Name(s): Lichen, Spanish Moss.

Usage: Material: **Pomo (Southwestern):** used for baby

diapers and other sanitary purposes.

Medicine: **Kawaiisu (Tehachapi):** used in weather manipulation to cause rain.

Citation(s): Zigmond 1981; Goodrich et al 1980; Strike 1994.

Terminology: **Kawaiisu (Tehachapi):** paazimo?ova. **Pomo (Southwestern):** qʰale qóci.

Citation(s): Zigmond 1981; Goodrich et al 1980.

1067 *Ranunculus* spp.

Family: *Ranunculaceae*

Usage: Medicine: **Costanoan:** decoction, used to wash wounds.

Citation(s): Rayburn 2012.

1068 *Ranunculus californicus*

Common Name(s): Yellow-Blossom, Crow-Foot, California Buttercup.

Usage: Food: **Maidu, Southern (Nisenan):** the seed is gathered by sweeping through the plant with a long-handled basket or gourd; the seed is parched a little then beaten into flour; eaten without further cooking, or, made into bread or mush. **Miwok:** seeds gathered in June; parched and pulverized.

Misc: **Pomo (Southwestern):** child puts flowers under chin; if reflected yellow, child will like butter.

Citation(s): Barrett and Gifford 1933; Powers 1877; Goodrich et al 1981.

Terminology: **Miwok (Central Sierra):** takalu. **Pomo (Southwestern):** ga?baia.

Citation(s): Barrett and Gifford 1933; Goodrich et al 1980.

1069 *Ranunculus cymbalaria*

Synonym(s): *Ranunculus cymbalaria* var. *saximontanus*
Common Name(s): Buttercup.

Usage: Medicine: **Kawaiisu (Tehachapi):** leaves ands flowers mashed into salve for sores and cuts; "it burns like pepper."
Citation(s): Zigmond 1981.

1070 *Ranunculus lobbii*

Synonym(s): *Ranunculus aquatilis*

Usage: Misc: **Tubatulabal:** considered weeds.
Citation(s): Voegelin 1938.

1071 *Ranunculus occidentalis* var. *occidentalis*

Synonym(s): *Ranunculus eisenii, Ranunculus occidentalis, Ranunculus occidentalis* var. *eisenii, Ranunculus occidentalis* var. *rattani*
Common Name(s): Summer Salmon's Eye.

Usage: Food: **Yuki:** gathered seeds in May for pinole; the acrid nature of the seeds is entirely destroyed in the parching of the seed. **Pomo:** used seeds for pinole.

Misc: **Shasta:** when this bloomed it was time to fish for summer salmon. **Karok:** named nut not used(?).

Citation(s): Chesnut 1902; Barrett 1952; Holt 1946; Schenck and Gifford 1952.
Terminology: **Shasta:** gitar itu'wi. **Karok:** mutmuut. **Pomo (Central):** mtca, tala'.
Citation(s): Holt 1946; Schenck and Gifford 1952; Barrett 1952.

1072 *Ranunculus uncinatus*

Common Name(s): Bongard's Buttercup.

Usage: Medicine: **Maidu:** used to soothe stiff, sore muscles.

Citation(s): Strike 1994.

1073 *Rhamnus crocea*

Common Name(s): Red-Berry.

Usage: Misc: **Tubatulabal:** rejected seeds as food because this is the "berries fox eats." **Pomo (Southwestern):** named but not used.

Citation(s): Gifford 1967; Voegelin 1938.

Terminology: **Pomo (Southwestern):** baxkum.

Citation(s): Gifford 1967.

1074 *Rhamnus ilicifolia*

Synonym(s): *Rhamnus crocea* ssp. *ilicifolia*

Common Name(s): Holly-Leaf Berry, Holly-Leaf Buckthorn.

Usage: Food: **Cahuilla:** berry eaten available August to October.

Medicine: **Pomo:** considered bark as "good medicine." **Kawaiisu (Tehachapi):** root and bark brewed, decoction drunk hot for relief of coughs, colds, internal soreness; root boiled for gonorrhea remedy.

Citation(s): Chesnut 1902; Zigmond 1981; Bean and Saubel 1972.

1075 *Rhododendron columbianum*

Family: *Ericaceae*
Synonym(s): *Ledum glandulosum*
Common Name(s): Labrador Tea.

Usage: Food: **Pomo (Southwestern):** beverage made from the leaves. **Maidu:** drank tea.

Material: **Maidu:** used externally to eliminate fleas; leaves smoked.

Citation(s): Goodrich et al 1980; Strike 1994.
Terminology: **Pomo (Southwestern):** naya mihše· qʰale.
Citation(s): Goodrich et al 1980.

1076 *Rhododendron macrophyllum*

Synonym(s): *Rhododendron californicum*
Common Name(s): California Rose Bay.

Usage: Material: **Pomo (Southwestern):** flowers used for dance wreaths for the Strawberry Festival.

Misc: **Karok:** used by the men as part of the luck-getting ceremony of the sweathouse. **Coast Yuki:** named but not used.

Citation(s): Gifford 1939; Schenck and Giofford 1952; Goodrich et al 1980.
Terminology: **Pomo (Southwestern):** michakawani, "digging stick" (kawani), bića?. **Yuki:** ulkelet. **Karok:** mahaxyamshurip, "mountain axyamshruip."
Citation(s): Gifford 1939, 1967; Schenck and Gifford 1952; Goodrich et al 1980.

1077 *Rhododendron occidentale*

Synonym(s): *Azalea occidentalis*
Common Name(s): Western Azalea.

Usage: Material: **Pomo (Southwestern):** flowers used for dance wreaths for the Strawberry Festival.

Medicine: **Achomawi:** regarded as a remedy for poisoning. **Maidu:** leaves used as a diuretic, as a narcotic.

Misc: **Karok:** named but not used; known for its pretty flowers.

Citation(s): Gifford 1967; Schenck and Gifford 1952; Goodrich et al 1980; Strike 1994.

Terminology: **Pomo (Southwestern):** bicha kale. **Karok:** kyamsurip.

Citation(s): Gifford 1967; Schenck and Gifford 1952; Goodrich et al 1980.

1078 *Rhus aromatica*

Family: *Anancardiaceae*
Synonym(s): *Rhus aromatica trilobata, Rhus trilobata*
Common Name(s): Sumach, Squaw Bush, Indian Lemonade, Aromatic Sumac, Skunk Berries, Basket-Weed.

Usage: Food: **Luiseño:** the berries are dried, crushed, and mixed with water. **Shasta:** the berries are eaten fresh or dried. **Cahuilla:** small red berry used fresh, or soaked in water to make a beverage, or dried, ground into flour used in soup. **Atsugewi:** berries gathered in midsummer; washed, dried, and stored; pounded into flour in a mortar basket, mixed with Manzanita (*Arctostaphylos* spp.) flour and water and drunk. **Wintun:** soaked berries in water to remove acidity then dried and pounded into flour which was mixed with cold water and eaten.

Material: **Achomawi:** used to enlarge the hole in the ear lobe of young girls about one month after the initial piercing. **Luiseño:** from the shrub, splints are obtained for use to wrap the coil in baskets; a seed-fan for beating the seeds from plants is made from the twigs; used as warp and woof in baskets. **Cahuilla, Diegueño, Panamint Shoshone (Koso):** used the stem as warp, woof, and wrapping in baskets. **Cahuilla:** sometimes dyed materials black by soaking for a week in solution made from stems of *Sambucus careulea.* **Panamint Shoshone (Koso):** used year-old shoots in basketry. **Chumash:** used stem as foundation and wrapping in baskets; used peeled shoots for sewing strands (whitish) in basketry, used in seedbeater baskets. **Wintun:** used a warp in storage baskets. **Yokuts:** used as warp and weft in baskets. **Kawaiisu (Tehachapi):** stems used in basketry.

Medicine: **Pomo:** berries dried and powder applied to smallpox spots as a cure. **Cahuilla:** used as restorative for inactive stomachs.

Citation(s): Barrows 1900; Chesnut 1902; Coville 1892; Curtis 1924; Dawson and Deetz 1965; Dixon 1907; Garth 1953; Merriam 1955, 1967; Merrill 1923; Sparkman 1908; Zigmond 1981; Bean and Saubel 1972; Timbrook 2007.

Terminology: **Luiseño:** shoval. **Miwok (Central Sierra):** ta'ma. **Atsugewi:** kópcĭr. **Cahuilla:** selet. **Chumash (Barbareño, Ynezeño):** shu'nay; **(Ventureño):** shuna'y.

Citation(s): Sparkman 1908; Barrett and Gifford 1933; Garth 1953; Bean and Saubel 1972; Timbrook 2007.

Note(s): berries contain:

Ash	1.8%
Protein	15.4%
Oil	13.4%

Citation(s): Earle and Jones 1962.

1079 *Rhus integrifolia*

Common Name(s): Lemonberry.

Usage: Food: **Cahuilla:** berries, slightly acidic, gathered in June to September, soaked in water to make beverage. **Chumash:** berries may have been eaten.

Citation(s): Bean and Saubel 1972; Timbrook 2007.

1080 *Rhus ovata*

Common Name(s): Sugar Bush.

Usage: Food: **Cahuilla:** flower clusters cooked in water and eaten; berries eaten fresh, or dried, ground into flour for mush; sugar exuded on the fruit used as sweetener. **Chumash:** berries may have been eaten.

Medicine: **Cahuilla, Diegueño:** tea of the leaves drunk for coughs, colds, pain in the chest.

Citation(s): Barrows 1900; Bean and Saubel 1972; Strike 1994; Timbrook 2007.

Terminology: **Cahuilla:** nakwet. **Chumash (Barbareño,**

Ynezeño): walqaqsh; **(Ventureño):** shtoyho'os.
Citation(s): Bean and Saubel 1972; Timbrook 2007.

1081 *Ribes* spp.

Family: *Grossulariaceae*
Common Name(s): Gooseberry, Currant.

Usage: Food: **Hupa:** the berries are eaten during the Spring and Summer months. **Shasta:** the berries are eaten either fresh or dried. **Atsugewi:** berries gathered and eaten when fresh. **Yokuts (North Hill):** picked when brownish but not quite black, eaten at once, never dried. **Luiseño:** eaten when found. **Salinan:** berries eaten. **Northern Paiute (Paviotso):** berries eaten fresh and uncooked. **Cahuilla:** berries eaten. **Chumash:** fruit eaten.
Citation(s): Dixon 1907; Garth 1953; Gayton 1948; Goddard 1903; Holt 1946; Kelly 1932; Mason 1912; Sparkman 1908; Bean and Saubel 1972; Timbrook 2007.
Terminology: **Chumash (Barbareño):** stɨmɨy, sqa'yi'nu; **(Ynezeño):** stɨmɨy, sqayi'nu; **(Ventureño):** chtɨmɨy.
Citation(s): Timbrook 2007.

1082 *Ribes aureum*

Common Name(s): Golden Currant.

Usage: Food: **Owens Valley Paiute:** eaten fresh or boiled in pots; usually cooked.
Citation(s): Steward 1933.
Terminology: **Owens Valley Paiute:** paɲwavū'hia.
Citation(s): Steward 1933.

1083 *Ribes californicum*

> Synonym(s): *Ribes occidentale*
> Common Name(s): Gooseberry.

Usage: Food: **Maidu (Northern):** berries eaten. **Yuki:** used berries for food, after singeing off the prickles in a basket with hot coals; children eat the fruit directly from the bush. **Kawaiisu (Tehachapi):** berries, ripe in June-July, eaten fresh, or dried and stored. **Pomo (Southwestern):** berries singed, eaten whole.

Citation(s): Chesnut 1902; Dixon 1905; Zigmond 1981; Goodrich et al 1980.

Terminology: **Kawaiisu (Tehachapi):** kɨzarabɨ. **Pomo (Southwestern):** butaga′ ʔlum?, "bear gooseberry."

Citation(s): Zigmond 1981; Goodrich et al 1980.

1084 *Ribes cereum*

Usage: Food: **Northern Paiute (Paviotso):** the berry was eaten fresh and uncooked.

Citation(s): Kelly 1932.

Terminology: **Northern Paiute (Paviotso):** atsa′pui.

Citation(s): Kelly 1932.

1085 *Ribes divaricatum*

> Common Name(s): Straggly Gooseberry.

Usage: Food: **Karok:** the berries are eaten raw, not preserved. **Yuki:** considered very delicious.

Medicine: **Maidu:** used to treat sore throat, a wash used to bathe sore eyes.

Citation(s): Chesnut 1902; Schenck and Gifford 1952; Strike

1994.
Terminology: **Karok:** yufĭvkunish, "crooked nose."
Citation(s): Schenck and Gifford 1952.

1086 *Ribes indecorum*

Usage: Medicine: **Luiseño:** the root is used to cure toothache. **Diegueño:** leaves, dried, used to treat toothache.
Citation(s): Sparkman 1908; Strike 1994.
Terminology: **Luiseño:** kawa'wal.
Citation(s): Sparkman 1908.

1087 *Ribes menziesii* var. *menziesii*
Synonym(s): *Ribes menziesii* var. *leptosum*

Usage: Misc: **Pomo (Southwestern):** named but not used.
Citation(s): Gifford 1967.
Terminology: **Pomo (Southwestern):** butaaka ilum, "bear gooseberry."
Citation(s): Gifford 1967.

1088 *Ribes nevadense*
Common Name(s): Wild Sierra Currant.

Usage: Food: **Miwok:** berries eaten raw.
Citation(s): Barrett and Gifford 1933.
Terminology: **Miwok (Central Sierra):** hemekine.
Citation(s): Barrett and Gifford 1933.

1089 *Ribes quercetorum*
Common Name(s): Gooseberry.

Usage: Food: **Tubatulabal:** used berries. **Kawaiisu**

(**Tehachapi**): berries, ripe in June-July, eaten fresh, or dried and stored.
Citation(s): Voegelin 1938; Zigmond 1981.
Terminology: **Kawaiisu (Tehachapi):** pahu?upivɨ.
Citation(s): Zigmond 1981.

1090 *Ribes roezlii*
Common Name(s): Gooseberry.

Usage: Food: **Atsugewi:** berries gathered and eaten when fresh. **Kawaiisu (Tehachapi):** berries, ripe in June-July, eaten fresh, or dried and stored.
Citation(s): Garth 1953; Zigmond 1981.
Terminology: **Miwok (Central Sierra):** kilili. **Atsugewi:** hEstĭkida. **Kawaiisu (Tehachapi):** kɨzarabɨ.
Citation(s): Barrett and Gifford 1933; Garth 1953; Zigmond 1981.

1091 *Ribes roezlii* var. *cruentum*
Common Name(s): Gooseberry.

Usage: Food: **Karok:** the berries eaten raw, not preserved. **Miwok:** the berries are eaten.
Citation(s): Barrett and Gifford 1933; Schenck and Gifford 1952.
Terminology: **Karok:** axrattip.
Citation(s): Schenck and Gifford 1952.

1092 *Ribes sanguineum*
Synonym(s): *Ribes sanguineum variegatum*
Common Name(s): Currant, Black Current.

Usage: Food: **Maidu (Northern):** berries used.
Citation(s): Dixon 1905.

1093 Ribes speciosum

Common Name(s): Fuchsia-Flowered Gooseberry.

Usage: Food: **Chumash:** berries eaten.
Citation(s): Timbrook 2007.
Terminology: **Chumash (Barbareño):** stɨmɨy 'iwɨ; **(Ynezeño):** tsiqun.
Citation(s): Timbrook 2007.

1094 Ribes velutinum

Synonym(s): *Ribes velutinum* var. *glanduliferum*
Common Name(s): Plateau Gooseberry.

Usage: Food: **Kawaiisu (Tehachapi):** berries, ripe in June-July, eaten fresh, or dried and stored.
Citation(s): Zigmond 1981.
Terminology: **Kawaiisu (Tehachapi):** kaayhi?apivɨ.
Citation(s): Zigmond 1981.

1095 Ricinus communis

Family: *Euphorbiaceae*
Common Name(s): Caster-Bean.

Usage: Medicine: **Cahuilla:** seed, when ripe, crushed into a greasy substance, said to have value in relieving sores.
Citation(s): Bean and Saubel 1972.
Terminology: **Cahuilla:** navish.
Note(s): seeds, if chewed, are toxic; leaves less so. Ricin, a toxin, causes burning of the mouth and throat, nausea, vomiting, sever stomach pains, diarrhea, excessive thirst, prostration, dullness of vision, convulsion, uremia, and death. Plant native of Asia and Africa.
Citation(s): Harden and Arena 1974; Munz and Keck 1965.

1096 *Romneya coulteri*

Family: *Papaveraceae*
Common Name(s): Matilija Poppy.

Usage: Food: **Cahuilla:** watery substance from stalk drunk.
Citation(s): Bean and Saubel 1972.
Terminology: **Cahuilla:** kulux'a.
Citation(s): Bean and Saubel 1972.

1097 *Romneya trichocalyx*

Common Name(s): Matilija Poppy.

Usage: Food: **Cahuilla:** watery substance from stalk drunk.
Citation(s): Strike 1994.

1098 *Rorippa* spp.

Family: *Brassicaceae*

Usage: Medicine: **Panamint Shoshone (Koso):** women ate while experiencing labor pains.
Costanoan: used to reduce fever.
Citation(s): Strike 1994.

1099 *Rorippa curvisiliqua*

Synonym(s): *Radicula curvisiliqua*
Common Name(s): Western Yellow Cress.
Usage: Food: **Mono Lake Paiute, Owens Valley Paiute:** small seeds eaten.
Citation(s): Steward 1933.
Terminology: **Mono Lake Paiute, Owens Valley Paiute:** ā'tsā.
Citation(s): Steward 1933.

1100 *Nasturtium officinale*

Synonym(s): *Rorippa nasturtium-aquaticum*
Common Name(s): Water Cress.

Usage: Food: **Luiseño:** used as greens; cooked, not eaten fresh. **Kawaiisu (Tehachapi):** leaves eaten raw, usually with salt; boiled and fried in grease and salt. **Cahuilla:** eaten fresh in Spring, or cooked like spinach and flavored before eating. **Chumash:** eaten raw, or boiled as cooked greens.

Medicine: **Cahuilla:** effective in treating low blood pressure; eaten each morning with salt would cure liver ailments with two months; patient expected to eat no other food before noon.

Citation(s): Sparkman 1908; Zigmond 1981; Bean and Saubel 1972; Timbrook 2007.

Terminology: **Kawaiisu (Tehachapi):** po?opaatooribɨ, "in the water." **Cahuilla:** pangasamat. **Chumash (Barbareño):** welu; **(Ventureño):** spe'ey he'so'o.

Citation(s): Zigmond 1981; Bean and Saubel 1972; Timbrook 2007.

Note(s): naturalized from Europe.

Citation(s): Munz and Keck 1965.

1101 *Rosa californica*

Family: *Rosaceae*
Common Name(s): Wild Rose, Rose.

Usage: Food: **Yuki:** rarely used for food. **Cahuilla:** capsules picked and eaten to a small extent; blossoms soaked in water to make beverage. **Kawaiisu (Tehachapi):**

fruit picked and eaten when ripe. **Pomo (Southwestern):** fresh fruit eaten; tastes sweetest after first light frost, or cold night of Fall. **Chumash:** fruit eaten raw. **Modoc (Lutuami):** hips (fruit) eaten.

Material: **Kawaiisu (Tehachapi):** stems used as rim in twined basketry.

Medicine: **Miwok:** leaves and berries steeped and drunk as medicine for pains, colic, etc. **Cahuilla:** tea made from blossoms used to relieve "clogged stomachs." **Costanoan:** rose "hips" decocted, used as a wash for scabs, sores, to treat indigestion, fever, sore throat, rheumatism, kidney ailments. **Chumash:** petals, dried, powdered, used like talcum to relieve chafing, skin rash in babies; petal, tea, drunk for stomach ache, colic, wash for sore eyes, teething, blockage in the stomach (for adults).

Citation(s): Barrows 1900; Chesnut 1902; Barrett and Gifford 1933; Zigmond 1981; Goodrich et al 1980; Bean and Saubel 1972; Strike 1994; Rayburn 2012; Timbrook 2007.

Terminology: **Miwok (Central Sierra):** mamute. **Kawaiisu (Tehachapi):** Čiyavipɨ, Čiyavipɨ?ivi "the fruit." **Pomo (Southwestern):** badú?den?. **Cahuilla:** ushal. **Chumash (Barbareño, Ynezeño):** washtiqʼoliqʼol; **(Ventureño):** watiqʼoniqʼon.

Citation(s): Barrett and Gifford 1933; Zigmond 1981; Goodrich et al 1980; Bean and Saubel 1972; Timbrook 2007.

1102 *Rosa gymnocarpa*

Common Name(s): Wood Rose, Wild Rose.

Usage: Food: **Pomo (Southwestern):** fresh fruit eaten, tastes sweetest after first light frost, or cold night of Fall.

Medicine: **Maidu:** stem decoction used as a tonic, as a treatment for indigestion; bark decoction used to bathe sore eyes.

Misc: **Pomo (Southwestern), Yuki:** named but used.

Citation(s): Curtin 1957; Gifford 1967; Goodrich et al 1980.

Terminology: **Yuki:** ka-li. **Pomo (Southwestern):** baduden.

Citation(s): Curtin 1957, Gifford 1967.

Note(s): note the difference of opinion for **Pomo (Southwestern)** which may be a temporal difference in the ethnographers visits [author].

1103 *Rosa pisocarpa*

Common Name(s): Wild Rose.

Usage: Food: **Northern Paiute (Paviotso):** the haws were pounded with deer tallow and eaten. **Mono Lake Paiute, Owens Valley Paiute:** only occasionally used. **Maidu (Northern):** rose hips eaten.

Material: **Maidu (Northern):** arrow shafts made of the wood.

Misc: **Karok:** believed that they could not touch the hips or the brush when it is in fruit; if they do they will drown.

Citation(s): Dixon 1905; Kelly 1932; Schenck and Gifford 1952; Steward 1933.

Terminology: **Mono Lake Paiute, Owens Valley Paiute:** tsia′va. **Northern Paiute (Paviotso):** tsia′b′. **Karok:** axanart sinvanahichkams, "drown-make-believe," "something with stickers only."

Citation(s): Steward 1933; Kelly 1932; Schenck and Gifford

1952.

1104 *Rosa spithamaea*
Common Name(s): Ground Rose.

Usage: Food: **Kawaiisu (Tehachapi):** fruit picked and eaten when ripe.

          Material: **Kawaiisu (Tehachapi):** stems used as rims in basketry.

          Misc: **Karok:** a belief regarding danger of drowning is held in regard with this plant and *Rosa pisocarpa*.

Citation(s): Schenck and Gifford 1952; Zigmond 1981.
Terminology: **Karok:** axanart sinvanahich. **Kawaiisu (Tehachapi):** Čiyavipɨ, Čiyavipɨʔivi "the fruit."
Citation(s): Schenck and Gifford 1952; Zigmond 1981.

1105 *Rosa woodsii* var. *ultramontana*
Synonym(s): *Rosa Woodsii*

Usage: Food: **Cahuilla, Chumash, Modoc (Lutuami) (Lutuami):** hip (fruit) eaten raw.

          Medicine: **Washo:** a tea is made from boiled roots or inner bark of the stems; used as a cure for colds.

Citation(s): Train et al 1941; Strike 1994.
Terminology: **Washo:** pet-som-a-lee, pet-su-mah-le.
Citation(s): Train et al 1941.

1106 *Rubus* spp.
Family: *Rosaceae*
Common Name(s): Blackberry, Thimbleberry.

Usage: Food: **Yokuts (North Hill):** picked when ripe and eaten at once; never dried. **Bear**

River: sun dried; packed in deep baskets between layers of leaves for storage. **Shasta:** berries eaten fresh. **Luiseño:** fruit eaten when found. **Huchnom; Salinan; Wintun, River (Patwin), Central (Nomlaki), Chumash:** berries eaten. **Sinkyone:** berries eaten raw. **Cahuilla:** berries eaten fresh, or dried and stored; dried boiled in small quantity of water for beverage.

Citation(s): Driver 1936; Foster 1944; Gayton 1948; Holt 1946; Goldschmidt 1951;Kroeber 1932; Mason 1912; Nomland 1935, 1938; Sparkman 1908; Bean and Saubel 1972; Landberg 1965.

Terminology: **Cahuilla:** pikwlyam.

Citation(s): Bean and Saubel 1972.

1107 *Rubus glaucifolius*
Common Name(s): Thimbleberry.

Usage: Food: **Shasta:** berries eaten either fresh or dried. **Maidu (Northern):** berries eaten.

Citation(s): Dixon 1905, 1907.

1108 *Rubus leucodermis*
Common Name(s): Raspberry, Western Raspberry.

Usage: Food: **Hupa:** the berries are eaten during the Spring and Summer months. **Huchnom:** ate berries. **Karok:** berries eaten raw, but not preserved. **Yuki:** berries eaten fresh or dried; gathered in July. **Coast Yuki:** berries eaten raw. **Pomo (Southwestern):** berries picked, eaten fresh; not stored for winter.

Medicine: **Pomo (Southwestern):** leaves or root boiled, tea drunk for diarrhea, weak bowels, or upset stomach.

Citation(s): Chesnut 1902; Curtin 1957; Foster 1944; Gifford 1967; Goddard 1903; Schenck and Gifford 1952; Goodrich et al 1980.

Terminology: **Pomo (Southwestern):** bashkût (loan word from **Pomo (Central)**). **Yuki:** kā-lā′-mäm, motsako. **Karok:** paturupven.

Citation(s): Gifford 1967; Curtin 1957; Gifford 1939; Schenck and Gifford 1952; Goodrich et al 1980.

1109 *Rubus parviflorus*

Common Name(s): Thimbleberry.

Usage: Food: **Luiseño:** the fruit is eaten. **Hupa:** the berries are eaten during the Spring and Summer months. **Karok:** berries eaten when ripe, but not preserved. **Maidu:** never gathered in quantity but eaten directly from bush by old and young alike. **Pomo (Southwestern), Coast Yuki:** berries eaten raw.

Material: **Pomo (Southwestern):** leaves used to wrap meat for baking.

Medicine: **Karok:** roots soaked in water; the water drunk for an appetizer or as a tonic for a person who is thin.

Citation(s): Chesnut 1902; Gifford 1939, 1967; Goddard 1903; Schenck and Gifford 1952; Sparkman 1908; Goodrich et al 1980.

Terminology: **Karok:** xapuxara. **Hupoa:** wûndauℰ. **Yuki:** hepella. **Luiseño:** paylash. **Pomo (Southwestern):** hemkolo kale, hemkolo (berry), kotolo (leaves). **Wiyot:** kiwaātchokwere.

Citation(s): Schenck and Gifford 1952; Goddard 1903, 1967; Gifford 1939; Sparkman 1908; Loud 1918; Goodrich et al 1980.

1110 *Rubus spectabilis*

Synonym(s): *Rubus menziesii, Rubus spectabilis* var. *franciscanus*

Common Name(s): Salmon Berry.

Usage: Food: **Pomo (Southwestern), Sinkyone:** berries eaten raw. **Bear River:** berries sun dried; stored in deep baskets between layers of leaves. **Wiyot:** used berries.

Citation(s): Gifford 1967; Nomland 1935, 1938; Goodrich et al 1980; Loud 1918.

Terminology: **Pomo (Southwestern):** kôtô kale, kôtô (berry). **Wiyot:** we'taw.

Citation(s): Gifford 1967; Goodrich et al 1980; Loud 1918.

1111 *Rubus ursinus*

Synonym(s): *Rubus macropetalus, Rubus vitifolius*

Common Name(s): California Blackberry, Large-Flowered Blackberry, Wild Blackberry.

Usage: Food: **Luiseño, Miwok, Coast Yuki:** the fruit is eaten. **Hupa:** the berries are eaten during the Spring and Summer months. **Karok:** the berries are eaten in season, but not preserved. **Shasta:** berries eaten either fresh or dried. **Yuki:** eaten directly from vine; occasionally dried for Winter use. **Pomo (Southwestern):** berries eaten fresh; whole or mashed and eaten with bread; mashed as topping for ice cream;

	cooked in pies; used a sauce fro dumplings. **Esselen:** berries, raw, eaten. **Chumash:** berries eaten. **Maidu, Modoc (Lutuami):** fruit eaten.
Material:	**Luiseño:** the juice of the berries is sometimes used to stain articles of wood (black dye).
Medicine:	**Yuki:** infusion of the root used for checking diarrhea. **Pomo (Southwestern):** root boiled; tea used to check diarrhea. **Costanoan:** roots used in treatment of infected sores; root, decoction, considered most effective remedy for diarrhea, dysentery. **Miwok (Coast):** root tea used to relieve menstrual pain. **Chumash:** roots, boiled, taken for diarrhea.
Misc:	**Pomo (Southwestern):** named but not used (?); berries not to be eaten by pregnant women or fathers-to-be else child will come out dark in color.

Citation(s): Barrett and Gifford 1933l; Chesnut 1902; Dixon 1907; Gifford 1939, 1967; Goddard 1903; Merrill 1923; Schenck and Gifford 1952; Sparkman 1908; Goodrich et al 1980; Rayburn 2012; Strike 1994; Breschini and Haversat 2004; Timbrook 2007.

Terminology: **Yuki:** motsumam. **Wiyot:** mī̄p. **Miwok (Central Sierra):** lututuya. **Pomo ((Southwestern):** tibakhai. **Luiseño:** pikwlax. **Karok:** attaichurip. **Chumash (Barbareño):** tɨq'ɨtɨq'; **(Ynezeño):** tɨqɨtɨq'; **(Ventureño):** tɨhɨ.

Citation(s): Gifford 1939, 1967; Loud 1918; Barrett and Gifford 1933; Sparkman 1908; Schenck and Gifford 1952; Timbrook 2007.

Note(s): note the difference of opinion for **Pomo (Southwestern)** which may be a temporal difference in the ethnographers visits [author].

1112 *Rudbeckia* spp.

Family: *Asteraceae*

Usage: Material: **Pomo (Southwestern):** children make
baskets of the blades; used to string
clamshell beads to hold them together
when being smoothed.

Misc: **Miwok:** named but not used.

Citation(s): Barrett and Gifford 1933; Goodrich et al 1980.

Terminology: **Miwok (Central Sierra):** mika. **Pomo
(Southwestern):** ci?ba.

Citation(s): Barrett and Gifford 1933; Goodrich et al 1980.

1113 *Rumex conglomeratus*

Family: *Polygonaceae*
Common Name(s): Green Dock.

Usage: Medicine: **Miwok:** root tea drank to cure boils; root
poultice used in boils.

Citation(s): Strike 1994.

Note(s): all species of Rumex contain oxalic acid. When
ingested it is corrosive to the tissues that it reaches.
Oxalate poisoning is complex and poorly understood.
This species is native to Europe.

Citation(s): Kingsbury 1964; Baldwin et al 2012.

1114 *Rumex crispus*

Common Name(s): Dock, Yellow Dock.

Usage: Food: **Yuki:** leaves used for greens; seeds used
for mush. **Kawaiisu (Tehachapi):** green
stems and seeds eaten. **Chumash:** leaves,
cooked as greens; stem, peeled, eaten

raw; seeds, pounded into dough, rolled
in balls, baked.

Medicine: **Yuki:** seeds, when ripe, made into tea for
dysentery; for sores a decoction of leaves
and seeds brewed and applied. **Owens
Valley Paiute:** roots peeled, eaten raw
for stomach disorders; or made into a
decoction, boiled for a long time.
Kawaiisu (Tehachapi): root mashed
with water; salve applied on sores, cuts,
and sore limbs. **Costanoan:** plant part,
decoction, used for urinary problems.
Chumash: root, boiled, tea used for
stomach trouble.

Misc: **Tubatulabal:** considered weeds.

Citation(s): Chesnut 1902; Curtin 1957; Steward 1933; Voegelin
1938; Zigmond 1981; Rayburn 2012; Timbrook 2007.

Terminology: **Owens Valley Paiute: ā′tsākān**[va]. **Kawaiisu
(Tehachapi):** avaaananɨbɨ. **Chumash (Barbareño):**
tsukhat'; **(Ynezeño):** ts'ukhat'; **(Ventureño):**
'alakhnipk'ɨsh.

Citation(s): Steward 1933; Zigmond 1981; Timbrook 2007.

Note(s): excessive handling of the leaves can cause dermatitis;
certain agents in the plant, rumicin, are slightly soluble
in hot water; the solution acts as a rubefacient and
discutient and can destroy parasites of the skin; if
drunk causes nausea, a dry brown feces, copious
urination, a dry spastic cough, and perspiration; can be
used as a stimulant and diuretic; all three species
(*Rumex acetosella, Rumex conglomeratus, Rumex crispus*)
are naturalized from Europe; native of Eurasia.

seeds contain:

Ash   2.1%
Protein   15.9%
Oil   5.0%

Citation(s): Meunscher 1951; Pammel 1911; Bailey 1950; Munz

and Keck 1965; Chevallier 1996; Earle and Jones 1962; Baldwin et al 2012.

1115 *Rumex hymenosepalus*

Common Name(s): Indian Rhubarb, Canaigre, Wild Rhubard.

Usage:	Food:	**Kawaiisu (Tehachapi):** green stems ands seeds eaten. **Cahuilla:** stalks, crisp and juicy, eaten as greens. **Chumash:** shoots, collected in March, boiled, or roasted, eaten, taste was sour but agreeable; seeds, made into thick mush, *qolowush*.
	Material:	**Cahuilla:** roots used in tanning hides. **Chumash:** root, used to make yellow dye for basketry materials.
	Medicine:	**Kawaiisu (Tehachapi):** root mashed with water; slave applied on sores, cuts, and sore limbs. **Chumash:** root, boiled, tea used for sore throat, dizziness, liver trouble.
	Misc:	**Tubatulabal:** considered weeds.

Citation(s): Voegelin 1938; Zigmond 1981; Bean and Saubel 1972; Timbrook 2007.

Terminology: **Kawaiisu (Tehachapi):** avaaananɨbɨ. **Cahuilla:** maalval. **Chumash (Barbareño):** shaw; **(Ynezeño):** sha'w; **(Ventureño):** 'alakhpɨy.

Citation(s): Zigmond 1981; Bean and Saubel 1972; Timbrook 2007.

Note(s): the dry roots contain as much as 35% tannin.

Citation(s): Munz and Keck 1965.

1116 *Rumex lacustris*

Common Name(s): Dock.

Usage: Food: **Modoc (Lutuami):** seeds eaten.
Citation(s): Strike 1994.

1117 *Rumex occidentalis*

Common Name(s): Western Dock.

Usage: Food: **Modoc (Lutuami):** seeds eaten.
Citation(s): Strike 1994.

1118 *Rumex salicifolius*

Common Name(s): Willow Bark.

Usage: Food: **Kawaiisu (Tehachapi):** green stems ands seeds eaten. **Maidu, Modoc (Lutuami):** seed eaten.

Medicine: **Kawaiisu (Tehachapi):** root mashed with water; slave applied on sores, cuts, and sore limbs.

Citation(s): Zigmond 1981; Strike 1994.
Terminology: paa?avaaananïbï, (paa-"water")
Citation(s): Zigmond 1981.

1119 *Ruta chalapensis*

Family: *Rutaceae*
Common Name(s): Rue.

Usage: Medicine: **Chumash:** drunk as a tea for nerves and heart palpitations; also used for ear problems. **Costanoan:** leaves, heated, placed inside ear for earache; used for stomach pain; used for paralysis; decoction, used for cough.

Citation(s): Timbrook 1987; Rayburn 2102.
Note(s): native of the Mediterranean region.
Citation(s): Munz and Keck 1965.

1120 *Sagittaria latifolia*

Family: *Alismataceae*
Common Name(s): Arrowleaf, Arrowhead.

Usage: Food: **Pomo:** tubers eaten. **Yokuts (Southern Valley):** boiled root and ate it.

Medicine: **Maidu:** root infusion used to clean and treat wounds.

Citation(s): Barrett 1952; Chesnut 1902; Gayton 1948; Strike 1994.
Terminology: **Yokuts:** ta'a˘na.
Citation(s): Gayton 1948.

1121 *Salix* spp.

Family: *Salicaceae*
Common Name(s): Willow.

Usage: Material: **Hupa:** a fiber is obtained from the root, used in basketry. **Hupa, Whilkut, Nongatl, Wailaki, Kato, Yurok, Wiyot, Yuki, Wappo, Pomo, Modoc (Lutuami), Achomawi, Yana, Wintun, Washo, Maidu, Yokuts, Monache (Western Mono), Northern Paiute (Paviotso), Panamint Shoshone (Koso), Chemehuevi, Kawaiisu (Tehachapi), Chumash, Cahuilla, Mohave:** used stems as warp in basketry. **Wailaki, Kato, Yuki, Pomo, Modoc (Lutuami), Achomawi, Yana, Washo, Maidu, Miwok, Yokuts, Monache (Western**

Mono), Northern Paiute (Paviotso), Chemehuevi, Kawaiisu (Tehachapi), Panamint Shoshone (Koso), Chumash, Cahuilla, Mohave: used stems as woof in basketry. Whilkut, Nongatl, Lassik, Wailaki, Wiyot, Yana, Pomo, Northern Paiute (Paviotso): used stems as rim hoop in basketry. Kato, Wappo, Pomo, Washo, Wintun, Maidu, Miwok, Yokuts, Monache (Western Mono), Panamint Shoshone (Koso), Kawaiisu (Tehachapi), Chemehuevi, Mohave: used stems as foundation in basketry. Hupa, Yurok, Wappo, Miwok, Yokuts, Mohave: used root as woof in basketry. Pomo: used root as wrapping in basketry. Wailaki, Modoc (Lutuami): used roots as rim binding in basketry. Pomo, Miwok, Panamint Shoshone (Koso) used roots as red pattern in basketry. Washo, Miwok, Yokuts, Kitanemuk (Tejon), Panamint Shoshone (Koso): used root as brown pattern in basketry. Washo: used bark (dyed) as black pattern in basketry. Yuki, Northern Paiute (Paviotso): used sapwood as rim binding in basketry. Washo, Maidu, Kitanemuk (Tejon), Panamint Shoshone (Koso), Chemehuevi, Kawaiisu (Tehachapi): used sapwood as wrapping in basketry. Wintun, Panamint Shoshone (Koso), Mohave: used sapwood as woof in basketry. Chemehuevi: used sapwood (dyed) as black pattern in basketry.

Chimariko: used some portion of the plant, although the part is not specified. Owens Valley Paiute: used in making bows. Shasta: used for ribs in basketry. Kamia: bark used by the women for skirts. Bear River: made weft in loose baskets from root fiber. Yokuts: used poles as framework for houses; made plain bows from wood. Luiseño: made apron from inner bark. Wintun, River (Patwin): made food bins of interlaced sticks. Miwok (Lake): used poles as frame of houses. Northern Paiute (Paviotso): made bowls for gruel or mush. Kawaiisu (Tehachapi): used in house construction. Esselen: used for construction, basketry. Chumash: poles used in house framework, sleeping platforms, storage platforms, ceremonial enclosures, ladder for sweathouse, seed beaters, thatching needles, firewood, dugout canoes.

Medicine: Costanoan: leaf, paste, rubbed into hair for treatment of falling hair; leaf, infusion, used as hair rinse.

Citation(s): Barrett 1917; Dixon 1907; Gayton 1948; Goddard 1903; Kelly 1932; Kroeber 1932; Merriam 1955, 1967; Merrill 1923; Nomland 1938; O'Neale 1932; Powers 1877; Sparkman 1908; Zigmond 1981; Rayburn 2012; Breschini and Haversat 2004; Timbrook 2007.

Terminology: Owens Valley Paiute: sagap. Mono Lake Paiute: sugupi. Miwok (Central Sierra): sisima. Yurok: paxkwo (stick). Karok: paruk, koovip (root).

Citation(s): Steward 1933; Barrett and Gifford 1933; O'Neale 1932.

1122 *Salix exigua*

Usage: Food: **Tubatulabal:** used to wrap fish which are roasted; used to wrap tobacco when it was to be cured, as this plant did not impart any odor to food.

Medicine: **Diegueño:** bark used to cure skin erutions, running sores; stems boiled, vapors inhaled to cure headache.

Citration(s): Voegelin 1938; Strike 1994.

1123 *Salix exigua* var. *hindsiana*

Synonym(s): *Salix fluviatilis argyrophylla, Salix sessilifolia, Salix sessilifolia hindsiana, Salix hindsiana, Salix argophylla, Salix hindsiana* var. *leucodendroides*

Common Name(s): Willow, Sandbar Willow.

Usage: Food: **Tubatulabal:** used to wrap fish which are roasted; used to wrap tobacco when it was to be cured, as this plant did not impart any odor to food.

Material: **Pomo:** used stem as warp; used sapwood and bark in baskets; straight wands made into arrows; the larger limbs are frequently used to make fish weirs. **Pomo (Southwestern):** used root in twined baskets; large branches used in framework for thatched summer homes. **Owens Valley Paiute:** used as foundation or warp and as weft in basketry; cut only in Winter. **Hupa:** the young shoots (stems) were used as warp and foundation in baskets. **Maidu (Northern):** most esteemed species of willow used in basketry construction. **Karok:** twigs gathered in April and August, used to make warp sticks for twined baskets; the roots are also basket material, gathered in the Winter after the river had receded and exposed the roots. **Kawaiisu (Tehachapi):** used in house construction; making cradles.

Medicine: **Pomo (Southwestern):** bark or leaves boiled; tea used for treating sore throat, laryngitis. **Salinan:** used to treat fever.

Citation(s): Dixon 1905; Goddard 1903; Steward 1933; Voegelin 1938; Zigmond 1981; Goodrich et al 1980; Strike 1994; Chesnut 1902; Merrill 1923.

Terminology: **Hupa:** kitdilmai "grey," tōxatawe "it grows by the water." **Owens Valley Paiute:** sūhū′va. **Karok:** pa'arak, ischa'sip (root fibers). **Kawaiisu (Tehachapi):** kahnav ɨ. **Pomo (Southwestern):** cuia·.

Citation(s): Goddard 1903; Steward 1933; Schenck and Gifford 1952; Zigmond 1981; Goodrich et al 1980.

1124 *Salix geyeriana*

Common Name(s): Geyer Willow.

Usage: Material: **Modoc (Lutuami):** use in basketry.
Citation(s): Strike 1994.

1125 *Salix gooddingii*

Synonym(s): *Salix nigra*
Common Name(s): Black Willow.

Usage: Material: **Panamint Shoshone (Koso):** used stem as woof in baskets. **Pomo:** used inner bark in baskets. **Cahuilla:** bows made from wood, cradle boards; used as warp in basketry; small branches used in weaving large storage and carrying baskets.

Misc: **Kawaiisu (Tehachapi):** named but not used.

Citation(s): Barrows 1900; Merrill 1923; Zigmond 1981, Bean and Saubel 1972.

Terminology: **Kawaiisu (Tehachapi):** kaawiidaavɨ?avɨ.

Cahuilla: avasily.

Citation(s): Zigmond 1981, Bean and Saubel 1972.

1126 *Salix laevigata*

Common Name(s): Red Willow.

Usage: Material: **Tubatulabal:** poles about 2 inches in diameter used for house frame; small trunks used as firewood; peeled shoots used as foundation in large baskets. **Karok:** used root fiber to make baskets; twigs for warp sticks. **Kawaiisu (Tehachapi):** stems, when young and green, satisfactory for basketry. **Chumash:** bark, fiber, basketry; two-ply cordage, belts, sandals, Fox Dance headdress tail, bags, nets, hunting slings, tumplines, tarring brushes, arrow quivers, saddle pads; wood, mush-stirring sticks, spoons, digging sticks, game ball for the hoop-and-pole game.

Medicine: **Kawaiisu (Tehachapi):** roots boiled, drunk for diarrhea. **Costanoan:** root infusion drank to stop diarrhea; bark, made fever remedy.

Misc: **Karok:** used as a protective charm by those ferrying turbulent waters.

Citation(s): Schenck and Gifford 1952; Voegelin 1938; Zigmond 1981; Strike 1994; Rayburn 2012; Timbrook 2007.
Terminology: **Karok:** kufǐp furak "red kufǐp." **Kawaiisu (Tehachapi):** wo?ovɨ. **Chumash (Barbareño, Island, Ynezeño, Ventureño):** wak.
Citation(s): Schenck and Gifford m1952; Zigmond 1981; Timbrook 2007.

1127 *Salix laevigata*

Synonym(s): *Salix araquipa, Salix laevigata* var. *araquipa*

Usage: Material: **Tubatulabal:** peeled shoots used as foundation in large baskets.
Citation(s): Voegelin 1938.

1128 *Salix lasiandra*

Common Name(s): Yellow Willow.

Usage: Material: **Panamint Shoshone (Koso):** used stems of year-old shoots in basketry.

Material: **Pomo (Southwestern):** branches used as warp in basketry and foundation in coiled baskets.

Medicine: **Pomo (Southwestern):** leaves boiled; tea used for treating colds and sore throats.
Citation(s): Coville 1892; Merrill 1923; Goodrich et al 1980.
Terminology: **Pomo (Southwestern):** cuia·.
Citation(s): Goodrich et al 1980.

1129 *Salix lasiolepsis*

Common Name(s): Mono Willow, Black Willow, Arroyo Willow.

Usage: Food: **Pomo:** portion of the inner bark used as a substitute for chewing tobacco.

Material: **Tubatulabal:** peeled shoots used as foundation in large baskets. **Pomo:** fibrous inner bark collected in the Spring and manufactured into rope and into a garment. **Yuki:** used to make basketry. **Kawaiisu (Tehachapi):** used extensively in basketry. **Chumash:** leaves, twigs, plug mouths of basketry water bottles; wood, fishing poles, switches or whips, firewood in sweat-house.

Medicine: **Pomo:** a strong decoction is used externally as a wash for the itch; internally as a tea to cure chills and fever, and a in large quantities to cause sweating in almost any disease; an infusion of leaves used to check diarrhea. **Costanoan:** bark, or young leaves, steeped in water, used for colds; flowers, decocted, used for cold remedy. **Miwok:** bark tea used to treat fever.

Citation(s): Chesnut 1902; Curtin 1957; Voegelin 1938; Zigmond 1981; Strike 1994; Rayburn 2012; Timbrook 2007.

Terminology: **Yuki:** shipa-nū. **Kawaiisu (Tehachapi):** puhisɨɨvibɨ. **Chumash (Barbareño, Ynezeño, Purisimeño):** shtayɨt; **(Island, Ventureño):** khaw; **(Obispeño):** tsaʼ.

Citation(s): Curtin 1957; Zigmond 1981; Timbrook 2007.

1130 *Salix lutea*

Usage: Material: **Tubatulabal:** peeled shoots used as foundation in large baskets.

Citation(s): Voegelin 1938.

1131 *Salix scouleriana*
Common Name(s): Nuttall Willow.

Usage: Material: **Karok:** best material for fire drill and fire hearth.
Citation(s): Schenck and Gifford 1952.
Terminology: **Karok:** kufǐp inara.
Citation(s): Schenck and Gifford 1952.

1132 *Salix sitchensis*
Synonym(s): *Salix sitchensis coulteri, Salix coulteri*
Common Name(s): Velvet Willow.

Usage: Material: **Karok:** roots used in basket making; twigs are used to string salmon; root used as fire dire and hearth. **Pomo:** used stem as warp and woof in baskets. **Pomo (Southwestern):** used for firewood.
        Misc: **Karok:** a fresh branch tied to the bow of a boat is a charm against danger when crossing the river at high water; a thicket of the plant is beat with a stick to make the wind blow on hot days.
Citation(s): Gifford 1967; Schenck and Gifford 1952; Merrill 1923.
Terminology: **Pomo (Southwestern):** chu'tah. **Karok:** kufip.
Citation(s): Gifford 1967; Schenck and Gifford 1952.

1133 *Salvia apiana*
Family: *Lamiaceae*
Synonym(s): *Ramona plystachya*
Common Name(s): White Sage.

Usage: Food: **Luiseño:** the tops of the stems, when tender, are peeled and eaten uncooked;

the seeds are eaten. **Kamia:** used for seasoning roasted seeds. **Cahuilla:** seeds, gathered July to September, parched, ground into flour for mush; sometimes leaves crushed and added as flavoring. **Chumash:** stem, tender young, peeled, eaten.

Material: **Cahuilla:** leaves crushed, mixed with water, used as hair shampoo, dye, and hair straightener, used by hunters to keep game from detecting human odor. **Chumash:** leaves, hunter would place in moth so deer would not detect presence, lined granaries.

Medicine: **Cahuilla:** leaves eaten, smoked, in sweat-house; seeds used as eye cleansers; leaves crushed, made into poultice, placed under armpits to cleanse sweat glands. **Chumash:** leaves, placed on head to treat headache, added to water, drunk to induce vomiting.

Citation(s): Merriam 1967; Sparkman 1908; Bean and Saubel 1972; Timbrook 2007.

Terminology: **Luiseño:** kashil. **Cahuilla:** qas'ily. **Chumash (Barbareño, Ynezeño, Ventureño):** khapshɨkh.

Citation(s): Sparkman 1908; Bean and Saubel 1972; Timbrook 2007.

1134 *Salvia carduacea*

Common Name(s): Thistle Sage.

Usage: Food: **Luiseño, Tubatulabal:** seeds were eaten. **Chumash:** seeds, sometimes eaten.

        Medicine: **Cahuilla:** used to prevent bad luck if a menstruating women accidently touched

a man's hunting equipment.
Citation(s): Sparkman 1908; Voegelin 1938; Bean and Saubel
1972; Timbrook 2007.
Terminology: **Luiseño:** palit. **Cahuilla:** palnat. **Chumash
(Ynezeño, Ventureño):** pakh.
Citation(s): Sparkman 1908; Bean and Saubel 1972; Timbrook
2007.

1135 *Salvia columbariae*
Common Name(s): Chia.

Usage:	Food:	**Luiseño:** the seeds are much esteemed as food. **Monache (Western Mono):** seeds eaten after parching and pulverizing as a thick, cold soup. **Owens Valley Paiute:** a favorite food for mush. **Pomo:** used seeds for pinole; used in making soup. **Tubatulabal, Salinan:** seeds gathered and eaten. **Diegueño:** used seeds in pinole. **Cahuilla:** seeds, harvested June to September, hulled, parched and ground into flour for mush or cakes; beverage made of unground seeds by soaking in water. **Kawaiisu (Tehachapi):** seeds parched, beaten, beverage of meal and water drunk. **Chumash:** seeds, collected in late Spring and Summer with a seedbeater, dried, toasted eaten, ground into fine flour, mixed with cold water, drunk.
	Material:	**Cahuilla:** burned over chia stands to facilitate next season's growth.
	Medicine:	**Kawaiisu (Tehachapi):** seeds used to clear irritated eyes. **Costanoan:** seeds, mixed in water, taken to reduce fever;

gelatinous seeds, placed in eye to remove foreign objects. **Chumash:** gelatinous seeds, placed in eye to remove foreign objects; gruel, treatment for inflammation of the stomach and bowels, dysentery, hemorroids; seeds, kept with charmstones.

Citation(s): Barrett 1952; Barrows 1900; Chesnut 1902; Curtis 1924; Gifford 1932; Mason 1912; Sparkman 1908; Steward 1933; Voegelin 1938; Zigmond 1981; Bean and Saubel 1972; Landberg 1965; Rayburn 2012; Timbrook 2007.

Terminology: **Owens Valley Paiute:** pāsī´da. **Pomo (Central):** pīce´l. **Luiseño:** pashal. **Cahuilla:** pasal. **Kawaiisu (Tehachapi):** pasidabɨ. **Chumash (Barbareño):** 'ilépesh; **(Ynezeño):** 'i'lepesh; **(Obispeño):** l'ɨpɨ; **(Ventureño):** 'itepesh.

Citation(s): Steward 1933; Barrett 1952; Sparkman 1908; Zigmond 1981; Bean and Saubel 1972; Timbrook 2007.

Note(s): seed contain:

protein	20.2%
oil	34.4%
ash	5.6%

Citation(s): Bean and Saubel 1972.

1136 *Salvia dorrii*

Synonym(s): *Salvia carnosa*
Common Name(s): Purple Sage.

Usage: Medicine: **Washo:** a tea made from leaves, or leaves and stems, taken for a cold remedy; to clear congested nasal passage the dried leaves are crushed and smoked in a pipe. **Kawaiisu (Tehachapi):** leaves soaked in water, decoction utilized as wash for

relief of headache; drunk for stomachache; plant thrown in fire at night to keep a spirit, ghost, away.
Citation(s): Train et al 1941; Zigmond 1981.
Terminology: **Washo:** poh-lo-<u>pee</u>-soh. **Kawaiisu (Tehachapi):** tugubasidabɨ.
Citation(s): Train et al 1941; Zigmond 1981.

1137 *Salvia dorrii* ssp. *incana*

Synonym(s): *Salvia dorrii* ssp. *carnosa*
Common Name(s): Desert Ramona.

Usage: Medicine: **Washo:** leaves, dried, smoke to clear congested nasal passages; leaves, stem decoction used to cure colds.
Citation(s): Strike 1994.

1138 *Salvia mellifera*

Synonym(s): *Ramona stachyoides*
Common Name(s): Black Sage.

Usage: Food: **Luiseño:** the seeds were used as food. **Cahuilla:** seeds, parched, ground into meal; leaves and stalks gathered, April to May, used as food flavoring.

Medicine: **Costanoan:** green leaves chewed for gas pain; soaked in water, taken for heart disorders; leaves, heated, held against ear for earache (to reduce pain), wrapped around neck for sore throat; decoction used in bathes for paralysis, taken internally for cough. **Esselen:** used for a number of medicinal and ceremonial purposes.
Citation(s): Sparkman 1908; Bean and Saubel 1972; Rayburn

2012; Breschini and Haversat 2004.
Terminology: **Luiseño:** kanavut. **Cahuilla:** qas'ily.
Citation(s): Sparkman 1908; Bean and Saubel 1972.

1139 *Salvia spathacea*

Common Name(s): Hummingbird Sage.

Usage: Material: **Esselen:** mixed with Madrone bark and Black Sage to be smoked ceremonially.

          Medicine: **Esselen:** used medicinally. **Chumash:** leaves, decoction, drunk or used as a bath to treat pulmonary ailments, rheumatism; leaves, rubbed on patient's body for treatment of sorcery.

Citation(s): Breschini and Haversat 2004; Timbrook 2007.
Terminology: **Chumash (Barbareño):** qimsh; **(Ventureño):** pakh.
Citation(s): Timbrook 2007.

1140 *Sambucus* spp.

Family: *Adoxaceae*
Common Name(s): Elder, Elderberry, Elderwood.

Usage: Food: **Maidu, Southern (Nisenan), Yurok, Yuki, Shasta, Salinan; Wintun, River (Patwin), Central (Nomlaki):** the berries are eaten. **Bear River:** sun dried berries are packed in baskets between leaves for storage. **Yokuts (Northern Hill):** berries gathered, dried, boiled, and eaten once a week in Winter; gathering generally done in August along with pine nuts. **Luiseño:** much like and gathered in large quantities. **Wappo:** made into nonintoxicating drink.

Material: **Yuki:** made arrows from shoots; also flutes. **Yokuts (Southern Valley):** flute made from wood; **(Northern Hill):** made pipe from wood. **Sinkyone:** made arrows from wood; whistle made the way as a bone whistle after pith is punched out. **Wappo:** made a Kuksu Whistle from wood. **Cahuilla, Diegueño, Luiseño, Cupeño:** used stem as black dye. **Kamia:** bark used by women for skirts.

Citation(s): Barrows 1900; Driver 1936; Foster 1944; Gayton 1948; Goldschmidt 1951; Holt 1946; Kroeber 1932; Mason 1912; Merriam 1967; Merrill 1923; Nomland 1935, 1938; Powers 1877; Sparkman 1908.

Terminology: **Owens Valley Paiute:** hūvū′via.

Citation(s): Steward 1933.

1141 *Sambucus nigra* ssp. *caerulea*

Synonym(s): *Sambucus coerula, Sambucus glauca*
Common Name(s): Elderberry, Blue Elderberry.

Usage: Food: **Luiseño:** used the fruit, both fresh and dried. **Hupa:** the berries are eaten during the Spring and Summer months. **Northern Paiute (Paviotso):** the berries were eaten fresh or dried; they were not cooked before being dried. **Karok:** ripe berries are eaten, sometimes mashed, but not mixed with anything. **Shasta, Yuki:** berries eaten, either fresh or dried. **Miwok:** berries always cooked; some dried for Winter use, these occasionally cooked, sometimes eaten without cooking. **Maidu (Northern):** berries eaten. **Kawaiisu (Tehachapi):** berries, ripe in July, dried, cooked into jelly. **Pomo (Southwestern):** tart berries eaten fresh in small quantities in the latter part of Summer.

Material: **Luiseño:** esteemed the wood for bows. **Hupa:** the shafts of arrows were sometimes made from this plant. **Karok:** a flute is made from lower branches. **Yuki:** made flutes, clappers, and small whistles from branches. **Kawaiisu (Tehachapi):** flute made; used as tobacco container. **Pomo (Southwestern):** branches made into whistles and clappers.

Medicine: **Luiseño:** used the flowers sometimes as a remedy for female complaints. **Karok:** a branch is used to sprinkle a sick child during the Brush Dance. **Yuki:** a decoction of blossoms used externally for sprains and bruises and in fevers; as an antiseptic wash for the itch; tea made from boiled flowers used for fever. **Pomo:** decoction of the blossoms drunk to stop bleeding and for consumption and to allay stomach troubles; a wash of leaves used as an antiseptic wash. **Pomo (Southwestern):** root boiled; solution used as a healing lotion on open sores and cuts; dried flower tea to break a fever. **Miwok:** decoction of blossoms drunk for ague, taken as soon as possible after shaking began, the patient was covered as profuse perspiration followed. **Kawaiisu (Tehachapi):**

flowers and leaves boiled, fumes inhaled to relieve headaches and colds; drunk to relieve fever, "good for measles." **Costanoan:** leaf decoction used as purgative, to cure new colds. **Diegueño:** stem, leaf, blossom decoction used as eyewash, especially for newborn babies, to reduce fever, clear chest congestion, diminish inflammations, treat female diseases, cure colds.

Misc: **Coast Yuki:** named but not used.

Citation(s): Barrett and Gifford 1933; Chesnut 1902; Curtin 1957; Dixon 1905, 1907; Gifford 1939; Goddard 1903; Kelly 1923; Schenck and Gifford 1952; Sparkman 1908; Zigmond 1981; Goodrich et al 1980; Strike 1994; Rayburn 2012.

Terminology: **Luiseño:** kutpat. **Northern Paiute (Paviotso):** hubu'. **Miwok (Central Sierra):** añta'iyu. **Yuki:** kē-wē-mäm, ki-wí-mäm, kluwa. **Karok:** yahuush. **Hupa:** tcūhɛūɛ. **Pomo (Southwestern):** tʰe gʰale', "feather tree."

Citation(s): Sparkman 1908; Kelly 1932; Barrett and Gifford 1933; Curtin 1957; Gifford 1939; Schenck and Gifford 1952; Goddard 1903; Goodrich et al 1980.

1142 *Sambucus nigra* ssp. *caerula*

Synonym(s): *Sambucus velutina, Sambucus mexicana*
Common Name(s): Blue Elderberry.

Usage: Food: **Tubatulabal:** used berries. **Atsugewi:** berries mashed and mixed with Manzanita (*Arctostaphylos* spp.) Flour and stored in dried cakes. **Owens Valley Paiute:** gathered in the Sierra or traded from the west; used berries. **Cahuilla:**

	berries gathered in large quantities in July and August and dried. **Esselen:** berries, ripe, eaten raw, or dried. **Chumash:** berries, eaten.
Material:	**Cahuilla:** berry juice made purplish or black coloring for dying basketry materials; stem used for yellow or orange dye; twigs used to make whistles. **Esselen:** made a split stick musical instrument. **Chumash:** wood, made bows, smoking pipes, dance wand handles, feathered pole erected at Winter Solstice, flutes, split-stick rattle, bullroarer, firesticks, pump-drill shaft; bark, woven bags.
Medicine:	**Cahuilla:** blossoms brewed into tea for during fevers, upset stomachs, cold, and the flu; considered beneficial to newborn babies, good for teeth; roots boiled, administered for constipation. **Chumash:** flowers, decoction, to treat colds, fevers, soaked injured parts; flowers, gathered, dried; root, pounded, boiled, tea drank hot as strong laxative, to close wounds.

Citation(s): Barrows 1900; Garth 1953; Merrill 1923; Steward 1933; Voegelin 1938; Bean and Saubel 1972; Breschini and Haversat 2004; Timbrook 2007.

Terminology: **Owens Valley Paiute:** pā′inoiya[a], sainō′wiyu′[u]. **Mono Lake Paiute:** hūbū′xia. **Atsugewi:** warakui. **Cahuilla:** hunqwat. **Chumash (Barbareño, Island, Ynezeño, Ventureño):** qayas.

Citation(s): Steward 1933; Garth 1953; Bean and Saubel 1972; Timbrook 2007.

Note(s): roots, stems, leaves, and much less the flowers and unripe berries, contain a poisonous alkaloid and

cyanogenic glycoside causing nausea, vomiting, and diarrhea. Children have been poisoned by making blowguns, whistles, popguns out of stems and having them in their mouths. The flowers and ripe fruit are edible without harm.
Citation(s): Hardin and Arena 1974.

1143 *Sambucus racemosa*

Synonym(s): *Sambucus callicarpa*
Common Name(s): Elderberry.

Usage: Material: **Pomo (Southwestern):** branches used for whistles and clappers.

  Medicine: **Pomo (Southwestern):** root was boiled and the decoction used as a healing lotion on open sores and cuts.

Citation(s): Gifford 1967.
Terminology: **Pomo (Southwestern):** tekale.
Citation(s): Gifford 1967.

1144 *Sambucus racemosa* var. *melanocarpa*

Synonym(s): *Sambucus melanocarpa*

Usage: Food: **Maidu (Northern):** berries eaten.

  Medicine: **Pomo:** root decoction used as a lotion for sores, cuts.

Citation(s): Dixon 1905; Strike 1994.

1145 *Sanicula* spp.

Family: *Apiaceae*

Usage: Food: **Maidu, Southern (Nisenan):** the root, long and slightly tuberous, is eaten.

  Misc: **Pomo (Southwestern):** bad medicinal use.

Citation(s): Powers 1877; Goodrich et al 1980.
Terminology: **Pomo (Southwestern):** cy?iá? Ci?do.
Citation(s): Goodrich et al 1980.

1146 *Sanicula arguta*
Common Name(s): Cow Parsley.

Usage: Food: **Diegueño:** roots, boiled and eaten.
         Medicine: **Diegueño:** leaf tea used to relieve
                      cramps.
Citation(s): Strike 1994.

1147 *Sanicula bipinnata*
Common Name(s): Poison Sanicle.

Usage: Food: **Karok, Maidu:** the young greens are
                      eaten. **Pomo:** roots eaten raw.
         Medicine: **Miwok:** the plant is boiled and applied
                      to snake bite.
Citation(s): Barrett and Gifford 1933; Schenck and Gifford 1952;
             Strike 1994.
Terminology: **Karok:** ikxash. **Miwok (Central Sierra):** wene.
Citation(s): Schenck and Gifford 1952; Barrett and Gifford 1933.

1148 *Sanicula bipinnatifida*

Usage: Medicine: **Miwok, Maidu:** used as a cure-all; root
                      boiled and decoction drunk; steeped
                      leaves applied to snake bites.
Citation(s): Barrett and Gifford 1933.
Terminology: **Miwok (Central Sierra):** wene.
Citation(s): Barrett and Gifford 1933.

1149 *Sanicula carssicaulis*
Synonym(s): *Sanicula Menziesii*

Common Name(s): Gamble Weed.

Usage: Food: **Miwok (Coast):** blossoms eaten.

Medicine: **Miwok:** pulverized leaves placed on rattlesnake bites and other wounds; never drunk as a tea as it caused sickness.

Misc: **Miwok:** named but not used (?). **Wailaki:** root is supposed to bring good luck in gambling if chewed and rubbed on the body.

Citation(s): Barrett and Gifford 1933; Chesnut 1902; Schenck and Gifford 1952; Strike 1994.

Terminology: **Miwok (Central Sierra):** lawati huzikus, "rattlesnake medicine." **Karok:** pufich'imkaanva, "deer imkaanva."

Citation(s): Barrett and Gifford 1933; Schenck and Gifford 1951.

Note(s): some difference in reporting for the **Miwok** [author].

1150 *Sanicula tuberosa*

Common Name(s): Turkey-Pea.

Usage: Food: **Maidu, Southern (Nisenan):** eaten raw, or roasted in ashes, or boiled. **Pomo:** tuberous roots eaten raw.

Citation(s): Barrett 1952; Chesnut 1902; Powers 1877.

1151 *Sarcodes sanguinea*

Family: *Ericaceae*

Common Name(s): Snow Plant.

Usage: Material: **Miwok:** plucked because of their beauty.

Misc: **Karok:** named but not used (?).

Citation(s): Barrett and Gifford 1933; Schenck and Gifford 1952.

Terminology: **Miwok (Central Sierra):** kokolpate. **Karok:**

kutannav, "itch medicine."
Citation(s): Barrett and Gifford 1933; Schenck and Gifford 1952.
Note(s): usage statement seems to disagree with plant name
used by **Karok** [author].

1152 *Sarcoscypha coccinea*
Family: *Sarcoscyphacae*
Common Name(s): Orange Peel Mushroom.

Usage: Food: **Pomo (Southwestern):** cooked on hot
stones; eaten fresh.
Citation(s): Goodrich et al 1980.
Terminology: **Pomo (Southwestern):** duwiĉe.
Citation(s): Goodrich et al 1980.

1153 *Schoenoplectus acutus* var. *occidentalis*
Family: Cyperaceae
Synonym(s): *Scirpus lacustrus occidentalis, Scirpus acutus*
Common Name(s): Tule, Common Tule, Circular-
Stemmed Tule.

Usage: Food: **Pomo:** the roots and tender young shoots
were eaten as greens. **Tubatulabal:** used
roots.

Material: **Owens Valley Paiute:** made into mats
for the covering of Winter valley house.
Karok, Chumash, Maidu (Northern):
used for making mats. **Modoc
(Lutuami):** mats made of the plant form
the outer three layers of the covering of
the Summer house; leaf used as the
warp, woof, and rim binding in basketry;
leaf (dyed) as black patten, leaf (aged) as
yellow pattern in baskets. **Achomawi:**
used leaf as warp, woof in baskets; leaf

(dyed) and root as black pattern in baskets. **Pomo, Yokuts, Northern Paiute (Paviotso), Chumash:** used leaf as warp and woof in baskets. **Pomo:** sometimes used as shredded to make tule skirt; also used for weaving mats and baskets and for house thatch as well as for balsa building. **Chumash:** made into 2-ply cordage for twining warp in basketry; balsa watercraft, caulking for canoes, thatching, mats. **Wintun:** used bundles of dry plants for fish seine.

Medicine: **Luiseño:** leaves used to poultice wounds, burns.

Citation(s): Barrett 1910, 1952; Curtis 1924; Dawson and Deetz 1965; Dixon 1905; Grant 1964; Merrill 1923; Schenck and Gifford 1952; Steward 1933; Voegelin 1938; Strike 1994; Timbrook 2007.

Terminology: **Karok:** taprara. **Pomo (Southwestern):** ta. **Chumash (Barbareño, Island):** stapan; **(Ynezeño):** swow; **(Ventureño):** kawɨyɨsh.

Citation(s): Schenck and Gifford 1952; Barrett 1952; Timbrook 2007.

Note(s):   Jepson (1925) states for this species that "It is our estimate that originally there were in California about 250,000 acres of tule lands; much of this area has now been reclaimed to cultivation."

1154 *Schoenoplectus americanus*
Synonym(s): Scirpus olneyi

Usage: Material: **Chumash:** used for making mats.
Citation(s): Grant 1964.

1155 *Schoenoplectus californicus*

Synonym(s): *Scirpus californicus*

Usage: Material: **Chumash:** used for making balsa watercraft, caulking for canoes, thatching, mats. **Esselen:** roots, used in basketry; stems, used in the construction of mats, thatch.

Citation(s): Grant 1964; Breschini and Haversat 2004; Timbrook 2007.

Terminology: **Esselen:** capanay "tule." **Chumash (Barbareño, Island, Ynezeño):** stapan; **(Ventureño):** kawɨyɨsh.

Citation(s): Breschini and Haversat 2004; Timbrook 2007.

1156 *Schoenoplectus pungens* var. *longispicatus*

Synonym(s): *Scripus pungens*
Common Name(s): Common Three Square Bulrush.

Usage: Material: **Chumash:** thatching, mats, oval baskets, containers for ceremonial paraphernalia, cradles for newborn babies.

Citation(s): Timbrook 2007.

Terminology: **Chumash (Barbareño, Ynezeño):** saw'; **(Ventureño):** tup'.

Citation(s): Timbrook 2007.

1157 *Schoenoplectus tabernaemontani*

Synonym(s): *Scirpus validus*
Common Name(s): Tule.

Usage: Food: **Yokuts:** the pollen is beaten off onto a cloth in great quantities and is made in pinole or mush; the bulbous root is also eaten. **Kawaiisu (Tehachapi):** tender lower portions eaten raw.

Material: **Kawaiisu (Tehachapi):** used to make mats, cradle board, saddles.

Medicine: **Kawaiisu (Tehachapi):** in the ceremony for the dead, images representing the deceased are made.

Citation(s): Powers 1877; Zigmond 1981.

Terminology: **Kawaiisu (Tehachapi):** seevibɨ.

Citation(s): Zigmond 1981.

1158 *Scirpus* spp.

Family: *Cyperaceae*

Common Name(s): Bulrush, Tule, Marsh Bulrush.

Usage: Food: **Luiseño:** the tender shoots are eaten raw. **Yokuts:** roots gathered, dried, pounded, and used as flour for mush; seeds, when ripe, are gathered. **Wintun, River (Patwin):** bulb eaten, also new shoots. **Wappo:** sprouts and inside of head eaten. **Cahuilla:** white, starchy root ground into sweet tasting flour; seeds eaten raw, or ground into mush; cake made from pollen.

Material: **Bear River:** made mats about 3 foot square. **Yuki:** made mats. **Yokuts:** made mats to cover houses; **(Southern Valley):** used as mats. **Wintun, River (Patwin):** roofed food bins with material; also used to separate layers of food in the bin. **Pomo:** used root (dyed) as black pattern in baskets. **Modoc (Lutuami):** used stem as warp and woof in baskets. **Cahuilla:** used stem as wrapping in baskets; used as house thatching; stalks used for bedding, mats. **Panamint Shoshone**

(Koso): used root as black pattern in baskets. **Owens Valley Paiute:** used bundles of the material to cover Winter house. **Chumash:** split root presumably dyed black by burial in mud used (rarely) for sewing strands in coiled basketry; stems used to make Water Bottles, waterproofed inside with asphaltum, used for thatching, mats, frameworks for the Fox Dance headdress, tied bundles for target practice.

Medicine: **Cahuilla:** ceremonial bundle and images for image-burning ceremony made from plant materials.

Citation(s): Barrows 1900; Dawson and Deetz 1965; Driver 1936; Foster 1944; Gayton 1948; Kroeber 1932; Mason 1912; Merrill 1923; Nomland 1938; Steward 1933; Sparkman 1908; Bean and Saubel 1972; Strike 1994; Hudson 1893; Timbrook 2007.

Terminology: **Luiseño:** pevesash. **Wiyot:** sòptik. **Yokuts:** tsos (seeds), pu'muk (root). **Cahuilla:** pa'al.

Citation(s): Sparkman 1908; Loud 1918; Gayton 1948; Bean and Saubel 1972.

1159 *Scirpus microcarpus*

Usage: Material: **Coast Yuki:** root used in basketry.

Misc: **Kawaiisu (Tehachapi):** named but not used.

Citation(s): Gifford 1939; Zigmond 1981.

Terminology: **Yuki:** sii. **Kawaiisu (Tehachapi):** nigo?odɨbɨ.

Citation(s): Gifford 1939; Zigmond 1981.

1160 *Scrophularia californica*

Family: *Scrophulariaceae*
Common Name(s): Figwort.

Usage: Food: **Wailaki, Yuki:** leaves rolled into small balls, dried, set on fire to produce salty ashes used to flavor foods.

Material: **Maidu:** roots produced a black dye.

Medicine: **Pomo (Southwestern):** leaves were warmed and used as a night poultice until boils came to a head. **Diegueño:** tea used to reduce fever. **Maidu, Pomo:** twigs, leaves used to poultice skin irritations, boils, infections, sores, as a disinfectant, treatment for eye problems. **Miwok (Coast):** leaf decoction used to wash sores, leaves, dried, pulverized sprinkled on sores. **Costanoan:** twigs, leaves used to poultice skin irritations, boils, infections, sores, swellings, as a disinfectant, treatment for eye problems; heated twigs tied over swollen sores; leaves, used in poultices over sore eyes; plant, juice, used in eyewash to treat poor vision.

Citation(s): Gifford 1967; Goodrich et al 1980; Strike 1994; Rayburn 2012.

Terminology: **Pomo (Southwestern):** hakacha kale, "hornet plant."

Citation(s): Gifford 1967; Goodrich et al 1980.

1161 *Scutellaria* spp.

Family: *Lamiaceae*
Common Name(s): Skullcap.

Usage: Medicine: **Miwok:** boiled; decoction drunk for coughs and colds.
Citation(s): Barrett and Gifford 1933.
Terminology: **Miwok (Central Sierra):** wenene.
Citation(s): Barrett and Gifford 1933.

1162 *Scutellaria bolanderi*

Usage: Misc: **Miwok:** named but not used.
Citation(s): Barrett and Gifford 1933.
Terminology: **Miwok (Central Sierra):** susupe.
Citation(s): Barrett and Gifford 1933.

1163 *Scutellaria californica*

Common Name(s): California Skullcap.

Usage: Medicine: **Maidu:** used to treat chills, fever.
         Misc: **Yuki:** named but not used.
Citation(s): Chesnut 1902; Strike 1994.

1164 *Scutellaria mexicana*

Family: *Lamiaceae*
Synonym(s): *Salazaris mexicana*
Common Name(s): Bladder Sage.

Usage: Misc: **Kawaiisu (Tehachapi):** named but not used.
Citation(s): Zigmond 1981.
Terminology: **Kawaiisu (Tehachapi):** paavo?orɨbɨ.
Citation(s): Zigmond 1981.

1165 *Scutellaria siphocamyploides*

Synonym(s): *Scutellaria augusttifolia*

Common Name(s): Grey-Leaved Skullcap.

Usage: Medicine: **Miwok, Maidu:** decoction used as a wash for sore eyes.

Citation(s): Barrett and Gifford 1933; Strike 1994.

1166 *Scutellaria tuberosa*

Common Name(s): Dannie's Skullcap.

Usage: Medicine: **Miwok:** tea used to cure colds. **Miwok, Maidu:** used as a tonic, or a sedative.

Citation(s): Strike 1994.

1167 *Sedum* spp.

Family: *Crassulaceae*

Usage: Medicine: **Costanoan:** decoction used to reduce fever, as a gargle for sore throat; decoction, or powder, used externally on wounds.

Citation(s): Rayburn 2012.

1168 *Sedum laxum* ssp. *heckneri*

Usage: Misc: **Karok:** named but not used (?).

Citation(s): Schenck and Gifford 1952.

Terminology: **Karok:** xanvathiich, "imitation clam."

Citation(s): Schenck and Gifford 1952.

1169 *Sedum spathulifolium*

Common Name(s): Orpin, Stonecrop.

Usage: Medicine: **Costanoan:** decoction used as a gargle to

reduce fever, relieve sore throat; plant powdered, sprinkled on wounds. **Maidu:** used to stop hemorrhages, diarrhea, sooth wounds, ulcerations; decoction drunk by women to facilitate childbirth.
Citation(s): Strike 1994.

1170 *Senecio serra*
Family: *Asteraceae*

Usage: Misc: **Kawaiisu (Tehachapi):** named but not used.
Citation(s): Zigmond 1981.
Terminology: **Kawaiisu (Tehachapi):** kamɨnagavivɨ keevowadɨ, "mountain rabbit ear."
Citation(s): Zigmond 1981.

1171 *Senecio flaccidus* var. *douglasii*
Synonym(s): *Senecio Douglasii*

Usage: Medicine: **Kawaiisu (Tehachapi):** leaves are brewed, drunk as "very strong medicine," given to women after childbirth, used to cure kidney ailments. **Costanoan:** used to treat dermatological, gynecological, kidney ailments, for a "cold in the kidneys,", as a disinfectant, for infected sores; taken internally for "lockjaw" in women after childbirth; externally, used for cuts.

Misc: **Tubatulabal:** considered weeds.
Citation(s): Voegelin 1938; Zigmond 1981; Strike 1994; Rayburn 2012.
Terminology: **Kawaiisu (Tehachapi):** cukʷa?nidɨbɨ.

Citation(s): Zigmond 1981.

1172 *Senecio triangularis*

Usage: Misc: **Miwok:** named but not used.
Citation(s): Barrett and Gifford 1933.
Terminology: **Miwok (Central Sierra):** yusuyi.
Citation(s): Barrett and Gifford 1933.

1173 *Sequoia sempervirens*
Family: *Cupressaceae*
Common Name(s): Redwood, Coast Redwood.

Usage: Material: **Hupa:** a fiber from the roots was imported from the coast for the woof in making baskets. **Wiyot:** planks were made and used for house-construction. **Wiyot, Yurok:** canoes were made. **Yurok, Sinkyone, Karok:** used root as woof in baskets. **Pomo:** tied driftwood logs together to make rafts. **Pomo (Southwestern):** bark used for house covering. **Bear River:** made canoes from logs floated down from Rainbow Ridge; also made rafts of logs tied together with rawhide thongs to skirt the cliffs along the beach and to visit the near-by rocks for gathering eggs or young shags. **Coast Yuki:** started fires using wood hearth and Buckeye (*Aesculus californica*) drill. **Chumash, Gabrielino:** used driftwood, splint into planks, to make seaworthy canoe, capable of carrying 30 people.

Medicine: **Pomo (Southwestern):** new foliage, warmed in fire, applied as a poultice for earache; gummy sap taken as medicine for rundown condition; soaked in water and liquid drunk as a tonic. **Yokuts (Southern Valley):** sap was shaved into pot of hot water; person with a wound which bled profusely steamed himself with this under a blanket; a little of the sap was chewed for a cold.

Misc: **Karok, Coast Yuki:** named but not used. **Costanoan:** considered large trees as sacred objects.

Citation(s): Curtis 1924; Gayton 1948; Gifford 1939, 1967; Goddard 1903; Loud 1918; Merrill 1923; Nomland 1938; O'Neale 1932; Schenck and Gifford 1952; Goodrich et al 1980; Strike 1994; Timbrook 2007.

Terminology: **Pomo (Northern):** kasil; **(Southwestern):** kasil, kasin. **Yurok:** qiL (roots), hape' (roots). **Yuki:** oloxtem. **Karok:** `uθkanpahip, "coast pahip." **Chumash (Barbareño):** wi'ma; **(Ynezeño):** wima'; **(Ventureño):** wima.

Citation(s): Barrett 1933; Gifford 1967; O'Neale 1932; Gifford 1939; Schenck and Gifford 1952; Goodrich et al 1980; Timbrook 2007.

Note(s): some difference in cited data about the **Coast Yuki** [author].

1174 *Sequoiadendron giganteum*

Family: *Cupressaceae*
Synonym(s): *Sequoia gigantea*
Common Name(s): Giant Sequoia.

Usage: Material: **Miwok:** slabs of bark used for covering their conical dwellings.

Medicine: **Monache (Western Mono), Yokuts:** bitter sap chewed to relieve cold symptoms.

Citation(s): Barrett and Gifford 1933; Strike 1994.

Terminology: **Monache (Western Mono):** woh-woh'-nau.
Miwok (Central Sierra): pusine.

Citation(s): Gifford 1932; Barrett and Gifford 1933.

1175 *Setaria pumila* ssp. *pumila*

Family: *Poaceae*
Synonym(s): *Setaria lutescens*
Common Name(s): Yellow Bristle Grass.

Usage" Misc: **Kawaiisu (Tehachapi):** named but not used.

Citation(s): Zigmond 1981.

Terminology: **Kawaiisu (Tehachapi):** pa?yig^{w}asivɨ, "kangaroo-rat tail."

Citation(s): Zigmond 1981.

Note(s): native of the Old World.

Citation(s): Munz and Keck 1965.

1176 *Sheperdia argentea*

Family: *Elaeagnaceae*
Synonym(s): *Eleagnus asgenta, Eleagnus asgente*
Common Name(s): Buffalo Berry.

Usage: Food: **Mono Lake Paiute:** seeds used. **Northern Paiute (Paviotso):** the berry was eaten fresh; if dried, they were boiled, drained, and crushed. **Owens Valley Paiute:** fruit eaten.

Citation(s): Kelly 1932; Steward 1933; Strike 1994.

Terminology: **Mono Lake Paiute:** wiyu'pi. **Northern Paiute (Paviotso):** wi'yüpui.

Citation(s): Steward 1933; Kelly 1932.

Note(s): berry

Water	71.28%
Nitrogenous substances	00.14%
Free Acid	2.45%
Total Sugar	5.47%
Other (Pectin, etc.)	19.79%
Ash	00.45%

Citation(s): Trimble 1881-1891.

1177 *Sidalcea asprella*

Family: *Malvaceae*
Synonym(s): *Sidalcea malviflora* ssp. *asprella*
Common Name(s): Checker Bloom.

Usage: Misc: **Karok:** named but not used.
Citation(s): Schenck and Gifford 1952.
Terminology: **Karok:** anuhich, "imitation thimble."
Citation(s): Schenck and Gifford 1952.

1178 *Sidalcea glaucescens*

Usage: Misc: **Miwok:** named but not used.
Citation(s): Barrett and Gifford 1933.
Terminology: **Miwok (Central Sierra):** lotowiye.
Citation(s): Barrett and Gifford 1933.

1179 *Sidalcea malviflora*

Common Name(s): Wild Hollyhock.

Usage: Food: **Luiseño:** used as greens. **Yana:** sometimes used dried and mashed leaves to mix with other mashed berries as a flavoring device. **Maidu:** stems, leaves, eaten raw, or cooked; dried, pulverized leaves used to flavor Manzanita (*Arctostaphylos*) berries in pinole.

Misc: **Tubatulabal:** considered weeds.

Citation(s): Sapir and Spier 1943; Sparkman 1908; Voegelin 1938; Strike 1994.

Terminology: **Luiseño:** pashangal. **Yana:** lakʻaʹtʹi (the fruit).

Citation(s): Sparkman 1908; Sapir and Spier 1943.

1180 *Silene antirrhina*

Family: *Caryophyllaceae*

Common Name(s): Sleepy Catchfly.

Usage: Material: **Miwok:** used juice that oozed from plant as a paint to decorate the face of young girls; brown color turns black after awhile.

Citation(s): Barrett and Gifford 1933.

1181 *Silene campanulata*

Usage: Medicine: **Karok:** used as a medicine for babies. **Maidu:** infusion used to poultice open sores.

Citation(s): Schenck and Gifford 1952; Strike 1994.

Terminology: **Karok:** yupshitanachpirish.

Citation(s): Schenck and Gifford 1952.

Note(s): plant only known from Red Mountain, Mendocino County.

Citation(s): Munz and Keck 1965.

1182 *Silene laciniata*

Common Name(s): Fringed Indian Pink.

Usage: Medicine: **Chumash:** plant, boiled, drank by women to make their menses flow.

Citation(s): Timbrook 2007.

Terminology: **Chumash (Barbareño):** s'akhtutu 'iyukhnuts'.
Citation(s): Timbrook 2007.

1183 *Silene laciniata* ssp. *californica*
Synonym(s): *Silene californica*
Common Name(s): Indian Pink.

Usage: Misc: **Karok, Miwok:** named but not used.
Citation(s): Barrett and Gifford 1933; Schenck and Gifford 1952.
Terminology: **Miwok:** pusulu. **Karok:** pinef ichishsrixa,
"Coyote flower."
Citation(s): Barrett and Gifford 1933; Schenck and Gifford 1952.

1184 *Simmondsia chinensis*
Family: *Simmondsiaceae*
Synonym(s): *Simmondsia californica*

Usage: Food: **Cahuilla:** nuts eaten fresh, or ground
and made a "coffee" by boiling the meal.
Citation(s): Barrows 1900; Bean and Saubel 1972.
Terminology: **Cahuilla:** qawnaxal.
Citation(s): Bean and Saubel 1972.

1185 *Sisymbrium irio*
Family: *Brassiaceae*
Common Name(s): London-Rocket.

Usage: Food: **Cahuilla:** immature leaves gathered,
fried, or boiled, for greens.
Citation(s): Bean and Saubel 1972.

1186 *Sisyrinchium bellum*
Family: *Iridaceae*
Common Name(s): Blue-Eyed Grass.

Usage: Medicine: **Luiseño, Maidu:** a purgative is made from the roots. **Pomo (Southwestern):** wash and boil roots, strain, drunk for upset stomach, heartburn, ulcers, and asthma. **Costanoan:** tea (decoction) drunk to treat stomach ache, to induce menstruation, to produce abortions, for chills.

Citation(s): Sparkman 1908; Goodrich et al 1980; Strike 1994; Rayburn 2012.

Terminology: **Luiseño:** patumkut. **Pomo (Southwestern):** šiwitá gawiyya, "little Iris." **Chumash (Barbareño):** sh'ichkɨ 'i'waqa.

Citation(s): Sparkman 1908; Goodrich et al 1980; Timbrook 2007.

1187 *Sium suave*

Family: *Apiaceae*
Synonym(s): *Sium cicutaefolium*

Usage: Food: **Northern Paiute (Paviotso):** root eaten fresh or dried. **Maidu:** roots eaten raw, or cooked. **Maidu, Modoc (Lutuami):** leaves, stems, eaten cooked.

Citation(s): Kelly 1932; Strike 1994.

Terminology: **Northern Paiute (Paviotso):** ya'pa'.

Citation(s): Kelly 1932.

1188 *Smilax californica*

Family: *Smilacaceae*
Common Name(s): Greenbriar.

Usage: Material: **Yuki:** used stem as black pattern in baskets.

Citation(s): Chesnut 1902; Merrill 1923.

1189 *Solanum* spp.

Family: *Solanaceae*

Usage: Material: **Costanoan:** fruit, used to make blue dye used in tattoos.

Citation(s): Rayburn 2012.

1190 *Solanum douglasii*

Common Name(s): Black Nightshade.

Usage: Food: **Luiseño:** the leaves used for greens. **Maidu:** leaves, stems eaten. **Chumash:** berries, eaten raw, or boiled.

Material: **Luiseño, Chumash:** the juice of the berries was used for tattooing.

Medicine: **Luiseño, Maidu, Miwok:** the juice of the berries was used for inflamed eyes. **Cahuilla:** the juice of the berries was used for inflamed eyes, pink eye, eye strain, said to improve night vision in the elderly. **Chumash:** leaves, crushed by hands and rubbed on rash from Poison Oak; fruit, used to treat sore eyes, possibly as a dye.

Citation(s): Sparkmen 1908; Bean and Saubel 1972; Strike 1994; Timbrook 2007.

Terminology: **Luiseño:** takovshish. **Cahuilla:** ayka'kal.
Chumash (Barbareño, Ynezeño, Purisimeño): 'aqulpop'; **(Island, Ventureño):** qolpo'op.

Citation(s): Sparkmen 1908; Bean and Saubel 1972; Timbrook 2007.

1191 *Solanum nigrum*

Common Name(s): Deadly Nightshade, Black Nightshade.

Usage: Food: **Tubatulabal:** used berries. **Wintun:** berries used for food, but only when ripe; green fruit considered poisonous.

Medicine: **Miwok:** decoction used as a wash for sore eyes. **Costanoan:** in cigarettes, smoke inhaled to relieve toothache; leaves, heated, applied to boils; decoction used as remedy for scarlet fever.

Misc: **Karok:** named but not used (known to be poisonous).

Citation(s): Barrett and Gifford 1933; Chesnut 1902; Schenck and Gifford 1952; Voegelin 1938; Rayburn 2012.

Terminology: **Miwok (Southern Sierra):** ma'nmantca. **Karok:** shishpuris, "dog huckleberry."

Citation(s): Barrett and Gifford 1933; Schenck and Gifford 1952.

Note(s): contain alkaloidal glucoside, solamine, the leaves and berries, especially in the unripe condition when it is poisonous; as the fruit ripens the solamine content decreases to non-toxic quantities; cooking apparently destroys the toxic principles; plant native to Europe.

seeds contain:

Protein 17.5%

Oil   27.2%

Native to Eurasia.

Citation(s): Muenscher 1951; Munz and Keck 1965; Earle and Jones 1962; Baldwin et al 2012.

1192 *Solanum umbelliferum*

Common Name(s): Blue Witch.

Usage: Medicine: **Maidu:** roots used to treat colic.

Citation(s): Strike 1994.

1193 *Solanum xantii*

Synonym(s): *Solanum xantii* var. *intermedium*
Common Name(s): Chaparral Nightshade, Purple
Nightshade.

Usage: Food: **Miwok, Maidu:** berries eaten raw.

Medicine: **Kawaiisu (Tehachapi):** leaves and
flowers placed on hot rocks with affected
part, swollen leg, sore swollen shoulder,
laid directly on heated plant material;
used on skin sores, as a rub on crippled
limbs.

Citation(s): Zigmond 1981; Barrett and Gifford 1933; Strike 1994.
Terminology: **Kawaiisu (Tehachapi):** cokoviyazigadɨbɨ. **Miwok
(Central Sierra):** watana.
Citation(s): Zigmond 1981; Barrett and Gifford 1933.

1194 *Solidago confinis*

Family: *Asteraceae*
Common Name(s): Southern Goldenrod.

Usage: Medicine: **Kawaiisu (Tehachapi):** leaves and
flowers boiled, infusion used to wash
sores, boils, skin irritations.

Citation(s): Zigmond 1981.
Terminology: **Kawaiisu (Tehachapi):** kamɨnagavivɨ, "rabbit
ear."
Citation(s): Zigmond 1981.

1195 *Solidago elongata*

Synonym(s): *Solidago canadensis*
Common Name(s): Meadow Goldenrod.

Usage: Medicine: **Maidu:** flowers, chewed to relieve sore
throat; plant used as poultice on the

neck.
Citation(s): Strike 1994.

1196 *Solidago velutina* ssp. *californica*
Synonym(s): *Solidago californica*
Common Name(s): Common Goldenrod.

Usage: Material: **Cahuilla:** made a hair rinse.

Medicine: **Miwok:** small quantity of a decoction was held in mouth to alleviate toothache; it was expectorated, never swallowed; pale green powder was made from the leaves and applied to open sores after washing with a decoction of *Euphorbia serpyllifolia*. **Kawaiisu (Tehachapi):** leaves and flowers boiled, infusion used to wash sores, boils, skin irritations. **Cahuilla:** used in feminine hygiene. **Kawaiisu (Tehachapi):** leaf, flower infusion used to treat skin irritations, open sores, boils. **Costanoan:** used as a skin medication, as a burn dressing; leaves used to wash sores, burns, leaves, toasted, crumbled, sprinkled on wounds. **Maidu:** used to bathes sores; leaves, dried, pulverized sprinkled on sores, wounds. **Chumash:** decoction, taken for coughs, colds, wash for sores, wounds, to treat sick horses as a drink or wash for bruises.

Citation(s): Barrett and Gifford 1933; Zigmond 1981; Bean and Saubel 1972; Strike 1994; Rayburn 2012; Timbrook 2007.

Terminology: **Miwok (Southern Sierra):** lo'yama. **Cahuilla:** pa'kily. **Chumash (Barbareño):** stu 'imá'; **(Yenezeño):**

shtu'ama'; **(Ventureño)**: chtu 'ima.

Citation(s): Barrett and Gifford 1933; Bean and Saubel 1972; Timbrook 2007.

1197 *Sonchus asper*

Family: *Asteraceae*

Common Name(s): Sow Thistle.

Usage: Food: **Luiseño:** used for greens.

         Misc: **Miwok:** named but not used.

Citation(s): Barrett and Gifford 1933; Sparkman 1908.

Terminology: **Miwok (Central Sierra):** tcittimpa. **Luiseño:** posi'kana.

Citation(s): Barrett and Gifford 1933; Sparkman 1908.

Note(s): naturalized from Europe.

Citation(s): Munz and Keck 1965.

1198 *Sonchus oleraceus*

Common Name(s): Sow Thistle.

Usage: Food: **Kamia:** leaves were boiled and eaten as greens. **Chumash:** may have used several kinds of thistles for food.

Citation(s): Gifford 1931; Timbrook 2007.

Terminology: **Chumash (Barbareño, Ynezeño):** qayish; **(Ventureño):** ts'aqsmi.

Citation(s): Timbrook 2007.

Note(s): naturalized from Europe.

Citation(s): Munz and Keck 1965.

1199 *Sparganium angustifolium*

Family: *Typhaceae*

Synonym(s): *Sparganium simplex, Sparganium miltipedunculatum*

Usage: Misc: **Karok:** named but not used.
Citation(s): Schenck and Gifford 1952.
Terminology: **Karok:** tapraratumnyaich.
Citation(s): Schenck and Gifford 1952.

1200 *Sparganium eurycarpum*
Common Name(s): Bur Reed.

Usage: Food: **Modoc (Lutuami):** stems, shoots, roots eaten. **Maidu:** root, base of plant cooked, eaten.
Citation(s): Strike 1994.

1201 *Spartina foliosa*
Family: *Poaceae*
Common Name(s): Cord Grass, Marsh Grass.

Usage: Material: **Diegueño:** grass bundles used to construct dwelling walls.
Medicine: **Diegueño:** roots, boiled, used as infant laxative, diuretic.
Citation(s): Strike 1994.

1202 *Sphaeralcea* spp.
Family: *Malvaceae*
Common Name(s): Desert Mallow.

Usage: Medicine: **Luiseño:** used as an emetic.
Citation(s): Strike 1994.

1203 *Sphenosciadium capitellatum*
Family: *Asteraceae*

Usage: Medicine: **Owens Valley Paiute:** cloth, soaked in a decoction of boiled roots, placed on

venereal sores. **Maidu:** root chewed to relieve sore throat, root decoction used to treat bronchial problems. **Owens Valley Paiute, Maidu:** root infusion used to repel lice.
Citation(s): Steward 1933; Strike 1994.
Terminology: **Owens Valley Paiute:** kosigonova. **Mono Lake Paiute:** paimüzü′pü.
Citation(s): Steward 1933.

1204 *Spiraea douglasii*

Family: *Rosaceae*
Common Name(s): Spiraea, Indian Lilac.

Usage: Medicine: **Maidu:** seed infusion used to treat diarrhea.

Misc: **Karok:** named but not used.
Citation(s): Schenck and Gifford 1952; Strike 1994.
Terminology: **Karok:** makchukinhish, "mountain-make-believe-crayon."
Citation(s): Schenck and Gifford 1952.

1205 *Spiranthes* spp.

Family: *Orchidaceae*

Usage: Food: **Mono Lake Paiute:** used roots.
Citation(s): Steward 1933.
Terminology: **Mono Lake Paiute:** ica′düma.
Citation(s): Steward 1933.

1206 *Sporobolus* spp.

Family: *Poaceae*

Usage: Material: **Yokuts:** grass bundles used to construct dwelling walls.

Citation(s): Merrill 1923; Powers 1877.

1207 *Sporobolus airoides*
Common Name(s): Akali Sacaton, Dropseed.

Usage: Material: **Yokuts:** stalks sometimes used as foundation for coiled baskets.
Misc: **Kawaiisu (Tehachapi):** seeds not eaten - "bad."
Citation(s): Zigmond 1981; Strike 1994.
Terminology: **Kawaiisu (Tehachapi):** hugʷišiivɨ.
Citation(s): Zigmond 1981.

1208 *Sporobolus flexuous*

Usage: Material: **Yokuts:** stalks sometimes used as foundation for coiled baskets.
Citation(s): Strike 1994.

1209 *Stachys albens*
Family: *Lamiaceae*
Common Name(s): White Hedge Nettle.

Usage: Material: **Kawaiisu (Tehachapi):** leaves, which have a pleasant odor, are bunched together and used as a "cork" for the basketry water bottle; "it gives a good taste to the water."
Medicine: **Costanoan:** decoction used to relieve skin irritations, ease stomach ache, poultice infected or swollen sores, as a gargle for sore throats, as a disinfectant.
Citation(s): Zigmond 1981; Strike 1994.
Terminology: **Kawaiisu (Tehachapi):** hugzɨbɨ.
Citation(s): Zigmond 1981.

1210 *Stachys bullata*

Common Name(s): Hedge Nettle.

Usage: Material: **Karok:** pick this up because it smells good.

Medicine: **Pomo (Southwestern):** leaves heated, used as poultice on boils to bring to head. **Maidu:** used to poultice cuts, bruises; leaf infusion used to cure cough, colds; leaves heated, placed on ear to cure earache. **Costanoan:** root, decoction, small amount drunk followed by compresses for infected, or swollen, sores, gargle for sore throat; leaves, heated, used externally for earache, sore throat.

Citation(s): Schenck and Gifford 1952; Goodrich et al 1980; Strike 1994; Rayburn 2012.

Terminology: **Karok:** saushixara, "go-down-the-hill-and-wash." **Pomo (Southwestern):** šima šúkʰle.

Citation(s): Schenck and Gifford 1952; Goodrich et al 1980.

1211 *Stachys rigida*

Usage: Medicine: **Miwok (Coast):** flowers, leaves brewed into tea for stomach ache; leaves used to poultice rheumatic pain.

Citation(s): Strike 1994.

1212 *Stanleya elata*

Family: *Brassicaceae*

Usage: Food: **Panamint Shoshone (Koso):** leaves and stems eaten. **Kawaiisu (Tehachapi):** leaves and flowers boiled, squeezed out

in cold water to remove bitterness, fried in grease and eaten; taste "like cabbage."
Citation(s): Coville 1892; Zigmond 1981.
Terminology: **Kawaiisu (Tehachapi):** tɨhmaribɨ.
Citation(s): Zigmond 1981.

1213 *Stanleya pinnata*

Usage: Food: **Panamint Shoshone (Koso):** leaves and stems eaten.
Citation(s): Coville 1892.

1214 *Stephanomeria* spp.
Family: *Asteraceae*

Usage: Medicine: **Owens Valley Paiute:** roots boiled into tea for various aliments, especially fever.
Citation(s): Steward 1933.

1215 *Stephanomeria exigua*

Usage: Misc: **Tubatulabal:** considered weeds.
Citation(s): Voegelin 1938.

1216 *Stephanomeria pauciflora*
Common Name(s): Wire Lettuce.

Usage: Food: **Kawaiisu (Tehachapi):** plant exudes a thick liquid that is collected and chewed as gum.
Citation(s): Zigmond 1981.
Terminology: **Kawaiisu (Tehachapi):** sanako?ogadɨbɨ.
Cuitation(s): Zigmond 1981.

1217 *Stephanomeria virgata*

Usage: Medicine: **Kawaiisu (Tehachapi):** sometimes used as an eye medicine.

          Misc: **Tubatulabal:** considered weeds.

Citation(s): Zigmond 1981; Voegelin 1938.

1218 *Stillingia paucidentata*

Family: *Euphorbiaceae*
Common Name(s): Mojave Stillingia.

Usage: Misc: **Kawaiisu (Tehachapi):** named but not used.

Citation(s): Zigmond 1981.
Terminology: **Kawaiisu (Tehachapi):** paahovɨ.
Citation(s): Zigmond 1981.

1219 *Stipa divaricata*

Family: *Poaceae*
Synonym(s): *Oryzopsis milliacea*
Common Name(s): Rice Grass.

Usage: Food: **Owens Valley Paiute:** harvested seeds in early Summer; beaten from grass; seeds generally roasted, ground into flour.

Citation(s): Steward 1933.
Terminology: **Owens Valley Paiute:** wai.
Citation(s): Steward 1933.
Note(s): introduced from the Mediterranean region.
Citation(s): Munz and Keck 1965.

1220 *Stipa hymenoides*

Synonym(s): *Oryzopsis membranacea, Oryzopsis hymenoides*

Common Name(s): Sand Bunch Grass, Indian
Mountain Rice.

Usage: Food: **Panamint Shoshone (Koso):** used seeds.
Owens Valley Paiute: harvested seeds
in early summer; beaten from grass;
seeds generally roasted, ground into
flour. **Kawaiisu (Tehachapi):** significant
food source; seeds freed by burning
plants in bunches; seeds pounded into
meal, eaten dry.
Citation(s): Coville 1892; Steward 1933; Zigmond 1981.
Terminology: **Owens Valley Paiute:** wai. **Kawaiisu**
(Tehachapi): we?evɨ.
Citation(s): Steward 1933; Zigmond 1981.

1221 *Stipa occidentalis*
Common Name(s): Western Neddlegrass.

Usage: Misc: **Kawaiisu (Tehachapi):** named but not
used.
Citation(s): Zigmond 1981.
Terminology: **Kawaiisu (Tehachapi):** hugʷišiivɨ.
Citation(s): Zigmond 1981.

1222 *Stipa speciosa*
Common Name(s): Porcupine Grass.

Usage: Food: **Owens Valley Paiute:** seeds favored for
mush; harvested in late Spring.
Kawaiisu (Tehachapi): grass, gathered
in June, dried; seeds boiled and eaten.
Citation(s): Steward 1933; Zigmond 1981.
Terminology: **Owens Valley Paiute:** hū́ki. **Kawaiisu**
(Tehachapi): hugʷivɨ.

Citation(s): Steward 1933; Zigmond 1981.

1223 *Styrax redivivus*

Family: *Styracaceae*

Synonym(s): *Styrax officinalis* var. *californicus*

Usage: Medicine: **Maidu:** sap used as an antiseptic, as an expectorant.

Citation(s): Strike 1994.

1224 *Suaeda calceoliformis*

Family: *Chenopodiaceae*

Synonym(s): *Suaeda depressa, Suaeda depressa* var. *erecta*

Usage: Food: **Northern Paiute (Paviotso):** the seeds were eaten; made seed gruel.

Citation()s): Kelly 1932.

Terminology: **Northern Paiute (Pavitoso):** wa'·da (seeds).

Citation(s): Kelly 1932.

1225 *Suaeda nigra*

Synonym(s): *Suaeda diffusa, Suaeda fruticosa, Dondia suffrutescens, Suaeda torreyana*

Common Name(s): Sea-Blight.

Usage: Food: **Cahuilla, Luiseño:** tiny dark leaves boiled for greens; seed ground into fine flour or mush or cakes.

Material: **Cahuilla, Chumash, Luiseño, Diegueño:** used whole plant as black dye. **Cahuilla:** leaves used for basket material black dye; leaves boiled, dye mixed with clay, applied to hair as a dye, left until dry

Citation(s): Barrows 1900; Merrill 1923; Bean and Saubel 1972;

Strike 1994.
Terminology: **Cahuilla:** ngayal.
Citation(s): Bean and Saubel 1972.

1226 *Symphoricarpos* spp.
Family: *Caprifoliaceae*

Usage: Material: **Chumash:** green sprigs, tied together around middle, used as broom. **Pomo:** slender stems use for pipe stems.

Medicine: **Maidu:** infusion used to bathe sores; root infusion used to treat colds, stomach ache.

Citation(s): Strike 1994.

1227 *Symphoricarpos albus* var. *rivularis*
Synonym(s): *Symphoricarpos rivularis*
Common Name(s): Snowberry.

Usage: Medicine: **Miwok:** root was pounded and steeped to make a decoction which was drunk to alleviate colds and stomachache. **Yuki:** a tea was mode of the plant to bathe sores, skin lesions.

Misc: **Pomo (Southwestern):** named but not used.

Citation(s): Barrett and Gifford 1933; Curtin 1957; Schenck and Gifford 1952; Goodrich et al 1980; Strike 1994.

Terminology: **Karok:** xanchifichpuris, "frog's huckleberry," ˋaxnatsinnihich, "shiny gooseberry." **Miwok (Central Sierra):** yutasena. **Yuki:** olumkópl. **Pomo (Southwestern):** bahqom?.

Citation(s): Schenck and Gifford 1952; Barrett and Gifford 1933; Curtin 1957; Goodrich et al 1980.

1228 *Symphoricarpos mollis*

> Synonym(s): *Symphoricarpos racemosus, Symphoricarpos acutus*
>
> Common Name(s): Snowberry.

Usage: Material: **Pomo:** slenderest twigs made into brooms; medium sized for arrows and pipe stems; branches used for revolving shaft of the drill used in making shell money.
Citation(s): Chesnut 1902.

1229 *Symphoricarpos rotundifolius*

> Synonym(s): *Symphoricarpos vaccinoides*

Usage: Misc: **Miwok:** named but not used.
Citation(s): Barrett and Gifford 1933.
Terminology: **Miwok (Central Sierra):** mokokine.
Citation(s): Barrett and Gifford 1933.

1230 *Symphyotrichum subulatum*

> Synonym(s): *Aster exilis*
>
> Common Name(s): Slender Aster, American Aster.

Usage: Food: **Maidu:** ate young leaves.
         Medicine: **Kawaiisu (Tehachapi):** to relieve toothache, root mashed and applied, or held between teeth; root boiled and mashed in liquid used to relieve headache. **Maidu:** drank an infusion of flowers to increase fertility.
Citation(s): Zigmond 1981; Strike 1994.

1231 *Taraxacum californicum*

Family: *Asteraceae*
Common Name(s): Dandelion.

Usage: Food: **Cahuilla:** stems and leaves eaten in Spring and early Summer.
Citation(s): Bean and Saubel 1972.

1232 *Taraxacum officinale*

Common Name(s): Dandelion.

Usage: Misc: **Miwok:** named but not used.
Citation(s): Barrett and Gifford 1933.
Terminology: **Miwok (Central Sierra):** natumnatumma.
Citation(s): Barrett and Gifford 1933.
Note(s): native to Europe.
Citation(s): Baldwin et al 2012.

1233 *Tauschia arguta*

Family: *Apiaceae*
Synonym(s): *Deweya arguta*

Usage: Medicine: **Luiseño:** the root was much esteemed for medicinal purposes.
Citation(s): Sparkman 1908.
Terminology: **Luiseño:** kaiyat.
Citation(s): Sparkman 1908.

1234 *Tauschia parishii*

Usage: Medicine: **Kawaiisu (Tehachapi):** root mashed, heated with cheek and placed against it for toothache; dried root burned, inhaled for relief of head pains and colds; applied as salve to aching limbs; drunk

as an infusion for relief of internal pain.
Citation(s): Zigmond 1981.
Terminology: **Kawaiisu (Tehachapi):** išaragɨbɨ.
Citation(s): Zigmond 1981.

1235 *Taxus brevifolia*

Family: *Taxaceae*
Common Name(s): Yew, Western Yew.

Usage: Food: **Yuki:** the berries are edible but the seeds are considered poisonous.

        Material: **Hupa:** the saplings were used in the construction of sinew-backed bows; used to make tobacco pipes. **Chimariko:** made flat sinew-backed bows. **Karok:** the wood is used to make bows, tobacco pipes; bark is used as a handle or covering for stone knife. **Shasta:** bows were made of this wood. **Pomo (Southwestern):** used for mush stirrers and bows, abalone pry; roots used in basketry as weft material. **Bear River:** made bows with small pieces of same wood glued on back as reinforcement; made long pipes for female shamans, short one for male shamans; square frames for snowshoes. **Yuki:** made bows. **Sinkyone:** made bows, backed with sinew glued on with cooked fish heads. **Maidu:** roots used in basketry.

        Medicine: **Karok:** the bark is scraped off twigs and boiled; the liquid drunk for stomachache.

Citation(s): Chesnut 1902; Dixon 1910; Foster 1944; Gifford 1967; Goddard 1903; Merriam 1967; Nomland 1935,

1938; Schenck and Gifford 1952; Goodrich et al 1980; Strike 1994.
Terminology: **Pomo (Southwestern):** kawani kale, "digging stick plant." **Karok:** xupari.
Citation(s): Gifford 1967; Schenck and Gifford 1952; Goodrich et al 1980.

1236 *Tellima grandiflora*

Family: *Saxifragaceae*
Common Name(s): Fringe Cups.

Usage: Medicine: **Maidu:** roots, chewed, to relieve colds, stomach aches; decoction used to stimulate poor appetites.
Citation(s): Strike 1994.

1237 *Tetradymia canescens*

Family: *Asteraceae*
Common Name(s): Spineless Horsebrush.

Usage: Medicine: **Modoc (Lutuami):** root infusion used to soothe sore eyes.
           Misc: **Kawaiisu (Tehachapi):** named but not used.
Citation(s): Zigmond 1981; Strike 1994.
Terminology: **Kawaiisu (Tehachapi):** sanaco?ovibɨ keevowadi, "mountain sanaco?ovibɨ."
Citation(s): Zigmond 1981.

1238 *Tetradymia spinosa*

Synonym(s): *Tetradymia spinosa* var. *longispina*
Common Name(s): Cotton Thorn.

Usage: Material: **Kawaiisu (Tehachapi):** leaves made into thorns for tattooing needles; pushed

under warts to cause warts to fall off.
Citation(s): Zigmond 1981.
Terminology: **Kawaiisu (Tehachapi):** taciipi.
Citation(s): Zigmond 1981.

1239 *Tetradymia stenolepis*

Common Name(s): Mojave Horsebrush.

Usage: Material: **Kawaiisu (Tehachapi):** leaves made into thorns for tattooing needles; pushed under warts to cause warts to fall off.
Citation(s): Zigmond 1981.
Terminology: **Kawaiisu (Tehachapi):** taciipi.
Citation(s): Zigmond 1981.

1240 *Tetrapteron palmeri*

Synonym(s): *Oenothera palmeri*

Usage: Misc: **Tubatulabal:** considered weeds..
Citation(s): Voegelin 1938.

1241 *Thalictrum fendleri*

Family: *Ranunculaceae*

Usage: Medicine: **Washo:** root decoction given for colds.
Citation(s): Train et al 1941.
Terminology: **Washo:** taba emlu.
Citation(s): Train et al 1941.

1242 *Thalictrum fendleri* var. *polycarpum*

Synonym(s): *Thalictrum polycarpum*
Common Name(s): Meadow Rue.

Usage: Medicine: **Wailaki:** cure headache by washing their heads with the juices from crushed stems and leaves. **Yuki:** for a sprain, a poultice of pounded plant covered with a hot stone is applied. **Maidu:** poultice applied to sprains; root decoction used to cure colic.

Misc: **Wailaki:** believe that the plant is capable of making dead Indians have bad dreams if allowed to grow on their graves; when living friends feel their conscience troubled, they go out to grave and, if they find the plant growing there, the dig it up, and, as a sort of propitiation for their neglect, wash their hands with juices from the crushed stems and leaves.

Citation(s): Chesnut 1902; Curtin 1957; Strike 1994.
Terminology: **Yuki:** kin-ku-hach.
Citation(s): Curtin 1957.

1243 *Thalictrum occidentale*

Synonym(s): *Thalictrum occidentale* var. *palousense*
Common Name(s): Western Meadow Rue.

Usage: Medicine: **Maidu:** root decoction used to treat headache, bronchial ailments, eye problems.

Citation(s): Strike 1994.

1244 *Thamnosma montana*

Family: *Rutaceae*
Common Name(s): Turpentine Broom.

Usage: Medicine: **Kawaiisu (Tehachapi):** stems are boiled; decoction drunk, hot for relief of colds and chest pains; powdered form kept in a bag, a pinch put in deer tracks to slow down the animal. **Panamint Shoshone (Koso):** crushed stems rubbed into sore to promote healing; decoction used as antiseptic, drunk as a tonic.

Citation(s): Zigmond 1981; Strike 1994.

Terminology: **Kawaiisu (Tehachapi):** muguruuvɨ.

Citation(s): Zigmond 1981.

1245 *Thelypodium* spp.

Family: *Brassicaceae*

Usage: Misc: **Mono Lake Paiute:** named but not used.

Citation(s): Steward 1933.

Terminology: **Mono Lake Paiute:** tsaʹava.

Citation(s): Steward 1933.

1246 *Thermopsis macrophylla*

Family: *Fabaceae*

Common Name(s): False Lupine.

Usage: Medicine: **Pomo (Southwestern):** leaves boiled and the cooled decoction used as an eyewash when the eyes are sore and vision was difficult; tea used to slow down menstrual flow.

Citation(s): Gifford 1967; Goodrich et al 1980.

Terminology: **Pomo (Southwestern):** amalashima, "jackrabbit" (amala) "ear" (shima).

Citation(s): Gifford 1967.

1247 Thuja plicata

Family: *Cupressaceae*
Common Name(s): Canoe Cedar, Giant Arborvitae, Western Red Cedar.

Usage: Material: **Tolowa:** used as twining material in basketry.
Citation(s): Strike 1994.

1248 Thysanocarpus curvipes ssp. elegans

Family: *Brassicaceae*
Synonym(s): *Thysanocarpus elegans*

Usage: Food: **Yuki:** seeds used in pinole.
        Medicine: **Yuki, Maidu:** decoction of the whole plant used to relieve stomach aches.
Citation(s): Chesnut 1902; Strike 1994.

1249 Tonella tenella

Family: *Plantaginaceae*

Usage: Misc: **Karok:** named but not used.
Citation(s): Schenck and Gifford 1952.
Terminology: **Karok:** maxaiyushich, "mountain-imitation-digger-pine-nuts."
Citation(s): Schenck and Gifford 1952.

1250 Torreya californica

Family: *Taxaceae*
Synonym(s): *Tumion californicum*
Common Name(s): California Nutmeg, Wild Nutmeg.

Usage: Food: **Pomo:** the nuts are highly esteemed; roasted. **Maidu (Northern):** nuts eaten.
        Material: **Pomo:** used root as woof in baskets; the

rigid, sharp-pointed leaves used a needles to prick pitch soot into skin in tattooing. **Miwok:** made bows from branches. **Maidu:** root used in basketry. **Maidu (Northern):** used charcoal to rub into cuts made from flint or obsidian for tattooing.

Medicine: **Pomo (Southwestern):** nut cracked and soaked in water overnight; drunk as remedy for tuberculosis. **Costanoan:** nuts, crushed, mixed with animal fat, rubbed on temples to soothe a headache; used to reduce chills, to cause sweating; nut, chewed, to relieve indigestion.

Citation(s): Chesnut 1902; Clark 1904; Dixon 1905; Gifford 1967; Merrill 1923; Goodrich et al 1980; Strike 1994; Rayburn 2012.

Terminology: **Pomo (Southwestern):** kabehe (nut).

Citation(s): Gifford 1967; Goodrich et al 1980.

1251 *Toxicodendron diversilobum*

Family: *Anacardiaceae*
Synonym(s): *Rhus diversiloba*
Common Name(s): Poison-Oak.

Usage: Material: **Panamint Shoshone (Koso):** used the stem as the warp and woof of baskets. **Wintun:** used the stem as warp in burden baskets. **Pomo:** used the stem as foundation in baskets; used the juice as a black dye, rubbed plant on fresh tattoo to irritate skin, make scarring more distinct; **Pomo (Southwestern), Miwok (Coast):** used charcoal or soot for tattoo pigment; **(Lake):** for a rash they bathed in Borax Lake, or applied moss from the lake to sores. **Karok:** twigs are used to split salmon steaks while they are being smoked; leaves used to cover *Chlorogalum pomeridianum* bulbs when they are cooked in the earth oven.

Medicine: **Maidu, Southern (Nisenan), Northwestern (Konkow):** the leaves were eaten both as a preventative and as a cure from the poisonous effects of the plant. **Karok:** used to swallow a little piece in the Spring as protection. **Pomo, Maidu, Yuki:** used to burn out warts and ringworms. **Yuki:** sometimes juice from root placed on warts to make them disappear. **Chumash:** formerly was an important medicinal plant; plasters of powdered material were very effective in healing wounds and lacerations; juice from stem and leaves freshly cut in early Spring was considered most effective remedy for sores, skin cancers, and other persistent sores; also used for calluses and corns on feet. **Wailaki:** used to poultice rattlesnake bites. **Diegueño:** roots, washed carefully, boiled, decoction used to bathe eyes, morning and night, made eyes itch, but said to improve vision. **Chumash:** juice, used to stem flow of blood from cut, plant cut in early Spring, juice, most effective remedy for warts, skin cancers, other persistent sores; root, boiled, cooled, treatment for severe dysentery, diarrhea.

Misc: **Miwok, Shasta:** named but not used.

Citation(s): Barrett and Gifford 1933; Chesnut 1902; Curtin

1957; Curtis 1924; Gifford 1967; Holt 1946; Merrill 1923; Schenck and Gifford 1952; Powers 1877; Timbrook 1987, 2007; Goodrich et al 1980; Strike 1994.

Terminology: **Miwok (Central Sierra):** nukusu. **Pomo (Southwestern):** matiho kale. **Yuki:** kin′-macho, "to cry" (kin). **Karok:** iyunawnval (twigs used to split salmon steaks), yunonwi (twigs used to split salmon steaks). **Chumash (Obispeño, Purisimeño):** wala.

Citation(s): Barrett and Gifford 1933; Gifford 1967; Curtin 1957; Schenck and Gifford 1952; Goodrich et al 1980; Timbrook 2007.

Note(s): after contact, the first symptoms appear: burning itching, redness, small blisters (within a few hours or up to 5 days later). Secondary infections may occur when blisters are broken. Combination of the poison with skin proteins is believed to be immediate. Washing thoroughly with soap only washes off excess poison.

Citation(s): Hardin and Arena 1974; Kinsbury 1964.

1252 *Toxicoscordion fremontii*

Family: *Melanthiaceae*
Synonym(s): *Anticlea Fremontii, Zygadenus Fremontii*
Common Name(s): Black-Bulb Grass-Nut.

Usage: Misc: **Maidu, Southern (Nisenan), Yuki:** considered to be poisonous.

Citation(s): Curtin 1957; Powers 1877.

Note(s): bulbs contain alkaloids; steroid alkaloids of the veratrum group; symptoms of poisoning are those of gastrointestinal irritation and vasomotor collapse.

Citation(s): Kingsbury 1964.

1253 *Toxicoscordion paniculatum*

Synonym(s): *Zigadenus paniculatus*

Usage: Medicine: **Washo:** raw bulb crushed to make wet dressing, or poultice, for rheumatism, sprain, lameness, neuralgia, toothache, or any sort of swelling.
Citation(s): Train et al 1941.
Terminology: **Washo:** koh-gah-des-ma.
Citation(s): Train et al 1941.

1254 *Toxicoscordion venenosum* var. *venenosum*

Synonym(s): *Zigadenus venenosus, Zigadenus venenosum* var. *venenosum*
Common Name(s): Poison Camas, Death Camas.

Usage: Food: **Modoc (Lutuami):** bulbs, dried, soaked in flowing stream for 3 days, re-dried, stored.

Medicine: **Yuki:** bulb cooked, mashed, and bound as a poultice to cure boils, rheumatism, and to alleviate pain caused by strains and bruises; tied to affect parts for about 12 hours. **Maidu:** used as a poultice on bruises, sprains, rheumatic pain, burns, snakebite. **Pomo, Wailaki, Yuki:** used as poultice on bruises, sprains, rheumatic pain. **Chumash:** bulbs, mashed, used as poultice for sores.

Misc: **Yuki:** known to be poisonous, not eaten. **Pomo (Southwestern):** considered poisonous.
Citation(s): Chesnut 1902; Goodrich et al 1980; Strike 1994; Timbrook 2007.
Terminology: **Yuki:** mās. **Pomo (Southwestern):** silom?.

Chumash (Barbareño): mo'yoq'; **(Ynezeño):** moyoq';
(Ventureño): moyoq.

Citation(s): Curtin 1957; Goodrich et al 1980; Timbrook 2007.

Note(s): alkaloids concentrated in the bulb cause muscular
weakness, slow heartbeat, subnormal temperature,
stomach upset with pain, vomiting, diarrhea, excessive
watering of the mouth; children have been poisoned
by eating bulb, or flowers.

Citation(s): Hardin and Arena 1974; Fuller and McClintok 1986.

1255 *Trametes* spp.

Family: *Polyporaceae*
Common Name(s): Tree Fungus.

Usage: Material: **Karok:** polished, smoothed buckskin
with white underside of fungus.

Citation(s): Strike 1994.

1256 *Trichostema lanatum*

Family: *Lamiaceae*
Common Name(s): Wooly Blue Curls, Romero.

Usage: Medicine: **Cahuilla:** leaves and flowers boiled into
tea for relief of stomach ailments; one
cup was "good for anything wrong in
your stomach." **Maidu Northwestern
(Konkow), Wailaki, Yuki:** leaf decoction
used as a general remedy. **Chumash:**
plant, considered as a powerful
disinfectant, humans and livestock,
effective cure for gangrene; leaves,
flowers, tea, drunk for stomachache,
nervous troubles, rheumatism, as a wash
for skin conditions or sores.

Citation(s): Bean and Saubel 1972; Timbrook 2007.

Terminology: **Chumash (Ventureño):** ʾakhiyeʾp.
Citation(s): Timbrook 2007.

1257 *Trichostema lanceolatum*

Common Name(s): Vinegar Weed, Camphor Weed, Tarweed.

Usage: Material: **Karok:** put in bedding to keep fleas away. **Maidu, Yokuts, Pomo:** used as a fish poison.

Medicine: **Maidu:** infusion of leaves used as head wash to cure feverish headaches; mixed with a decoction of the leaves of *Eremocarpus setigerus*, the extract of the leaves is valued to some extent as a wash in cases of typhoid fever; leaves boiled, inhaling steam was used to treat uterine problems. **Miwok:** decoction of the leaves and flowers drunk for colds, malaria, headache, ague, general debility, and the structure of the bladder; bath of the decoction was a preventative measure against ague and smallpox; sitting over a steaming decoction was a cure for uterine trouble, should not be used during pregnancy; leaves were chewed and placed in cavity of or around an aching tooth. **Karok, Maidu Northwestern (Konkow), Miwok:** used as a general cure to relieve pain, cure colds, soothe skin sores, ease stomach ache, reduce fevers, as an insecticide, disinfectant. **Kawaiisu (Tehachapi):** plants brewed to make a tea drunk for relief of colds, stomach ache, soothe skin sores, ease stomach ache, reduce fevers, as an insecticide, disinfectant. **Tubatulabal:** infusion sniffed to stop headache, nosebleed. **Costanoan:** used as a general cure to relieve pain, cure colds, soothe skin sores, ease stomach ache, reduce fevers, as an insecticide, disinfectant; leaves, ground, rubbed on skin for pain remedy, rubbed on face and chest to relieve colds; decoction used on infected sores; leaves, steeped in cold water, used a cold remedy.

Citation(s): Barrett and Gifford 1933; Chesnut 1902; Heizer 1953; Schenck and Gifford 1952; Zigmond 1981; Strike 1994; Rayburn 2012.

Terminology: **Miwok (Central Sierra):** tcūkū′tcū. **Karok:** yufivmatnakvanna, "stinging the nose" (pungent, burning). **Kawaiisu (Tehachapi):** ponoho(m)dɨbɨ.

Citation(s): Barrett and Gifford 1933; Schenck and Gifford 1952; Zigmond 1981.

1258 *Trientalis latifolia*

Family: *Myrsinaceae*
Synonym(s): *Trientalis europaea latifolia*
Common Name(s): Star Flower.

Usage: Medicine: **Maidu:** sap, mixed with water, used as eye wash.

Misc: **Karok:** named but not used.

Citation(s): Schenck and Gifford 1952; Strike 1994.

Terminology: **Karok:** konyepxrichtahitihan, "that-which-grows-in-the-oaks."

Citation(s): Schenck and Gifford 1952.

1259 *Trifolium* spp.

Family: *Fabaceae*
Common Name(s): Wild Clover, Hairy Clover, Clover.

Usage: Food:

Maidu, Southern (Nisenan): eaten raw when young and tender, or boiled for greens; gathered green, soaked, dried for Winter. **Yokuts, Wappo:** eaten raw. **Wiyot:** blossoms and leaves were eaten raw. **Miwok:** eaten raw or steamed; steamed clover dried for later use. **Pomo (Southwestern):** new foliage eaten raw as greens. **Yuki:** eaten as greens; used seeds in pinole. **Bear River:** ate seeds and raw leaves with roasted pepper nuts; leaves eaten raw with meat and other foods. **Huchnom:** used as greens. **Luiseño:** eaten fresh and cooked. **Salinan:** greatly relished and eaten raw. **Wintun, Northern (Wintu):** eaten raw in Spring, when tender, often sprinkled with salt, sometimes steamed; **Central (Nomlaki):** used as greens. **Sinkyone:** always eaten raw. **Cahuilla:** leaves and seeds gathered from February through July; seeds ground for mush, leaves eaten as greens. **Chumash:** leaves, eaten raw like lettuce; seeds, eaten "quite a bit."

Material:

Maidu: while gathering the plant, young girls often put the flowers in their hair.

Medicine:

Pomo (Southwestern): blossoms were boiled, the decoction drunk as a medicine to check vomiting. **Costanoan:**

used as a cathartic; foliage eaten, or decocted for a purgative. **Shasta:** bulbs, fresh, mashed, used as a poultice to bring boils to a head.

Citation(s): Barrett and Gifford 1933; Beals 1933; Curtin 1957; Driver 1936; Du Bois 1935; Foster 1944; Gayton 1948; Gifford 1967; Goldschmidt 1951; Loud 1918; Mason 1912; Nomland 1935, 1938; Powers 1877; Sparkman 1908; Bean and Saubel 1972; Strike 1994; Rayburn 2012; Timbrook 2007.

Terminology: **Miwok (Central Sierra):** pumusayu. **Wiyot:** rokoiyi. **Cahuilla:** tre'evula (derived from Spanish *trebol*). **Chumash (Barbareño):** sha'puk'; **(Ynezeño):** shapuk'; **(Ventureño):** shapuk.

Citation(s): Barrett and Gifford 1933; Loud 1918; Bean and Saubel 1972; Timbrook 2007.

1260 *Trifolium albopurpureum*

Common Name(s): Rancheria Clover.

Usage: Food: **Pomo (Southwestern):** leaves eaten alone or with salt or peppernut cakes.

Citation(s): Goodrich et al 1980.

Terminology: **Pomo (Southwestern):** ?méhso.

Citation(s): Goodrich et al 1980.

1261 *Trifolium bifidum* var. *decipens*

Usage: Food: **Yuki:** eaten very sparingly and only when very young.

Citation(s): Chesnut 1902.

1262 *Trifolium ciliolatum*

Synonym(s): *Trifolium ciliatum*
Common Name(s): Tree Clover.

Usage: Food: **Luiseño:** eaten both cooked and raw; the seeds are also used. **Wailaki:** eaten as greens. **Miwok:** steamed and eaten, or dried and stored.

Misc: **Yuki, Pomo, Maidu:** considered as not fit to eat.

Citation(s): Barrett and Gifford 1933; Chesnut 1902; Sparkman 1908.

Terminology: **Miwok (Central Sierra):** patcuku, olisa; **(Southern Sierra):** pa'tcūkū. **Luiseño:** mukalwut.

Citation(s): Barrett and Gifford 1933; Sparkman 1908.

1263 *Trifolium cyathiferum*

Usage: Food: **Yuki:** little used.
Citation(s): Chesnut 1902.

1264 *Trifolium dichotomum*

Usage: Food: **Pomo:** leaves eaten sparingly when very young; seeds sometimes gathered for pinole.
Citation(s): Chesnut 1902.

1265 *Trifolium facatum*

Synonym(s): *Trifolium virescens, Trifolium facatum* var. *virescens*

Usage: Food: **Yuki:** eaten when plant is in flower; flower and seeds pods also used for food; seeds are not separated but eaten

raw. **Pomo (Southwestern):** sweet
flowers and leaves eaten alone or with
salt and peppernut cakes.
Citation(s): Chesnut 1902; Goodrich et al 1980.
Terminology: **Pomo (Southwestern):** pʰa?am?so.
Citation(s): Goodrich et al 1980.

1266 *Trifolium gracilentum*

Usage: Food: **Luiseño:** eaten both cooked and raw.
Citation(s): Sparkman 1908.
Terminology: **Luiseño:** ke'kesh.
Citation(s): Sparkman 1908.

1267 *Trifolium microcephalum*

Usage: Food: **Luiseño:** eaten cooked.
Citation(s): Sparkman 1908.
Terminology: **Luiseño:** pehevi.
Citation(s): Sparkman 1908.

1268 *Trifolium obtusiflorum*

Usage: Food: **Luiseño:** eaten cooked. **Pomo:** eaten raw;
sometimes dipped in salt water to add
relish to taste.
Citation(s): Chesnut 1902; Sparkman 1908.
Terminology: **Luiseño:** shoo'kut.
Citation(s): Sparkman 1908.

1269 *Trifolium pratense*

Usage: Food: **Miwok:** named but not used.
Citation(s): Barrett and Gifford 1933.
Terminology: **Miwok (Central Sierra):** yenwa.

Citation(s): Barrett and Gifford 1933.
Note(s): native of Eurasia.
Citation(s): Munz and Keck 1965.

1270 *Trifolium variegatum*

Usage: Food: **Yuki:** used greens.
Citation(s): Chesnut 1902.

1271 *Trifolium willdenovii*
Synonym(s): *Trifolium tridentatum*
Common Name(s): Tomcat Clover.

Usage: Food: **Luiseño:** eaten both raw and cooked; the seeds are also used. **Owens Valley Paiute:** seeds eaten; the entire plant was eaten uncooked as greens. **Miwok:** eaten steamed or raw before it bloomed; leaves, stems, buds were eaten; for storage the steamed leaves were spread on *Wyethia helenioides* leaves to dry in the sun; dried leaves either soaked in water or boiled before eaten.
Citation(s): Barrett and Gifford 1933; Sparkman 1908; Steward 1933.
Terminology: **Owens Valley Paiute:** posi´da. **Miwok (Central Sierra):** wilamü. **Luiseño:** chokat.
Citation(s): Steward 1933; Barrett and Gifford 1933; Sparkman 1908.

1272 *Trifolium wormskioldii*
Synonym(s): *Trifolium involucratum Fendlari*
Common Name(s): Cow Clover.

Usage: Food: **Yuki:** toward last of June only this species is gathered and eaten; flowers are eaten as well. **Owens Valley Paiute:** eaten uncooked as greens. **Miwok:** eaten raw, never cooked or dried; both leaves and flowers eaten; if wilted or dry, the leaves are soaked and stirred in basket of cold water for 10 minutes, making a sour drink. **Kawaiisu (Tehachapi):** green leaves eaten raw with salt. **Pomo (Southwestern):** flowers and leaves are sweet; gathered in large quantities in late Spring or early Summer; eaten alone or with salt and peppernut cakes.

Citation(s): Barrett and Gifford 1933; Chesnut 1902; Steward 1933; Voegelin 1938; Zigmond 1981; Goodrich et al 1980.

Terminology: **Miwok (Central Sierra):** saksamö. **Owens Valley Paiute:** po′tsidava. **Kawaiisu (Tehachapi):** paaziwaatïbï. **Pomo (Southwestern):** boho.

Citation(s): Barrett and Gifford 1933; Steward 1933; Zigmond 1981; Goodrich et al 1980.

1273 *Trillium* spp.

Family: *Melanthiaceae*

Usage: Medicine: **Shasta:** the bulbs were grated, with a drop of water if dry, without water if green, and put on a boil to bring it to a head.

Misc: **Pomo (Southwestern):** said to be poisonous.

Citation(s): Holt 1946; Goodrich et al 1980.

Terminology: **Pomo (Southwestern):** silom?.

Citation(s): Goodrich et al 1980.

1274 *Trillium chloropetalum*

Common Name(s): Wakerobin, Giant Trillium.

Usage: Medicine: **Costanoan:** used poultice to relieve chest pains; ate leaves, drank a decoction, as a purgative; hot compresses used for chest pain. **Pomo:** used decoction to arrest vomiting. **Maidu:** used to treat sores, wounds, boils, toothache, sore throat.

Citation(s): Strike 1994; Rayburn 2012.

1275 *Trillium chloropetalum*

Synonym(s): *Trillium sessile* ssp. *giganteum*, *Trillium chloropetalum* var. *giganteum*

Common Name(s): Giant Wakerobin.

Usage: Medicine: **Yuki, Wailaki:** drink a decoction for "any kind of sick;" both the bruised leaves and crushed root used as a poultice for boils.

Citation(s): Chesnut 1902.

1276 *Trillium ovatum*

Common Name(s): Coast Trillium.

Usage: Food: **Pomo (Southwestern):** sweet flowers and leaves eaten alone or with salt or peppernut cakes.

Medicine: **Karok:** the root is scraped until juices come and then is placed immediately over a boil and allowed to stay; when it is taken off, it has brought the boil to a head. **Yuki, Wailaki:** leaves, root, crushed, used as a poultice to treat most illness, to prevent deep sleep.

Citation(s): Schenck and Gifford 1952; Goodrich et al 1980; Strike 1994.

Terminology: **Karok:** anupuhich, "imitation navel" (root resembles that of *Smilacina racemosa* var. *amplexicaulis*). **Pomo (Southwestern):** gʰabóhso.

Citation(s): Schenck and Gifford 1952; Goodrich et al 1980.

1277 *Triphysaria versicolor* ssp. *faucibarbata*

Family: *Orobanchaceae*

Synonym(s): *Orthocarpus faucibarbatus*

Usag: Misc: **Pomo (Southwestern):** flowers used in dance wreaths at Strawberry Festival.

Citation(s): Goodrich et al 1980.

Terminology: **Pomo (Southwestern):** s'amo· hu?uy, "fly eyes."

Citation(s): Goodrich et al 1980.

1278 *Triteleia grandiflora*

Family: Themidaceae

Synonym(s): *Brodiaea douglasii* var. *howellii*

Usage: Food: **Maidu (Northern):** bulbs eaten.

Citation(s): Dixon 1905.

1279 *Triteleia hyacinthina*

Synonym(s): *Brodiaea lactea, Hesperocordium lacteum, Hookera nyocintlima lactea, Brodiaea hyacinthina*

Common Name(s): White Brodiaea, White-Flowered Grass Nut.

Usage: Food: **Pomo:** bulbs eaten raw or cooked. **Maidu (Northern):** bulbs eaten; **Southern (Nisenan):** bulbs eaten raw, or roasted in ashes, or boiled. **Northern**

Paiute (Paviotso): the bulbs were eaten. **Miwok:** bulbs dug about the first of May, cooked in earth oven. **Atsugewi:** cooked in earth oven all night, mashed and made into cakes, cooked again in pit, dried, stored; soaked for use.

Citation(s): Barrett 1952; Barrett and Gifford 1933; Chesnut 1902; Dixon 1905; Garth 1953; Kelly 1932; Powers 1877.

Terminology: **Miwok (Central Sierra):** wüsumayü. **Northern Paiute (Paviotsu):** wini'da. **Atsugewi:** boskĭtira. **Pomo (Central):** stala'i bū.

Citation(s): Barrett and Gifford 1933; Kelly 1932; Garth 1953; Barrett 1952.

1280 *Triteleia ixioides*

Synonym(s): *Brodiaea ixioides, Brodiaea lutea*
Common Name(s): Golden Brodiaea, Yellow-Blossom Grass-Nut.

Usage: Food: **Miwok:** bulbs were eaten.
Citation(s): Barrett and Gifford 1933.
Terminology: **Miwok (Central Sierra):** silūwū.
Citation(s): Barrett and Gifford 1933.

1281 *Triteleia laxa*

Synonym(s): *Hookera laxa, Triteleia laxa, Brodiaea laxa*
Common Name(s): Grass Nut, Blue-Flowered Brodiaea.

Usage: Food: **Pomo:** bulbs eaten. **Yuki:** bulbs were dug and cooked; esteemed as food. **Hupa:** bulbs were roasted in ashes or boiled in baskets. **Karok:** bulbs cooked overnight in an earthen oven. **Pomo**

(**Southwestern**): bulbs cooked in hot
ashes or boiled for food.

Material: **Pomo:** used flowers in dance wreaths.

Misc: **Kawaiisu (Tehachapi):** named but not
used.

Citation(s): Barrett 1952; Chesnut 1902; Curtin 1957; Gifford
1967; Goddard 1903; Schenck and Gifford 1952;
Zigmond 1981; Goodrich et al 1980; Strike 1994.

Terminology: **Pomo (Central):** ba'ba; **(Southwestern):**
hibula; **(Southeastern):** būn. **Yuki:** hut. **Karok:** pufish
tayish, "deer potato." **Kawaiisu (Tehachapi):**
kogosivɨ.

Citation(s): Barrett 1952; Gifford 1967; Curtin 1957;
Schenck and Gifford 1952; Zigmond 1981;
Goodrich et al 1980.

1282 *Triteleia peduncularis*

Synonym(s): *Brodiaea peduncularis*
Common Name(s): White-Flowered Brodiaea.

Usage: Food: **Pomo:** bulbs eaten. **Yuki:** bulbs used to
some extent.

Citation(s): Barrett 1952; Chesnut 1902.

1283 *Triticum aestivum*

Family: *Poaceae*
Common Name(s): Wheat Grass, Wheat.

Usage: Food: **Pomo (Southwestern):** seed used in
mush. **Cahuilla:** parched and ground
into flour; often mixed with other wild
seeds, such as chia, to form mush.

Citation(s): Goodrich et al 1980; Bean and Saubel 1972

Terminology: **Cahuilla:** pachesal (?).

Citation(s): Bean and Saubel 1972.

Note(s): native of the Old World; first introduced to the **Pomo (Southwestern)** area by Russians at Fort Ross.
Citation(s): Munz and Keck 1965; Goodrich et al 1980.

1284 *Tropidocarpum gracile*
Family: *Brassicaceae*

Usage: Misc: **Tubatulabal:** considered weeds.
Citation(s): Voegelin 1938.

1285 *Typha* spp.
Family: *Typhaceae*
Common Name(s): Cat-Tail.

Usage: Food: **Yokuts:** roots made into flour; the blossoms into yellow bread. **Northern Paiute (Paviotso):** ate seeds, but considered root more important food. **Yuki:** the white base of the young stem was eaten raw. **Pomo (Southwestern):** young shoots eaten. **Chumash:** thick starchy rhizomes, dug, pulverized, dried, balls of dough baked in ashes; spikes, eaten like corn; pollen, before fully rip, stirred in water to make thun, uncooked mush.

Material: **Chumash:** stems, leaves, thatching in houses, mats.

Misc: **Pomo (Southwestern):** young girls and women not supposed to walk near areas where these plants grow.

Citation(s): Curtin 1957; Gayton 1948; Kelly 1932; Goodrich et al 1980; Timbrook 2007.

Terminology: **Yuki:** sŌn. **Pomo (Southwestern):** baco.
Chumash (Barbareño, Ynezeño): taqsh; **(Ventureño):**

khap.
Citation(s): Curtin 1957; Goodrich et al 1980; Timbrook 2007.

1286 *Typha augustifolia*

Common Name(s): Cat-Tail.

Usage: Food: **Kawaiisu (Tehachapi):** white portions at base of stems and brown flowers eaten raw; when green seeds are eaten.

Material: **Owens Valley Paiute:** bundles of the material were used to cover the Winter cook house.

Citation(s): Steward 1933; Zigmond 1981.
Terminology: **Kawaiisu (Tehachapi):** to?ivɨ.
Citation(s): Zigmond 1981.

1287 *Typha latifolia*

Common Name(s): Cat-Tail, Cattail Rush.

Usage: Food: **Pomo:** the roots and tender young shoots eaten as greens. **Tubatulabal:** used roots. **Kawaiisu (Tehachapi):** white portions at base of stems and brown flowers eaten raw; when green, seeds are eaten. **Cahuilla:** gathered from June through July; root dried, ground into meal; pollen used to make cakes and mush.

Material: **Modoc (Lutuami):** used leaf as warp, woof and white pattern in baskets; leaves dyed black by burying in mud. **Owens Valley Paiute:** bundles of the material were used to cover the Winter cook house. **Pomo:** used as house thatching. **Maidu (Northern):** made mats. **Cahuilla:** stalks used as matting, bedding, in constructing ceremonial bundles.

Medicine: **Washo:** the young flowering heads sometimes are eaten to stop diarrhea. **Cahuilla:** roots used to heal bleeding wounds.

Citation(s): Barrett 192o; Dixon 1905; Merrill 1923; Steward 1933; Train et al 1941; Voegelin 1938; Zigmond 1981; Bean and Saubel 1972; Strike 1994.

Terminology: **Pomo (Eastern):** hal; **(Southwestern):** la. **Washo:** mah-ha-tahl-<u>lahl</u>. **Kawaiisu (Tehachapi):** to?ivɨ. **Cahuilla:** ku'ut.

Citation(s): Barrett 1952; Train et al 1941; Zigmond 1981; Bean and Saubel 1972.

1288 *Ulmus* spp.

Family: *Ulmaceae*
Common Name(s): Elm.

Usage: Medicine: **Diegueño:** bark, cut and molded to hold broken limb in place, tied with wet buckskin straps, sap from bark soothed the limb.

Citation(s): Strike 1994.

1289 *Ulva californica*

Family: *Ulvaceae*
Synonym(s): *Ulva augusta*
Common Name(s): Sea Weed, Sea Lettuce.

Usage: Misc: **Coast Yuki:** named but not used.
Citation(s): Gifford 1939.
Terminology: **Yuki:** sielilbal, "sandpiper rock leaves."
Citation(s): Gifford 1939.
Note(s): species nomenclature recently changed.

Citation(s): Tanner 1986.

1290 *Ulva lactuca*
Common Name(s): Sea Lettuce.

Usage: Food: **Pomo (Southwestern):** light-green peppery seaweed used as flavoring with other seaweeds.
Citation(s): Goodrich et al 1980.
Terminology: **Pomo (Southwestern):** s´ihtaʧʰóno.
Citation(s): Goodrich et al 1980.

1291 *Umbellularia californica*
Family: *Lauraceae*
Common Name(s): Pepperwood, California Laurel, Peppernut Tree, Laurel, Mountain Laurel, Balm of Heaven, Spice Bush.

Usage: Food: **Hupa, Miwok:** the nuts are roasted in ashes and eaten. **Miwok (Coast):** fruit made into beverage. **Karok:** nuts are parched in the ashes of a fire, cracked open, and eaten. **Maidu:** bark of the root used to make a beverage. **Pomo (Southwestern):** nuts gathered from ground, roast, hulled on stone slab, pounded in mortar into meal, rolled into balls, sun dried, and stored in grass-lined baskets. **Yuki:** nuts gathered in November; the green flesh and yellow butt-tip are "real good to eat;" nuts roasted and eaten with clover, or roasted and pounded into small mass. **Coast Yuki:** nuts were eaten. **Bear River:** nuts roasted in ashes; packed in small baskets for Winter; used for pepper, eaten with meat, vegetables. **Huchnom:** used nuts. **Wappo:** obtained "pepper" from the nuts; prepared the same as **Pomo**. **Esselen:** fruits, kernels, gathered in the Fall, prepared for food.

Material: **Shasta:** mush paddles made from this wood because the wood is smooth and fine grained. **Karok:** children throw leaves into fire to hear them crack like firecrackers. **Monache (Western Mono):** used leaves as fish poison. **Yuki:** leafy branches scattered about to keep fleas away. **Chumash:** wood, made bowls; twigs, burned, hunters stood in smoke, to hide human smell.

Medicine: **Wappo:** leaves used by doctors for purification after a case; tea from leaves drunk for sores and boils; also used to cure colds. **Yuki:** leaves soaked in hot water; the hot water used tp bathes in for 2-3 days, twice a day to cure rheumatism; vapor from leaves in fire esteemed as a cure for many diseases; leaves good for driving fleas away; leaves crushed and inhaled vapors to cure a headache; also applied leaves externally to cure headache; applied as a strong decoction to cure headaches; also to kill vermin; sometimes taken internally for both headaches and stomach ache. **Miwok:** leaves and twigs bound on forehead as a cure for headache. **Pomo (Southwestern):** leaves heated and used as poultice for rheumatism and neuralgic pains; small, leafy branches hung in homes for protection against harm; leaves rubbed on hunters to take body smell away. **Karok:** boughs are buried in coals of a fire to fumigate the house when colds or other kinds of sickness are prevalent [see *Artemisia Douglasiana* for other uses.] **Cahuilla:** leaves rubbed on hands and inhaled, or rubbed against face which produces a distressing headache; used to cure headaches. **Costanoan:** leaves, burned, smoke used to rid one of fleas; decoction used as wash to treat Poison Oak dermatitis, rubbed on body for rheumatism, damped and stuck on forehead for headache. **Esselen:** aromatic leaves used as medicine. **Chumash:** leaves, boiled, drunk for colds, treat diarrhea, bath water to relieve rheumatism.

Citation(s): Barrett and Gifford 1933; Chesnut 1902; Curtin 1957; Driver 1936; Foster 1944; Gifford 1939, 1967; Goddard 1903; Heizer 1953; Holt 1946; Nomland 1938; Schenck and Gifford 1952; Goodrich et al 1980; Bean and Saubel 1972; Strike 1994; Rayburn 2012; Breschini and Haversat 2004; Timbrook 2007.

Terminology: **Miwok (Central Sierra):** loko. **Yuki:** bokumol, bokum (nut), po´-kum ōl, pō´kum. **Pomo (Eastern, Southeastern):** behe´; **(Central)** bahe´; **(Southwestern):** hem, behem, behe (nut). **Karok:** pahip, pah (nut). **Chumash (Barbareño, Ynezeño):** psha'n; **(Ventureño):** psha'an.

Citation(s): Barrett and Gifford 1933; Gifford 1939 1967; Curtin 1957; Barrett 1952; Schenck and Gifford 1952; Timbrook 2007.

Note(s): fruit persistent on the branches after ripening; falls late

in the Autumn; the kernel contains starch and from 40-
60 percent of a peculiar fatty oil which may have some
food value.

Citation(s): Sargent 1922; Chesnut 1902.

1292 *Urtica* spp.

Family: *Urticaceae*
Common Name(s): Nettle.

Usage: Material: **Yokuts (Southern Valley):** string made
from fibers. **Luiseño:** made 2-ply
cordage for nets from fibers.

Medicine: **Maidu, Kawaiisu (Tehachapi):** leaf
poultice used for headache. **Costanoan:**
used as disinfectant; decoction used for
hives, infected sores. **Gabrielino:** stalks,
burned on skin to relieve rheumatic
pain; stalks struck on body to treat
sluggish blood. **Miwok (Coast):** root tea
eased menstrual pain. **Costanoan:**
decoction used for hives, infected sores;
aching joints beaten lightly with leaves
"to reduce the pain.".

Citation(s): Curtis 1924; Gayton 1948; Strike 1994; Rayburn
2012.

1293 *Urtica dioica* spp. *gracilis*

Synonym(s): *Urtica Lyallii, Urtica breweri, Urtica serra*
Common Name(s): Nettle, Sierra Nettle.

Usage: Medicine: **Pomo (Southwestern):** used as counter-
irritant for rheumatism and other pains.

Citation(s): Goodrich et al 1980.
Terminology: **Pomo (Southwestern):** ?ohom?.
Citation(s): Goodrich et al 1980.

1294 *Urtica dioica* spp. *holosericea*

Synonym(s): *Urtica californica, Urtica holosericea, Urtica gracilis, Urtica gracilis* var. *holosericea*

Common Name(s): Coast Nettle.

Usage: Food: **Cahuilla:** leaves eaten raw or boiled as greens.

Material: **Luiseño:** a fiber is obtained, bit is not much esteemed. **Kawaiisu (Tehachapi):** produce cordage from fibers; made rabbit and carrying nets. **Cahuilla:** fiber made into cordage, used in bowstrings, and basketry.

Medicine: **Pomo (Southwestern):** used as counter-irritant for rheumatism and other pains. **Costanoan:** used as counter-irritant for rheumatism; root tea drunk to cure consumption. **Miwok:** to cure rheumatism, a decoction was made of the roots and the affected part was bathed; the powdered leaves were sometimes rubbed on the affected part; to relieve pains the affected part was struck with a branch. **Kawaiisu (Tehachapi), Cahuilla:** used leaves as counter-irritant for sore limbs. **Modoc (Lutuami):** used bark as woof and rim binding in baskets; bark fiber made into cordage. **Northern Paiute (Paviotso):** used bark as woof in baskets.

Misc: **Karok:** named but not used (?). **Coast Yuki, Owens Valley Paiute:** named but not used.

Citation(s): Schenck and Gifford 1952; Goodrich et al 1980;

Merrill 1923; Strike 1994.

Terminology: **Karok:** anievxaat, "smells like under arm." **Pomo (Southwestern):** ?ohom?, oʼhom. **Owens Valley Paiute:** tuʹnünūʹvüva. **Miwok:** sosoloʹyū. **Luiseño:** shakishla. **Yuki:** yistot. **Karok:** ʾakviin. **Kawaiisu (Tehachapi):** kʷiči?atabɨ. **Cahuilla:** chikishyam.

Citation(s): Schenck and Gifford 1952; Goodrich et al 1980; Steward 1933; Barrett and Gifford 1933; Sparkman1908; Gifford 1939, 1967; Zigmond 1981; Bean and Saubel 1972.

1295 *Utricularia macrorhiza*

Family: *Lentibulariaceae*
Synonym(s): *Utricularia vulgaris*
Common Name(s): Bladderwort.

Usage: Medicine: **Maidu:** used as diuretic, to treat dysentery.
Citation(s): Strike 1994.

1296 *Usnea barbara*

Family: *Usneaceae*
Common Name(s): Tree Moss.

Usage: Misc: **Karok:** named but not used.
Citation(s): Schenck and Gifford 1952.
Terminology: **Karok:** ashaxaeme.
Citation(s): Schenck and Gifford 1952.

1297 *Usnea californica*

Common Name(s): Spanish Moss.

Usage: Material: **Pomo (Southwestern):** used as diapers for babies.
Citation(s): Gifford 1967.

Terminology: **Pomo (Southwestern):** kôchih.
Citation(s): Gifford 1967.

1298 *Usnea lacunosa*
Common Name(s): Lichen.

Usage: Material: **Yuki:** used as bedding material.
Citation(s): Chesnut 1902.

1299 *Vaccaria hispanica*
Family: *Caryophyllaceae*
Synonym(s): *Vaccaria vulgaris, Vaccaria segetalis*

Usage: Misc: **Tubatulabal:** considered weeds.
Citation(s): Voegelin 1938.
Note(s): naturalized from Europe, Eurasia
Citation(s): Munz and Keck 1965; Baldwin et al 2012.

1300 *Vaccinium ovatum*
Family: *Ericaceae*
Common Name(s): Huckleberry, California
Huckleberry.

Usage: Food: **Hupa:** the berries are eaten; remain on bushes until Christmas. **Wiyot:** the berries were eaten. **Karok:** berries gathered in the Fall, preferably after the first frost; stored in baskets. **Pomo:** gathered berries in July and August; eaten immediately. **Pomo (Southwestern):** berries eaten either fresh or dried; the dried were boiled before using; fresh eaten raw or were stone-boiled in a coiled basket. **Yurok:** berries used extensively for food. **Coast**

Yuki: berries eaten raw.

Material: Hupa: small arrows for use by boys were made. Karok: leaves used to cover *Chlorogalum pomeridianum* bulbs in the earth oven

Medicine: Pomo (Southwestern): leaves boiled, tea drunk as a medicine for diabetes.

Citation(s): Chesnut 1902; Gifford 1939, 1967; Goddard 1903; Loud 1918; Merriam 1967; Schenck and Gifford 1952; Goodrich et al 1980.

Terminology: Wiyot: mò´kel. Yuki: shimalo, shiman (ripe berries). Hupa: tcwiltc. Pomo (Southwestern): sununu, s´u?nú´nu qʰale, "huckleberry plant." Karok: purissipam, puriss (berry).

Citation(s): Loud 1918; Gifford 1939, 1967; Goddard 1903; Schenck and Gifford 1952.

1301 *Vaccinium parvifolium*

Common Name(s): Red Bilberry.

Usage: Food: Karok: the ripe berries are eaten raw. Yurok: berries used extensively for food. Pomo (Southwestern): berry eaten fresh.

Material: Karok: the best brooms are made from twigs and stems because they are pliable and do not break. Yurok: the long, straight, green branches are used for brooms.

Citation(s): Merriam 1967; Schenck and Gifford 1952; Goodrich et al 1980.

Terminology: Karok: meisiipar. Pomo (Southwestern): sabíto·yi?.

Citation(s): Schenck and Gifford 1952; Goodrich et al 1980.

1302 *Valeriana* spp.

Family: *Valerianaceae*

Usage: Food: **Northern Paiute (Paviotso):** the roots
were left overnight in an earth oven
without a fire, in the morning the oven
was uncovered and smelled bad but the
roots were sweet to the taste. **Maidu:**
roots eaten.
Citation(s): Kelly 1932; Strike 1994.
Terminology: **Northern Paiute (Paviotso):** kuyi.
Citation(s): Kelly 1932.

1303 *Vancouveria planipetala*

Family: *Berberidaceae*
Synonym(s): *Vancouveria parviflora*
Common Name(s): Inside-Out Flower.

Usage: Medicine: **Hupa:** tea drunk during pregnancy to
insure that baby would be small and
easily delivered.
         Misc: **Karok, Pomo (Southwestern):** named
but not used.
Citation(s): Schenck and Gifford 1952; Goodrich et al 1980;
Strike 1994.
Terminology: **Karok:** absikkinayachis. **Pomo (Southwestern):**
šihpʰa šálu? qʰale, "leaf glisten plant."
Citation(s): Schenck and Gifford 1952; Goodrich et al 1980.

1304 *Veratrum californicum*

Family: *Melanthiaceae*
Common Name(s): Corn Lily, False Hellebore, Skunk
Cabbage.

Usage: Food: **Miwok:** bulb roasted in ashes, peeled and eaten; not stored.

Material: **Karok:** inner white stem is torn into ribbons And braided into girl's hair for ornament.

Medicine: **Washo:** half-cupful of concentrated decoction of root used as emetic; pulped, mashed raw root applied with friction serves as a liniment; dry, powdered root sometimes sprinkled on sores to promote healing. **Maidu:** root used to poultice neck and throat problems, sores, rheumatic pain, sore nipples, infections; root decoction taken as contraceptive. **Tubatulabal:** used to stun fish.

Citation(s): Barrett and Gifford 1933; Schenck and Gifford 1952; Train et al 1941; Strike 1994.

Terminology: **Northern Paiute (Paviotso):** tüba´sup. **Miwok (Central Sierra):** sülumta. **Karok:** oxorupan, "shred the stem." **Washo:** bah-<u>do</u>-po.

Citation(s): Kelly 1932; Barrett and Gifford 1933; Schenck and Gifford 1952.

1305 *Verbascum blattaria*

Family: *Scrophulariaceae*

Usage: Misc: **Miwok:** named but not used.

Citation(s): Barrett and Gifford 1933.

Terminology: **Miwok (Central Sierra):** yotowina.

Citation(s): Barrett and Gifford 1933.

Note(s): naturalized from Eurasia.

Citation(s): Munz and Keck 1965.

1306 *Verbascum thapsus*

   Common Name(s): Mullein.

Usage: Medicine: **Atsugewi:** leaves boiled and drunk for bad colds and rheumatism; leaves pounded and raw as a poultice on cuts; during sweat bath the crushed leaves might be rubbed over the body.

Citation(s): Garth 1953.
Note(s): naturalized from Eurasia.
Citation(s): Munz and Keck 1865.

1307 *Verbena hastata*

   Family: *Verbenaceae*
   Common Name(s): Wild Vervain.

Usage: Medicine: **Costanoan:** root used to relieve stomach pain, as a general remedy. **Maidu Northwestern (Konkow):** root decoction taken to treat coughs, colds, applied to wounds to accelerate healing.

Citation(s): Strike 1994.

1308 *Verbena lasiostachys*

   Common Name(s): Western Verbena.

Usage: Medicine: **Costanoan:** tea used as remedy for "fever of the stomach," remedy for typhoid fever.

Citation(s): Rayburn 2012.

1309 *Verbena lasiostachys* var. *lasiostachys*
Common Name(s): Western Vervain.

Usage: Medicine: **Chumash:** root, boiled, decoction drunk for fever; leaves, crushed, soaked in water, strained, used to wash hair, to freshen scalp.
Citation(s): Timbrook 2007.
Terminology: **Chumash (Barbareño):** shikhwapsh 'i'ashk'á'; **(Ynezeño):** s'uwmo' 'oyoso; **(Ventureño):** 'also'o' shikhwapsh 'i'ashk'á'.
Citation(s): Timbrook 2007.

1310 *Viburnum ellipticum*
Family: *Caprifiliaceae*
Common Name(s): Viburnum.

Usage: Material: **Hupa:** the small black fruits were strung along with dentalium shells, shells of *Olivella biplicata* and Pine Nuts (*Pinus spp.*) shells. **Wiyot:** nuts were used as beads.
Citation(s): Goddard 1903; Loud 1918.

1311 *Vicia* spp.
Family: *Leguminosae*

Usage: Misc: **Pomo (Southwestern):** named but not used.
Citation(s): Goodrich et al 1980.
Terminology: **Pomo (Southwestern):** ?ama·la má?a, "rabbit food."
Citation(s): Goodrich et al 1980.

1312 *Vicia americana* ssp. *americana*

Synonym(s): *Vicia americana, Vicia californica*

Usage: Food: **Yuki:** when young, cooked and eaten as greens.

Misc: **Yuki:** small bunch of the roots kept in pocket for luck while gambling. **Karok:** named but not used.

Citation(s): Chesnut 1902, Schenck and Gifford 1952.

Terminology: **Karok:** 'iknitiptiipanach, "fringed."

Citation(s): Schenck and Gifford 1952.

1313 *Vicia gigantea*

Common Name(s): Giant Vetch.

Usage: Medicine: **Costanoan:** root, decocted for a laxative.

Citation(s): Rayburn 2012.

1314 *Viola* spp.

Family: *Violaceae*

Usage: Medicine: **Costanoan:** plant, juice, used as slave on sores; plants, boiled, used as poultice the following day.

Citation(s): Rayburn 2012.

1315 *Viola adunca*

Common Name(s): Dog Violet.

Usage: Food: **Maidu:** leaves eaten.

Medicine: **Maidu:** poultice used to relieve chest pain; decoction used to relieve stomach ache; root chewed to relieve labor pains.

Misc: **Miwok:** named but not used.

Citation(s): Barrett and Gifford 1933; Strike 1994

Terminology: **Miwok (Central Sierra):** sitimpa.
Citation(s): Barrett and Gifford 1933.

1316 *Viola pedunculata*
Common Name(s): Viola, Violet.

Usage: Food: **Luiseño:** the leaves were used as greens.
Citation(s): Sparkman 1908.
Terminology: **Luiseño:** ashla.
Citation(s): Sparkman 1908.

1317 *Viola sempervirens*
Synonym(s): *Viola sarmentosa*
Common Name(s): Wood Violet.

Usage: Misc: **Karok:** the children play with the blossoms.
Citation(s): Schenck and Gifford 1952.
Terminology: **Karok:** ichniahich, "flowers" (blossoms).
Citation(s): Schenck and Gifford 1952.

1318 *Vitis* spp.
Family: *Vitaceae*
Common Name(s): Grape, Wild Grape.

Usage: Food: **Shasta, Salinan:** ate berries. **Yokuts (Northern Hill):** eaten, never dried; sometimes crushed for juice which was drunk. **Luiseño:** cooked and eaten; never dried and preserved. **Wintun, River (Patwin):** berries occasionally gathered; **Hill (Patwin):** eaten raw; **Central (Nomlaki):** used berries; **Northern (Wintu):** placed in small basket and mashed; if too sour Sweet Manzanita

(*Arctostaphylos* spp.) flour added; eaten just before fermentation. **Wappo:** ate berries raw. **Chumash:** fruit, eaten.

Medicine: **Wintun, Northern (Wintu):** roots steeped in water; decoction drunk as blood purifier. **Shasta:** root was broken up and pounded and boiled to make a deep green tea to be drunk lukewarm as an emetic and cathartic.

Citation(s): Driver 1936; Du Bois 1935; Gayton 1948; Goldschmidt 1951; Holt 1946; Kroeber 1932; Mason 1912; Sparkman 1908; Timbrook 2007.

1319 *Vitis californica*

Common Name(s): Wild Grape, California Grape, California Wild Grape.

Usage: Food: **Hupa:** the leaves are used to line fire pit used to cook *Chlorogalum pomeridianum*; also mixed with *C. Pomeridianum*, said to improve flavor. **Tubatulabal:** used berries. **Karok:** berries gathered and eaten when ripe, but not preserved. **Pomo:** fruit eaten. **Pomo (Southwestern):** grapes eaten raw. **Miwok:** fruit mashed with hands in basket; eaten raw. **Yuki:** ate berries. **Luiseño:** reportedly ate fruit cooked, never dried it. **Cahuilla:** fruit dried, boiled in water to eat; made fermented drink from fruit; tender shoots eaten

Material: **Hupa:** a fiber was obtained from the roots for the woof in making baskets; used unsplit roots to make loose, open-work baskets. **Whilkut (Chilula),**

Nongatl, Lassik, Wailaki, Yurok, Wiyot, Yuki, Pomo, Maidu: stem used as rim binding in baskets; split stem, or sapwood, used as twining in basketry. **Yokuts:** stem used as woof in baskets; split stem, or sapwood, used as twining in basketry. **Yuki:** root used as woof in baskets. **Yurok:** root used as warp in baskets; used split root to start a basket. **Karok:** small roots used for baskets; leaves used to put over bulbs while they cooked; vines sometimes used to moor boats. **Miwok:** used leaves to line earth ovens; split stems, or sapwood, used as twining in basketry. **Pomo (Southwestern):** vines used for withes to tie things when hunting or traveling; to make hoop on baby basket.

Citation(s): Barrett and Gifford 1933; Chesnut 1902; Curtin 1957; Gifford 1967; Goddard 1903; Merrill 1923; O'Neale 1932; Schenck and Gifford 1952; Voegelin 1938; Goodrich et al 1980; Strike 1994; Barrett 1908.

Terminology: **Pomo (Southwestern):** shitim. **Yuki:** mö´-mäm. **Miwok (Plains:** mū´te; **(Central Sierra):** tolmesu, kimisu; **(Northern Sierra):** mü´te; **(Southern Sierra):** mū´te, mü´te, tolmesu. **Karok:** aiyi¹pa, ai (fruit), ahip'aha (roots).

Citation(s): Gifford 1967; Curtin 1957; Barrett and Gifford 1933; Schenck and Gifford 1952; O'Neale 1932.

1320 *Vitis girdiana*

Common Name(s): Wild Grape.

Usage: Food: **Maidu, Southern (Nisenan):** the berries are eaten. **Luiseño:** the fruit is cooked.

Citation(s): Powers 1877; Sparkman 1908.
Terminology: **Luiseño:** makwit. **Chumash (Barbareño):** nu'nit'; **(Ynezeño):** nunit'; **(Ventureño):** nunit.
Citation(s): Sparkman 1908; Timbrook 2007.

1321 *Vitia vinifera*

Common Name(s): Cultivated Grape.

Usage: Food: **Cahuilla:** as early as 1852 had "producing vineyards" at Agua Caliente.
Citation(s): Bean and Saubel 1972.
Note(s): native of Europe; brought into California by the Spanish.
Citation(s): Munz and Keck 1965; Bean and Saubel 1972.

1322 *Washingtonia filifera*

Family: *Arecaceae*
Synonym(s): *Neowashingtonia filamentosa*
Common Name(s): Desert Palm, Fan Palm, California Fan Palm.

Usage: Food: **Cahuilla:** fruit gathered late June to early November; eaten fresh, or dried and stored; ground into flour that contained flesh and seeds, mixed with other flours, water, eaten as mush; drank beverage from fruits soaked in water; the pith, center of the plant, sometimes eaten as famine food.

       Material: **Luiseño, Cahuilla:** used leaf as wrapping in baskets; used fronds as house thatching, to make cooking utensils, spoons and stirring implements; seeds used for gourd rattles; made sandals or foot pads (wakutem); stems used in fire-making process; plants burned to improve fruit yield and as pest control.

Citation(s): Barrows 1900; Merrill 1923; Bean and Saubel 1963; 1972.

Terminology: **Cahuilla:** maul.

Citation(s): Bean and Saubel 1972.

1323 *Woodwardia fimbriata*

Family: *Blechnaceae*

Synonym(s): *Woodwardia radicans, Woodwardia spinulosa*

Common Name(s): Giant Chain Fern, Brake Fern, Giant Fern, Chain Fern.

Usage: Material: **Yurok:** stem (dyed) used a red pattern; stems used as yellow pattern in baskets. **Sinkyone:** stem (dyed) used as red pattern in baskets. **Karok:** stem used in basketry, dyed in Alder bark dye. **Pomo (Southwestern):** fronds used in lining top and bottom of earth oven in baking acorn bread and other foods. **Wiyot, Hupa:** stem used as white pattern; stems furnished a reddish brown color picked up from bark of Alder (*Alnus* spp.) Boiled in water, or chewed in the mouth, the final dyed material being used to decorate baskets.

Medicine: **Luiseño:** a decoction of the root is used both externally and internally to relieve pain from juries to the body. **Maidu:** root decoction used to treat wounds, sores.

Citation(s): Gifford 1967; Goddard 1903; Merriam 1967; Merrill 1923; O'Neale 1932; Schenck and Gifford 1952;

Sparkman 1908; Goodrich et al 1980; Strike 1994.
Terminology: **Wiyot:** tigwanetsha-wèl. **Luiseño:** mashla.
Yurok: paap. **Karok:** tiptip. **Hupa:** mēme. **Pomo (Southwestern):** kahpa.
Citation(s): Curtis 1924; Loud 1918; Sparkman 1908; O'Neale 1932; Schenck and Gifford 1952; Goddard 1903; Gifford 1967; Goodrich et al 1980.

1324 *Wyethia* spp.

Family: *Astteraceae*
Common Name(s): Mountain Mule Ears, Narrow-Leaved Mule Ears, Mule Ears, Sunflower.

Usage: Food: **Wintun:** roast and pound seeds.
Citation(s): Merriam 1955.

1325 *Wyethia augustifolia*

Common Name(s): Sunflower, Mule Ears.

Usage: Food: **Hupa:** the fresh shoots are eaten raw. **Pomo:** used seeds in pinole; **(Southwestern):** fresh seeds eaten, taste like sunflower seeds; dried for Winter use, ground for pinole. **Maidu (Northern):** seeds eaten. **Yuki, Miwok (Coast):** in Autumn the ripe seeds were dried, pounded, winnowed in a flat basket and mixed with Wild Oats for pinole. **Northern Paiute (Paviotso):** pulled entire plant from ground; stems, seed eaten raw.

Medicine: **Miwok:** decoction of the leaves used as a bath for fever patients and produced a profuse perspiration, so that the patient must be covered immediately; never taken internally as it was considered poisonous. **Yuki:** roots boiled a little, tea used as an emetic; used to treat sore eyes. **Costanoan:** used as a skin medication; root, pounded, lather used in poultices to draw blisters, to treat lung problems. **Maidu:** decoction of the leaves used as a bath for fever patients and produced a profuse perspiration, used an emetic.

Citation(s): Barrett 1952; Barrett and Gifford 1933; Curtin 1957; Dixon 1905; Goddard 1903; Goodrich et al 1980; Strike 1994; Rayburn 2012.

Terminology: **Miwok (Northrn Sierra):** hū´ssūpu; **(Central Sierra):** hotcotca. **Yuki:** bish´-ki. **Hupa:** tcalatdûñ. **Pomo:** calam.

Citation(s): Barrett and Gifford 1933; Curtin 1957; Goddard 1903; Barrett 1952.

1326 *Wyethia glabra*

Common Name(s): Mule Ears.

Usage: Food: **Pomo (Southwestern):** seeds used in pinole, eaten fresh.

Citation(s): Goodrich et al1980.

Terminology: **Pomo (Southwestern):** cʰalam?.

Citation(s): Goodrich et al 1980.

1327 *Wyethia helenioides*

Usage: Food: **Miwok:** young shoots were eaten raw after peeling off the outer coating; had a sweetish taste.

Material: **Miwok:** used leaves to line earth oven.

Citation(s): Barrett and Gifford 1933.
Terminology: **Miwok (Central Sierra, Southern Sierra):**
notopayu.
Citation(s): Barrett and Gifford 1933.

1328 *Wyethia longicaulis*

Common Name(s): Sunflower.

Usage: Food: **Wiyot:** gathered seeds, generally from burned-off forest areas deliberately cleared so that these plants would grow. **Yuki:** lower part of fresh leaves and the stem, taken before flowering time, eaten in the field and occasionally at home; seed used for pinole.

Medicine: **Yuki:** root decoction used as an emetic; used for stomach complaints; root baked in hot ashes and applied as a poultice to cure rheumatism; root dried and powdered, the moistened and used for poultice for running sores and burns; decoction used as a wash to relieve headache, and to allay inflammation in sore eyes. **Wailaki:** root decoction used to soothe headache, inflamed eyes, as an emetic, for stomach problems; root, dried, pulverized sprinkled on running sores, burns.

Citation(s): Chesnut 1902; Loud 1918; Strike 1994.

1329 *Wyethia mollis*

Usage: Food: **Northern Paiute (Paviotso):** the seeds were eaten; the stems were eaten raw.

Medicine: **Washo:** half-cupful of root decoction

used as a physic or emetic, should be boiled sufficiently to become quite concentrated. **Maidu, Modoc (Lutuami):** poultice used to heal bruises, reduce swellings.

Citation(s): Kelly 1932; Train et al 1941; Strike 1994.

Terminology: **Northern Paiute (Paviotso):** a'gü´ (seed). **Washo:** shu-gil.

Citation(s): Kelly 1932; Train et al 1941.

1330 *Wyethia ovata*

Usage: Food: **Mono Lake Paiute:** used as food source.

        Misc: **Kawaiisu (Tehachapi):** named but not used.

Citation(s): Steward 1933; Zigmond 1981.

Terminology: **Mono Lake Paiute:** akü. **Kawaiisu (Tehachapi):** tɨhɨyanagavivi, "dear ear."

Citation(s): Steward 1933; Zigmond 1981.

1331 *Xanthium X spinosum*

Family: *Asteraceae*

Common Name(s): Spiny Clotbur, Spanish Thistle.

Usage: Misc: **Kawaiisu (Tehachapi):** named but not used.

Citation(s): Zigmond 1981.

Terminology: **Kawaiisu (Tehachapi):** wiyarɨbɨ.

Citation(s): Zigmond 1981.

1332 *Xanthium X strumarium*

Synonym(s): *Xanathium strumarium* var. *canadense*

Common Name(s): Cockleburr

Usage: Medicine: **Costanoan:** seed, decocted, used to cure

bladder ailments. **Chumash:** leaves,
boiled, used as a wash for cuts, drank for
bladder trouble (people and horses);
leaves, crushed, rubbed on skin for
ringworm.

Citation(s): Rayburn 2012; Timbrook 2007.

Terminology: **Chumash (Barbareño):** sho'moy; **(Ynezeño):**
mokoksh; **(Ventureño):** shomoy.

Citation(s): Timbrook 2007.

1333 *Xerophyllum tenax*

Family: *Melanthiaceae*

Common Name(s): Bear-Grass, Squaw Grass, Elk
Grass, Fire Lily.

Usage: Material: **Atsugewi:** woven into fringe at base of
arrow quiver. **Tolowa:** baskets more or
less covered with an overlay of this
grass. **Hupa:** braided the leaves on twine
upon which pine nut shell was strung;
used for clear white decoration of
baskets; made cord for use in knee-
length fringe apron for women. **Tolowa,
Whilkut (Chilula), Nongatl, Lassik,
Wailaki, Sinkyone, Yurok, Wiyot,
Chimariko, Shasta, Yuki,Yana, Karok,
Wintun, Yokuts, Achomawi, Atsugewi:**
used leaves as white pattern in baskets.
Hupa, Yurok, Karok, Yokuts: used
leaves (dyed) as yellow pattern in
baskets. **Yana:** the women often
wrapped the front of their skirts with the
white "grass;" twined around eight
buckskin strings which formed part of
girl's brow band during puberty

ceremonies. **Shasta:** the buckskin fringe on women's skirts are wrapped with the stem for a greater part of their length. **Yurok:** women wear a tight-fitting necklace of two to several strands braided together. **Maidu (Northern):** used in basketry construction. **Achomawi:** women wore a kilt of the grass. **Hupa, Karok:** fiber, shredded, used for women's skirts, cordage.

Medicine: **Hupa, Karok:** members of the household of a deceased person wear braided strands around their necks; the strands are never removed, but worn until they fall off. **Pomo (Southwestern):** roots washed, rubbed to make lather to wash sores.

Citation(s): Baumhoff 1958; Curtis 1924; Dixon 1905, 1907; Garth 1953; Goddard 1903; Merriam 1967; Merrill 1923; O'Neale 1932; Sapir and Spier 1943; Schenck and Gifford 1952; Goodrich et al 1980; Strike 1994.

Terminology: **Wintun:** pili. **Yurok:** häämo. **Karok:** panura, panyura, "wild grass." **Wiyot:** himene-wèl. **Hupa:** Lōtel (leaves). **Pomo (Southwestern):** šuċum qá?di, "grape grass."

Citation(s): Du Bois 1935; O'Neale 1932; Schenck and Gifford 1952; Loud 1918; Goddard 1903; Goodrich et al 1980.

1334 *Xylococcus bicolor*

Family: *Ericaceae*
Common Name(s): Manzanita.

Usage: Food: **Diegueño:** ripe berries, soaked, made a refreshing drink.
Citation(s): Strike 1994.

1335 *Yucca* spp.

Family: *Agavaceae*
Common Name(s): Spanish Bayonet.

Usage: Food: **Monache (Western Mono):** roots dug up in Fall and roasted; in the Spring collected the tall blossoms, cooked stems; often stored for Winter; cooked food which was dried and pounded up. **Cahuilla:** root grated and used for soap.

Citation(s): Gayton 1948; Barrows 1900.
Terminology: **Monache (Western Mono):** o·ˊpa·dra.
Citation(s): Gayton 1948.

1336 *Yucca baccata*

Common Name(s): Spanish Bayonet.

Usage: Material: **Cahuilla:** root grated and used for soap.
        Medicine: **Salinan:** roots, mashed, used to treat sunstroke, to poultice eyes as a cure for blindness, cataracts. **Diegueño:** seeds used to treat minor skin irritations.

Citation(s): Barrows 1900; Strike 1994.

1337 *Yucca brevifolia*

Synonym(s): *Yucca arborescens*
Common Name(s): Tree Yucca. Joshua Tree.

Usage: Food: **Tubatulabal:** pods eaten. **Panamint Shoshone (Koso):** in April the flower buds are gathered, roasted, and eaten. **Kawaiisu (Tehachapi):** in the Spring the fruit is prepared and eaten.
        Material: **Monache (Western Mono), Kitanemuk (Tejon), Panamint Shoshone (Koso):**

used root bark as red pattern in baskets. **Kawaiisu (Tehachapi):** used root bark as brown pattern in baskets. **Panamint Shoshone (Koso):** used long red roots in basketry. **Chemehuevi (Southern Paiute), Serrano, Tubatulabal:** roots used in coiled basketry.

Citation(s): Coville 1892; Merrill 1923; Voegelin 1938; Zigmond 1981; Strike 1994.

Terminology: **Kawaiisu (Tehachapi):** cawarabɨ (the tree), pakaibi (the fruit), sɨɨtɨɨvi (core of root stock).

Citation(s): Zigmond 1981.

1338 *Yucca schidigera*

Synonym(s): *Yucca filamentosa, Yucca mohavensis*
Common Name(s): Spanish Bayonet.

Usage: Food: **Luiseño:** the flowers are boiled and eaten; the pods are roasted and eaten. **Cahuilla:** fruit picked, April to May, when green and roasted in hot coals. **Diegueño:** flower petals from young flowers eaten raw; older petals boiled, eaten; seeds, made tea, or pounded, cooked as mush.

Material: **Kamia:** bark is used for soap. **Cahuilla:** used fronds to bind framework of house together; furnished a valuable material (fiber) for weaving; root is grated and used for soap; fiber was used in making ropes, strings, mats, coiled soles for sandals, saddle blankets, tying beams and poles in house construction; seeds in necklaces for women, seeds pods for toys for children. **Diegueño:** leaves used to make rough, throw-away baskets; seeds, strung on cords, used as beads.. **Kamia, Panamint Shoshone (Koso), Tubatulabal:** roots, used occasionally in basketry.

Citation(s): Barrows 1900; Merriam 1967; Sparkman 1908; Bean and Saubel 1972; Strike 1994.

Terminology: **Luiseño, Cahuilla:** hunuvut.

Citation(s): Sparkman 1908.

1339 *Zea mays*

Family: *Poaceae*

Common Name(s): Maise, Indian Corn.

Usage: Food: **Cahuilla:** ground into meal, boiled and eaten.

Material: **Cahuilla:** corn sprinkled on images of the dead during mourning ceremonies.

Citation(s): Bean and Saubel 1972.

Terminology: **Cahuilla:** pahavoshlum (?).

Citation(s): Bean and Saubel 1972.

Note(s): in 1823 it was noted by Don José Maria Estudillo, of the Romero expedition, that the Cahuilla were planting corn, pumpkins, melons, watermelons; introduced from Mesoamerica in prehistoric times.

Citation(s): Bean and Saubel 1972.

1340 *Zeltnera exaltata*

Family: *Gentianaceae*

Synonym(s): *Centaurium exaltatum*

Usage: Medicine: **Miwok:** a decoction of stems and leaves was drunk for toothache, stomach ache, other internal pains, and consumption.

Citation(s): Barrett and Gifford 1933.

Note(s): some *Centaurium* spp. are considered bitter tonic.
Citation(s): Stuhr 1933.

1341 *Zeltnera venusta*

Synonym(s): *Erythraea venusta, Centaurium venustum*
Common Name(s): Canchalagua (Spanish).

Usage: Medicine: **Luiseño:** tea made from this is used as a remedy for fever. **Miwok:** decoction made with water drunk to abate fever and ague. **Cahuilla:** tea used as a remedy for ague. **Chumash:** boiled, tea drunk as a tonic, as a blood purifier.

Misc: **Tubatulabal:** considered as weeds.
Citation(s): Barrett and Gifford 1933; Sparkman 1908; Voegelin 1938; Bean and Saubel 1972.
Terminology: **Luiseño:** ashoshkit.
Citation(s): Bean and Saubel 1972.

1342 *Ziziphus parryi* var. *parryi*

Family: *Rhamnaceae*
Synonym(s): *Zizyphus parryi, Condalia parryi*
Common Name(s): Crucillo, Wild Plum.

Usage: Food: **Cahuilla:** yellowish-red berry eaten fresh or dried, pounded into meal for atole; the edible nutlet of the drupe was ground and leached to produce tasty flour.

Citation(s): Barrows 1900; Bean and Saubel 1972.
Terminology: **Cahuilla:** o-ot [not recognized as proper term by modern Cahuilla].
Citation(s): Barrows 1900; Bean and Saubel 1972.

1343 *Zostera marina*

Family: Zosteraceae
Synonym(s): *Zostera marina* var. *latifolia*
Common Name(s): Eel Grass.

Usage: Material: **Chumash:** made nets and skirts.
         Misc: **Pomo (Southwestern):** named but not
                       used.
Citation(s): Goodrich et al 1980; Strike 1994.
Terminology: **Pomo (Southwestern):** qʰasˊáqaˑ.
Citation(s) Goodrich et al 1980.

References

BAE   Bureau of American Ethnology.

MPMB Milwaukee Public Museum Bulletin.

UCAR University of California Anthropological Records.

UCPAAE University of California Publications in American Archaeology and Ethnology.

Abrams, Leroy, and Roxana Stinchfield Ferris
  1940-1960 Illustrated Flora of the Pacific States: Washington, Oregon, and California, 4 vols. Stanford: Stanford University Press.

Anderson, M. Kat
  1991 California Indian Horticulture: Management and Use of Redbud By the Southern Sierra Miwok, *Journal of Ethnobotany* 11 (1): 145-157.

Bailey, L. H.
  1950 The Standard Cyclopedia of Horticulture, 3 vols. New York: The MacMillan Company.

Baldwin, Bruce C., Douglas Goldman, and David J. Keil, Robert Patterson, Thomas J. Tosatti (editors)
  2012 The Jepson Manual. Vascular Plants of

California, Thoroughly Revised and Expanded. Berkeley: University of California Press.

Barrett, S. A.
1908   Pomo Indian Basketry. UCPAAE, 7 (3): 134-276. (Reproduced in whole, *in*, Pomo Indian Basketry, Samuel Barrett, Phoebe Hearst Museum Reprint series, Phoebe Hearst Museum of Anthropology, University of California, Berkeley, 1996).
1910   The Material Culture of the Klamath Lake and Modoc Indians of Northeastern California and Oregon. UCPAAE, 5 (4): 239-292.
1917   The Washo Indians. MPMB, 2 (1): 1-52.
1933   Pomo Myths. MPMB, 15: 1-608.
1952   Material Aspects of Pomo Culture, 2 parts. MPMB, 20 (1); 1-260; (2): 261-508.

Barrett, S.A., and E. W. Gifford
1933   Miwok Material Culture. MPMB, 2 (4): 117-376.

Barrows, David Prescott
1900   The Ethnobotany of the Coahuilla Indians of Southern California. The University of Chicago Press.

Baumhoff, Martin A.
    1958   California Athabascan Groups. UCAR, 16 (5).

Beals, Ralph L.
    1933   Ethnology of the Nisenan. UCPAAE, 31 (6):
            335-414.

Breschini, Gary S., and Trudy Hoversat.
    2004   The Esselen Indians of the Big Sur Country.
            Salinas, CA: Coyote Press.

Bean, John Lowell, and Katherine Siva Saubel
    1961   *Cahuilla Ethnobotanical Notes: The Aboriginal
            Uses of the Oak. In,* Archaeological Survey
            Annual Report 1960-1961: 237-245. University
            of California, Los Angeles.
    1963   Cahuilla   Ethnobotanical   Notes: The
            Aboriginal   Uses   of   the   Mesquite   and
            Screwbean. Archaeological Survey Annual
            Report 1962-1963: 55-76.
    1972   Temalpakh. Cahuilla Indian Knowledge and
            Usage   of   Plants.   Morongo   Indian
            Reservation: Malki Museum Press.

Blanco, O., A. Crepo, R. H. Ree, and H. T. Lumbsch
    2006   Major   clades   of   parmelioid   lichens
            (Parmeliaceae,   Ascomycota)   and   the
            evolution of their morphological and
            chemical diversity. *Mol Hyolgenet Evol,* 39 (1):

52-69.

Bolton, Eileen M.
    1960   Lichens   For   Vegetable   Dyeing.
            Massachusetts: Charles   T.   Branford   Co.,
            Newton Centre.

Chesnut, V. K.
    1902   Plants Used by the Indians of Mendocino
            County, California. *Contributions of the U.S.
            National   Herbarium*,   7   (3):295-422.
            Washington, D.C.

Clark, Galen
    1904   Indians of the Yosemite Valley and Vicinity.
            Yosemite Valley, California: Galen Clark.

Chevallier, Andrew
    1996   The Encyclopedia of Medical Plants. New
            York: DK Publishing, Inc.

Coville, Frederick V.
    1892   The Panamint Indians of California. *American
            Anthropologist*, 5: 351-361.

Crespo, A., H.T. Lumbsch, J. E. Mattsson, O. Blanco, P.
        K. Divakar, K. Articus, E. Wiklund, P. A. Bawingan,
        and M. Wedin
    2006   Testing   morphology-based   hypotheses   of

phylogenetic relationships in Parmelaiceae (Ascomycota) using three ribosomal markers and the nuclear RPB1 gene. *Mol Phylogenet Evol*, 44 (2): 812-824.

Curtin, Lenore Scott Muse
  1957   Some Plants Used by the Yuki Indians of Round Valley, Northern California. *Masterkey*, 31 (2): 40-48; 31 (3): 85-94.

Curtis, Edward S.
  1924   The North American Indians, 20 vols. Norwood, Massachusetts: The Plimpton Press.

Dawson, Lawrence, and James Deetz
  1965   A Corpus of Chumash Basketry. *Archaeological Survey Annual Report*: 197-280. University of California, Los Angeles.

Demes, Kyle, Michael H. Graham, and Thew S. Suskiewicz
  2009   Phenotypic Plasticity Reconciles Incongruous Molecular and Morphological Taxonomies: The Giant Kelp, *Macrocystis* (Laminariales, Phaeophyceae), Is A Monospecific Genus. *Journal of Phycology*, 45 (6): 1266-1269.

Dixon, R. B.
  1905   The Northern Miadu. *American Museum of*

Natural History, Bulletin, 17 (3): 119-346. New York.

1907   The Shasta. *American Museum of Natural History, Bulletin*, 17 (5): 381-498. New York.

1910   The Chimariko Indians and Language. UCPAAE, 5 (5): 293-380.

Driver, Harold E.

1936   Wappo Ethnography. UCPAAE, 36 (3): 179-220.

Drucker, Philip

1937   The Tolowa and Their Southwest Kin. UCPAAE, 36 (4): 221-300.

Du Bois, Cora

1935   Wintu Ethnography. UCPAAE, 36 (1): 1-148.

Duncan, Ursala K.

1959   A Guide to the Study of Lichens. New York: Scholars Library, Publishers.

Dutcher, B. H.

1893   Piñon Gathering Among the Panamint Indians. *American Anthropologist*, 6: 377-380.

Earle, F. R., and Quentin Jones

1972   Analysis of Seed Samples from 113 Plant Families. *Economic Botany*, 16: 221-250

Foster, George M.
    1944   A Summary of Yuki Culture. UCAR, 5 (3): 155-244.

Fuller, Thomas C, and Elizabeth McClintock
    1986   Poisonous Plants of California. California Natural History Guides: 53. Berkeley: University of California Press.

Garth, Thomas
    1953   Atsugewi Ethnography, UCAR, 14 (2).

Gayton, A. H.
    1948   Yokuts and Western Mono Ethnography. UCAR, 10 (1, 2).

Gentry, Howard Scott
    1998   Agaves of Continental North America. University of Arizona Press.

Gifford, E. W.
    1914   Notes On the Chilula Indians of Northwestern California. UCPAAE, 10 (6): 265-288.
    1931   The Kamia of Imperial Valley. BAE, *Bulletin*, 97.
    1932   The Northfork Mono. UCPAAE, 31 (2): 15-65.
    1939   The Coast Yuki. *Anthropos*, 34: 292-375.
    1967   Ethnographic Notes on the Southwestern Pomo. UCAR, 25.

Goddard, Pliny Earl
    1903   Life and Culture of the Hupa. UCPAAE, 1 (1): 1-88.

Goldschmidt, Walter
    1951   Nomlaki Ethnography. UCPAAE, 42 (4): 303-443.

Goodrich, Jennie, Claudia Lawson, and Vanna Parrish Lawson.
    1980   Kashaya Pomo Plants. American Indians Study Center, University of California, Los Angeles.

Grant, Campbell
    1964   *Chumash Artifact Collected in Santa Barbara County, In,* California. Archaeological Survey Reports, 63: 1-44. University of California, Berkeley.

Grout, A. J.
    1928-1940 Moss Flora of North America North of Mexico. New York and Vermont: A. J.

Guberlet, Muriel Lewin
    1956   Seaweeds at Ebb Tide. Seattle: University of Washington Press.

Haber, Erich
    1983   Morphological Variability and Flavonol Chemistry of the Pyrola asarifolia Complex (Ericaceae) in North America. *Systematic Botany*, 8 (3): 277-298.

Hardin, James W., and Jay. M. Arena, M.D.
    1974   Human Poisoning from Native and Cultivated Plants (2nd edition). Duke University Press.

Harrington, John P.
    1932   Tobacco Among the Karok Indians of California. BAE, *Bulletin*, 94.

Heizer, Robert F.
    1953   Aboriginal Fish Poisons. BAE, *Bulletin*, 151, *Anthropological Paper*, 38: 225-284.
    1966   Languages, Territories, and Names of California Indian Tribes. Berkeley: University of California Press.

Holt, Catherine
    1946   Shasta Ethnography. UCAR, 3 (4): 299-350.

Huson, J. W.
    1893   Pomo Basket Makers. *Overland Monthly* XXI (Second Series), (126): 561-578.

Hwan Su Yoon, Ju Teon Lee, Sung Min Boo, and Debashish Bhattacharya

    2001   Phylogeny of Alariaceae, Laminariaceae, and Lessoniaceae (Phaeophyceae) Based on Plastid-Encoded RuBisCo Spacer and Nuclear-Encoded ITS Sequence Comparisons. *Molecular Phylogenetics and Evolution*, 21 (2): 231-243.

Iovin, June

    1963   *A Summary Description of Luiseño Material Culture. In*, Archaeological Survey Annual Report 1962-1963: 79-134. University of California, Los Angeles.

Jepson, Willis Linn

    1925   A Manual of the Flowering Plants of California. Berkeley: University of California Press.

Jones, Henry Albert, and Louis K. Mann

    1963   Onions and Their Allies: Botany, Cultivation, and Utilization. New York: Interscience Publishers, Inc.

Jordon, W. P.

    1973   The genus Lobartia in North America north of Mexico. *The Bryologist*, 76: 225-251.

Justo, A., and D. S. Hibbett.
>2011 Phylogenetic classification of *Trametes* (Basidiomycota, Polyporales) based on a five-marker dataset. *Taxon*, 60 (6): 1567-1583.

Kelly, Isabel T.
>1932 Ethnography of the Surprise Valley Paiute. UCPAAE, 31 (3): 67-210.

Kingsbury, John M.
>1964 Poisonous Plants of the United States and Canada. New Jersey: Prentice-Hall, Inc.

Kniffen, Fred B.
>1939 Pomo Geography. UCPAAE, 36 (6): 353-400.

Krieger, Louis C. C.
>1936 The Mushroom Handbook. New York: The MacMillan Company.

Kroeber, A. L.
>1932 The Patwin and Their Neighbors. UCPAAE, 29 (4): 252-423.

Kyung Mo Kim, Yuh-Gang Yoon, and Hack Sung Jung
>2005 Evaluation of the monophyly of *Fomitopsis* using parsimony and MCMC methods. *Mycologia*, 97 (4): 812-822.

Landberg, Leif C.W.
  1965   The Chumash Indians of Southern California. Southwest Museum Papers, 19. Higland Park, Los Angeles, CA: Southwest Musem.

Lane, Christopher E., Charlene Mayes, Louis D. Druehl, and Gary W. Saunders
  2006   A Multi-Gene Molecular Investigation of the Kelp (Laminariales, Phaeophyceae) Supports Substantial Taxonomic Re-Organization. *Journal of Phycology*, 42: 493-512.

Lichvar, Robert, and John T. Kartesz
  2009   Nomenclature and the National Wetland Plant List. ERDC/CRREL TN, 9 (1). US Army Corps of Engineers, Engineer Research and Development Center, Cold Regions Research and Engineering Laboratory.

Liener, Irwin E.
  1980   Toxic Constituents of Plant Foodstuffs. Academic Press.

Loeb, E. M.
  1932   The Western Kuksu Cult. UCPAAE, 33 (1): 1-137.

Loud, Llewellyn
  1918   Ethnography and Archaeology of the Wiyot

Territory. UCPAAE, 14 (3): 221-436.

Mason, J. Alden
    1912   The Ethnology of the Salinan Indians. UCPAAE, 10 (4): 97-240.

Mason, Otis
    1889   *The Ray Collection From Hupa Reservation. In,* Annual Report of the Board of Regents of the Smithsonian Institution for the Year Ending June 30, 1886: 205-239. Washington, D.C.

Mattar, Davicino R., Y. Casali, C. Porporatto, S. G. Correa, and B. Micalizzi
    2007   In vivo immunomodulatory effects of aqueous extracts of Larra divaricata Cav. *Immunopharmacol immunotoxicaol,* 29 (3-4): 351-366.

Merriam, Clinton Hart
    1955   Studies of California Indians. Berkeley: University of California Press.
    1966   Ethnographic Notes of the California Indian Tribes, compiled and edited by Robert F. Heizer. University of California Archaeological Survey, *Reports,* 68 (1).
    1967   Ethnographic Notes on California Indian Tribes. II. Ethnographic Notes on Northern and Southern California Indian Tribes,

compiled and edited by Robert F. Heizer. University of California Survey, *Reports*, 68 (2).

Merrill, Ruth Earl
  1923   Plants Used in Basketry by California Indians. UCPAAE, 20: 215-242.

Meunscher, W. C.
  1951   Poisonous Plants of the United States, 2nd Edition. New York: The Macmillan Company.

Miller, K. A.
  2012   Seaweeds of California: Updates of California Seaweed Species List, pp. 1-59. Berkeley: University of California Jepson Herbarium.

Munz, Philip A.
  1968   Supplement To A California Flora. Berkeley: University of California Press.

Munz, Philip A., and David D. Keck
  1965   A California Flora, 3rd Printing. Berkeley: University of California Press.

Nash, T. H., B. D. Ryan, C. Gries, and F. Bugartz (eds.)
  2001   Flora of the Greater Sonoran Desert Region, Vol. 1. Tempe, AZ

Nomland, Gladys Ayer
    1935   Sinkyone Notes. UCPAAE, 36 (2): 149-178.
    1938   Bear River Ethnography. UCAR, 2 (2).

O'Neale, Lila M.
    1932   Yurok-Karok Basket Weavers. UCPAAE, 32 (1): 1-184.

Pammel, L. H.
    1911   A Manual of Poisonous Plants. Cedar Rapids: The Torch Press.

Peck, Morton E.
    1961   A Manual of the Higher Plants of Oregon 2nd edition. Binford & Mort, Publishers.

Powers, Stephen
    1877   Tribes of California. Contributions to North American Ethnology, Vol. 3. GPO: Washington, D.C.

Rayburn, Keith
    2012   Native Medicinal Ethnobotany of Central Coastal California; The Healing Herbs used by Ohlone Native Americans. Saarbrücken, Germany: LAP Lambert Academic Publising.

St. John, Harold
    1963   Flora of Southeastern Washington and Adjacent Idaho, 3rd Edition. Escondido,

California: Outdoor Pictures.

Sandhar, Harleen Kaur, Billesh Kumar, Sunil Prasher,
   Prashant Tiwari, Manoj Salhan, and Pardeep Sharma
   2011   A   Review   of   Phytochemistry   and
           Pharmacology of Flavonoides. *Internationale
           Pharmaceutica Sciencia*, 1 (1).

Sapir, Edward, and Leslie Spier
   1943   Notes of the Culture of the Yana. UCAR, 3
           (3): 239-298.

Sargent, Charles Sprague
   1922   Manual of the Trees of North America, 2nd
           Corrected Edition, Reprinted 1961, 2 Vols.
           New York: Dover Publications, Inc.

Schenck, Sara M., and E. W. Gifford
   1952   Karok Ethnobotany. UCAR, 13 (6): 377-392.

Schneider, Albert
   1898   A Guide to the Study of Lichens. Boston:
           Bradlee Whidden.

Schultes, R. E.
   1969   *Hallucinogens of Plant Origin*. Science, 163
           (3864): 245-254.

Smith, Gilbert M.
    1944   Marine Algae of the Montery Peninsula, California. Stanford: Stanford University Press.

Sparkman, Philip Stedman
    1908   The Culture of the Luiseño Indians. UCPAAE, 8 (4): 187-234.

Steward, Julian Haynes
    1933   Ethnography of the Owens Valley Pauite, UCPAAE, 33 (3): 233-350.

Stille, Alfred, and John M. Maisch
    1880   The National Dispensatory, 2nd Edition. Philadelphia: Henry C. Lea's Son and Co.

Strike, Sandra S.
    1994   Ethnobotany of the California Indians. Vol. 2: Aboriginal Uses of California's Indigenous Plants. Koeltz Scientific Books USA.

Tanner, Christopher E.
    1986   Investigations of the taxonomy and morphological variation of *Ulva* (Chlorophyta): *Ulva californica* Willie. *Phycologia*, 25 (4): 510-520.

Timbrook, Jan
  1982   Use of Wild Cherry Pits as Food by the California Indians. *Journal of Ethnobotany*, 2 (2): 162-176.
  1987   Virtuous Herbs: Plants in Chumash Medicine. *Journal of Ethnobotany*, 7 (2): 171-180.
  2007   Chumash Ethnobotany. Plant Knowledge Among the Chumash People of Southern California. Santa Barbara Museum of Natural History, Santa Barbara, California. Berkeley, California: Heyday Books.

Train, Percy, James R. Henrichs, and W. Andrew Archer
  1941   *Medicinal Uses of Plants by Indian Tribes of Nevada, 3 Parts. In*, Contributions Toward A Flora of Nevada, 33. Division of Plant Exploration and Introduction, Bureau of Plant Industry. U.S. Department of Agriculture.

Trimble, H.
  1888-1891 *American Journal of Pharmacy* 8, (860): 593.

Trudell, Steve, and Joe Ammirati
  2009   Mushrooms of the Pacific Northwest: Timber Press Field Guide. Portland, OR: Timber Press.

Tsukada, Matsuo
>1967 Chenopod and Amaranth Pollens: Electron-Microscopic Identification. *Science*, 157 (3784): 80-82.

Tuckerman, Edward
>1882 A Synopsis of the North American Lichens. Boston: S. E. Cassino, Publisher.

Voegelin, Ermine Wheeler
>1938 Tubatulabal Ethnography. UCAR, 2 (1): 1-84.

Waterman, T. T.
>1911 Phonetic Elements of the Northern Pauite Language. UCPAAE, 10 (2): 13-44.
>1920 Yurok Geography. UCPAAE, 16 (5): 177-314.

Weber, William A.
>1953 Balsamorhiza Terebinthacea and Other Hybrid Balsom-roots. *Madroño*, 12 (2): 47-49. Calfironia Botanical Society.

Yanovsky, Elias, and R. M. Kingsbury
>1938 Analysis of Some Indian Food Plants. *Journal of The Association of Official Agricultural Chemists*, 21 (4).

Zigmond, Maurice L.
  1981   Kawaiisu Ethnobotany. University of Utah
          Press.

Appendix A

Tribal Listing of Plants

Achomawi

12, 30, 33, 34, 64, 77, 140, 142, 182, 279, 287, 295, 296, 369, 374, 380, 468, 549, 573, 691, 694, 947, 975, 1029, 1030, 1061, 1077, 1078, 1121, 1153, 1333.

Atsugewi

34, 49, 78, 83, 119, 130,, 142, 173, 174, 175, 176, 206, 276, 295, 335, 374, 418, 424, 441, 502, 569, 573, 655, 691, 694, 736, 738, 760, 761, 762, 930, 966, 971, 1029, 1030, 1043, 1054, 1078, 1081, 1090, 1142, 1279, 1306, 1333.

Bear River

1, 14, 33, 34, 49, 62, 88, 106, 108, 110, 163, 205, 305, 314, 317, 374, 380, 564, 795, 797, 963, 969, 1054, 1106, 1110, 1121, 1140, 1158, 1173, 1235, 1259, 1291.

Cahuilla

7, 8, 13, 20, 25, 27, 39, 48, 61, 71, 78, 87, 96, 97, 99, 106, 110, 114, 130, 131, 135, 137, 138, 146, 147, 149, 156, 158, 160, 165, 172, 183, 192, 193, 205, 216, 217, 223, 224, 249, 269, 276, 299, 308, 309, 311, 317, 328, 332, 335, 338, 352, 364, 379, 385, 386, 387, 388, 389, 391, 400, 418, 427, 434, 439, 442, 443, 448, 454, 458, 468, 490, 492, 496, 502, 506, 507, 513, 527, 530, 536, 540, 542, 547, 552, 563, 566, 567, 610, 612, 616, 621, 634, 640, 641, 648, 649, 653, 654, 658, 660, 664, 672, 680, 681, 682, 683, 688, 689, 692, 694, 696, 705, 707, 708, 713, 715, 717, 722, 782, 784, 809, 820, 821,

822, 832, 854, 858, 860, 868, 869, 872, 885, 886, 887, 888, 889, 891, 903, 905, 907, 908, 919, 926, 931, 935, 954, 955, 958, 960, 966, 972, 976, 987, 989, 991, 1006, 1016, 1020, 1021, 1022, 1024, 1028, 1030, 1040, 1042, 1043, 1054, 1055, 1056, 1058, 1061, 1074, 1078, 1079, 1078, 1079, 1080, 1081, 1095, 1096, 1097, 1100, 1101, 1105, 1106, 1115, 1125, 1133, 1134, 1135, 1138, 1140, 1142, 1158, 1184, 1185, 1190, 1196,1225, 1231, 1256, 1259, 1283, 1287, 1291, 1294, 1319, 1321, 1322, 1335, 1336, 1338, 1339, 1341, 1342.

Chemehuevi (Southern Paiute)
39, 295, 705, 860, 1017, 1018, 1121, 1337.

Chilula (see Whilkut)

Chimariko
1, 374, 948, 952, 966, 971, 1054, 1078, 1121, 1235.

Chumash
1, 15, 20, 25, 28, 29, 49, 64, 68, 83, 86, 87, 98, 99, 110, 131, 133, 139, 142, 147, 155, 163, 169, 171, 172, 201, 205, 219, 220, 221, 224, 262, 263, 282, 283, 285, 296, 307, 308, 317, 331, 348, 349, 369, 373, 379, 381, 395, 399, 400, 403, 404, 421, 427, 428, 432, 448, 432, 448, 457, 459, 461, 468, 496, 503, 513, 518, 531, 532, 537, 542, 573, 574, 604, 610, 624, 638, 640, 641, 642, 680, 681, 683, 686, 692, 699, 717, 722, 755,, 769, 771, 791, 806, 808, 811, 814, 817, 820, 821, 828, 858, 860, 869, 870, 871, 886, 901, 957, 959, 960, 962, 966, 972, 982, 983, 987, 1000, 1008, 1028, 1033, 1038, 1040, 1043, 1053, 1055, 1057, 1058, 1062, 1080, 1081, 1093, 1100, 1101, 1105, 1106, 1111, 1114, 1115, 1119, 1121, 1126, 1129, 1133, 1134, 1135, 1139, 1142, 1153, 1154, 1155, 1156, 1158, 1173, 1182, 1190, 1196, 1198, 1225, 1226, 1251, 1254, 1256, 1259, 1285, 1291, 1309, 1318, 1320, 1332, 1341, 1343.

Chumash Tribal Labels

(Heizer 1966)	(Timbrook 2007)
Obispeño	Obispeño
Purisimeño	Purisimeño
Ynezeño	Ineseño
Barbareño	Barbareño
Ventureño	Ventureño
Island	Cruzeño

Costanoan (Ohlone)

14, 15, 20, 32, 33, 34, 76, 81, 87, 88, 106, 110, 131, 133, 134, 145, 168, 169, 172, 249, 264, 266, 276, 308, 314, 317, 331, 349, 352, 372, 380, 399, 400, 403, 433, 462, 469, 471, 502, 507, 513, 516, 531, 537, 545, 568, 573, 575, 583, 585, 610, 613, 624, 636, 639, 641, 650, 658, 680, 691, 692, 704, 712, 743, 744, 763, 767, 795, 803, 810, 815, 817, 820, 821, 823, 827, 844, 854, 863, 871, 876, 886, 895, 903, 916, 919, 941, 945, 977, 986, 987, 1005, 1019, 1032, 1043, 1054, 1055, 1062, 1067, 1098, 1101, 1111, 1114, 1119, 1126, 1129, 1135, 1138, 1141, 1160, 1167, 1169, 1171, 1173, 1186, 1189, 1191, 1196, 1209, 1210, 1250, 1257, 1259, 1274, 1291, 1292, 1294, 1307, 1308, 1313, 1314, 1325, 1332.

Diegueño

6, 27, 39, 75, 95, 105, 137, 142, 156, 180, 296, 350, 367, 379, 401, 404, 427, 438, 457, 488, 491, 547, 570, 584, 629, 641, 642, 681, 683, 692, 696, 729, 744, 771, 805, 817, 821, 844, 860, 904, 914, 958, 959, 978, 992, 1000, 1025, 1031, 1040, 1078, 1080, 1086, 1122, 1135, 1140, 1141, 1146, 1160, 1201, 1225, 1251, 1288, 1334, 1336, 1338.

Esselen

2, 28, 34, 98, 106, 110, 133, 142, 254, 286, 352, 471, 484, 502, 573, 640, 1028, 1055, 1056, 1061, 1062, 1111, 1121, 1138, 1142, 1155, 1291.

Gabrielino

1, 370, 871, 1173, 1292.

Huchnom

34, 49, 86, 110, 114, 314, 402, 530, 532, 626, 797, 801, 971, 1106, 1108, 1259, 1291.

Hupa

33, 63, 64, 78, 88, 91, 96, 106, 110, 115, 182, 183, 186, 230, 280, 304, 317, 322, 352, 374, 422, 424, 470, 551, 632, 641, 653, 675, 678, 755, 797, 871, 902, 927, 948, 965, 967, 971, 975, 977, 1008, 1039, 1051, 1056, 1060, 1081, 1108, 1109, 1111, 1121, 1123, 1141, 1174, 1235, 1291, 1300, 1303, 1310, 1319, 1323, 1325, 1333.

Kamia

39, 162, 394, 640, 1021, 1022, 1121, 1133, 1140, 1198, 1338.

Karok

1, 2, 12, 14, 15, 17, 19, 30, 33, 40, 51, 52, 63, 64, 71, 78, 82, 88, 91, 91, 94, 100, 103, 105, 106, 110, 111, 115, 117, 119, 133, 142, 145, 166, 183, 186, 194, 204, 211, 213, 223, 229, 259, 271, 279, 280, 287, 295, 296, 303, 313, 315, 317, 322, 325, 347, 438, 352, 370, 371, 372, 373, 374, 377, 380, 398, 399, 406, 416, 421, 423, 430, 433, 450, 463, 467, 469, 470, 502, 518, 519, 531, 551, 559, 560, 566, 568, 572, 573, 575, 585, 589, 592, 596, 606, 611, 620, 632, 641, 644, 653, 658, 659, 670, 678, 679, 687, 711, 723, 735, 737, 738, 740, 755, 766, 773, 777, 792, 803, 818, 845, 848, 853, 857, 859, 866, 871, 876, 877, 879, 897, 902, 910, 911, 923, 927, 931, 939, 948, 959, 963, 967, 970, 971, 975, 977, 993, 997, 1003, 1008, 1013, 1019, 1029, 1030, 1032, 1034, 1035, 1037, 1039, 1043, 1049, 1050, 1060, 1061, 1071, 1076, 1077, 1085, 1091, 1103, 1104, 1108, 1109, 1111, 1123, 1126, 1131, 1132, 1140, 1147, 1149, 1151, 1168, 1173, 1176, 1181, 1183, 1191, 1199, 1204, 1210, 1235, 1249, 1251, 1255, 1257, 1258, 1276, 1291, 1296, 1300, 1301, 1303, 1304, 1312, 1317, 1319, 1323, 1333.

Kato

295, 317, 1121.

Kawaiisu (Tehachapi)

2, 9, 10, 15, 17, 22, 34, 41, 44, 49, 64, 72, 78, 83, 84, 87, 100, 109, 110, 114, 119, 120, 125, 133, 134, 137, 143, 146, 147, 150, 151, 152, 153, 156, 157, 159, 161, 165, 172, 174, 185, 192, 193, 195, 215, 237, 240, 258, 263, 274, 275, 279, 281, 288, 290, 291, 295, 296, 297, 298, 300, 304, 306, 308, 317, 320, 323, 330, 332, 333, 336, 337, 349, 356, 367, 380, 386, 387, 390, 400, 410, 413, 415, 418, 419, 421, 427, 443, 445, 451, 452, 455, 457, 463, 477, 479, 489, 490, 491, 495, 498, 502, 504, 507, 508, 510, 511, 513, 515, 518, 519, 520, 521, 524, 525, 526, 539, 543, 554, 555, 562, 573, 575, 576, 586, 588, 601, 605, 608, 610, 615, 617, 626, 637, 640, 662, 684, 690, 691, 692, 694, 697, 700, 705, 721, 724, 728, 730, 745, 755, 757, 759, 762, 767, 782, 783, 793, 804, 813, 815, 816, 820, 825, 829, 832, 833, 834, 835, 836, 838, 840, 852, 853, 856, 860, 861, 866, 868, 871, 887, 898, 916, 923, 925, 934, 935, 940, 943, 945, 949, 950, 953, 956, 958, 959, 960, 971, 972, 975, 981, 982, 985, 986, 987, 995, 999, 1006, 1011, 1013, 1017, 1018, 1021, 1022, 1026, 1030, 1035, 1041, 1043, 1044, 1047, 1056, 1057, 1058, 1060, 1061, 1062, 1065, 1066, 1069, 1074, 1078, 1083, 1089, 1090, 1094, 1100, 1101, 1104, 1114, 1115, 1118, 1121, 1123, 1125, 1126, 1129, 1135, 1136, 1141, 1157, 1159, 1164, 1170, 1171, 1175, 1193, 1194, 1196, 1207, 1209, 1212, 1216, 1217, 1218, 1220, 1221, 1222, 1230, 1234, 1237, 1238, 1239, 1244, 1257, 1272, 1281, 1286, 1287, 1292, 1294, 1330, 1331, 1337.

Kashaya (see Pomo Southwestern)

Kitanemuk (Tejon)

256, 860, 1018, 1028, 1043, 1121, 1337.

Konkow (see Maidu, Northwestern)

Koso (see Panamint Shoshone)

Kumeyaay (see Diegueño)

Lassik
33, 63, 295, 374, 965, 971, 1121, 1319, 1333.

Luiseño
23, 28, 73, 96, 97, 98, 99, 118, 126, 130, 133, 134, 137, 142, 145, 163, 165, 168, 192, 193, 202, 211, 221, 253, 262, 305, 306, 308, 316, 317, 352, 355, 366, 379, 386, 400, 434, 448, 474, 493, 503, 504, 505, 537, 547, 582, 600, 626, 634, 640, 641, 642, 649, 650, 681, 683, 685, 692, 702, 703, 706, 715, 717, 722, 771, 805, 807, 817, 849, 850, 858, 860, 868, 869, 886, 916, 945, 991, 992, 1005, 1006, 1009, 1021, 1028, 1030, 1040, 1055, 1056, 1058, 1059, 1061, 1078, 1081, 1086, 1100, 1109, 1111, 1121, 1133, 1134, 1135, 1138, 1140, 1141, 1153, 1158, 1179, 1186, 1190, 1197, 1202, 1225, 1233, 1259, 1262, 1265, 1267, 1268, 1271, 1292, 1294, 1316, 1318, 1319, 1320, 1322, 1338, 1341.

Lutuami (see Modoc)

Maidu
42, 46, 47, 64, 65, 70, 76, 77, 80, 99, 102, 104, 105, 106, 108, 110, 115, 129, 140, 144, 154, 168, 172, 174, 178, 184, 185, 186, 190, 200, 203, 205, 222, 223, 233, 234, 244, 245, 250, 251, 252, 254, 263, 280, 292, 294, 295, 296, 298, 312, 323, 330, 332, 349, 352, 354, 357, 361, 365, 370, 371, 373, 376, 380, 387, 392, 398, 412, 416, 418, 429, 431, 432, 433, 436, 445, 447, 449, 462, 464, 465, 467, 470, 476, 490, 499, 500, 501, 516, 524, 533, 550, 553, 557, 569, 571, 573, 575, 577, 583, 591, 593, 594, 595, 605, 610, 618, 629, 631, 632, 639, 641, 643, 645, 646, 649, 650, 657, 665, 666, 667, 668, 669, 671, 673, 675, 681, 691, 698, 706, 709, 714, 717, 718, 719, 720, 727, 731, 742, 746, 747, 748, 749, 750, 756, 758, 760, 764, 765, 766, 767, 768, 770, 771, 773, 790, 794, 798, 803, 804, 807, 812, 815, 819, 821, 829, 839, 840, 847, 850, 853, 854, 855, 865, 878, 880, 881, 890, 892, 893, 894, 897, 911, 912, 913, 916, 918, 920, 921, 923, 927, 936, 939, 944, 948, 958, 959, 973, 975, 984, 987, 992, 996, 997, 1000, 1003, 1004, 1007, 1008, 1015, 1023,

1025, 1029, 1030, 1034, 1043, 1044, 1050, 1072, 1057, 1075, 1077, 1085, 1102, 1109, 1111, 1118, 1120, 1121, 1147, 1148, 1160, 1163, 1165, 1166, 1169, 1179, 1181, 1186, 1187, 1190, 1192, 1193, 1195, 1196, 1200, 1203, 1204, 1210, 1223, 1226, 1230, 1235, 1236, 1242, 1243, 1248, 1250, 1254, 1257, 1258, 1259, 1262, 1274, 1291, 1292, 1295, 1302, 1304, 1315, 1319, 1323, 1325, 1329.

Maidu, Southern (Nisenan)

1, 5, 13, 14, 19, 24, 30, 31, 34, 49, 64, 88, 107, 110, 114, 142, 163, 165, 191, 205, 209, 224, 295, 317, 318, 374, 380, 422, 425, 437, 530, 532, 537, 573, 583, 641, 681, 797, 846, 858, 868, 871, 909, 969, 971, 977, 1057, 1060, 1061, 1068, 1140, 1145, 1150, 1251, 1252, 1259, 1279, 1320.

Maidu (Northern)

33, 34, 56, 78, 98, 102, 121, 148, 278, 280, 287, 297, 317, 374, 564, 620, 732, 741, 797, 799, 828, 869, 948, 955, 968, 971, 975, 977, 1029, 1030, 1056, 1057, 1061, 1065, 1083, 1092, 1103, 1107, 1123, 1141, 1144, 1153, 1250, 1278, 1279, 1287, 1325, 1333.

Maidu, Northwestern (Konkow)

380, 1251, 1256, 1257, 1307.

Maidu (Plains)

148.

Miwok

1, 13, 14, 15, 17, 23, 34, 38, 64, 78, 89, 90, 92, 99, 102, 103, 110, 114, 115, 116, 119, 122, 124, 127, 133, 135, 142, 144, 147, 148, 164, 176, 179, 187, 190, 201, 204, 205, 206, 207, 209, 214, 219, 221, 223, 226, 231, 232, 236, 238, 243, 253, 254, 256, 257, 263, 265, 266, 27`, 278, 279, 280, 289, 295, 296, 302, 306, 317, 333, 339, 40, 341, 342, 343, 345, 348, 358, 360, 364, 374, 381, 382, 393, 396, 398, 399, 400, 401, 403, 405, 408, 421, 425, 432, 433, 440, 447, 460, 462, 464, 465, 468, 469, 484, 486, 489, 496, 498, 502, 518, 528, 546, 561, 569, 573, 575, 576, 581, 583, 585, 597, 598, 610, 619, 623,

624, 632, 635, 641, 643, 651, 652, 662, 669, 671, 675, 677, 681, 694, 698, 701, 712, 719, 725, 738, 748, 753, 766, 767, 775, 777, 778, 787, 788, 798, 800, 801, 811, 820, 831, 832, 841, 843, 846, 847, 850, 853, 858, 860, 862, 864, 868, 869, 871, 881, 884, 896, 897, 900, 916, 920, 921, 922, 927, 929, 930, 931, 932, 936, 937, 944, 948, 963, 972, 975, 977, 985, 994, 997, 1007, 1029, 1030, 1032, 1043, 1048, 1054, 1057, 1061, 1068, 1088, 1091, 1101, 1111, 1112, 1113, 1121, 1129, 1141, 1147, 1148, 1149, 1151, 1161, 1162, 1165, 1166, 1172, 1174, 1178, 1180, 1183, 1190, 1191, 1193, 1196, 1197, 1227, 1229, 1232, 1250, 1251, 1257, 1259, 1262, 1269, 1271, 1272, 1279, 1279, 1280, 1291, 1294, 1304, 1305, 1315, 1319, 1325, 1327, 1340, 1341.

Miwok (Northern Sierra)

13, 136, 221, 223, 256, 295, 302, 317, 374, 569, 738, 916, 971, 977, 1057, 1061, 1062, 1065, 1319, 1325.

Miwok (Central Sierra)

13, 15, 17, 23, 34, 78, 92, 103, 106, 110, 115, 119, 122, 124, 127, 136, 148, 164, 176, 179, 187, 190, 201, 204, 206, 209, 212, 214, 223, 226, 231, 236, 238, 253, 254, 256, 263, 265, 266, 271, 278, 279, 280, 295, 302, 306, 317, 333, 340, 341, 343, 345, 348, 358, 359, 360, 374, 382, 393, 398, 399, 400, 403, 405, 408, 421, 425, 432, 440, 447, 465, 466, 469, 496, 581, 585, 597, 598, 610, 619, 623, 624, 641, 643, 652, 662, 669, 671, 694, 698, 701, 712, 719, 725, 738, 748, 753, 766, 767, 777, 778, 788, 798, 800, 811, 820, 831, 832, 843, 846, 847, 858, 862, 879, 881, 884, 896, 897, 922, 927, 930, 931, 932, 937, 944, 963, 971, 975, 985, 994, 997, 1007, 1029, 1030, 1054, 1057, 1061, 1062, 1065, 1068, 1078, 1088, 1090, 1101, 1111, 1112, 1121, 1141, 1148, 1149, 1151, 1161, 1162, 1172, 1174, 1178, 1191, 1193, 1196, 1197, 1227, 1229, 1232, 1251, 1257, 1259, 1262, 1269, 1271, 1272, 1279, 1280, 1291, 1304, 1305, 1315, 1319, 1325, 1327.

Miwok (Southern Sierra)

15, 147, 221, 223, 243, 280, 295, 374, 400, 465, 712, 777, 798, 816, 977, 1030, 1262, 1319, 1327.

Miwok (Coast)

83, 90, 133, 169, 219, 362, 406, 420, 426, 432, 487, 527, 632, 676, 706, 717, 776, 795, 821, 842, 963, 941, 1012, 1014, 1051, 1111, 1149, 1160, 1211, 1251, 1191, 1292, 1325.

Miwok (Lake)

1121, 1251.

Modoc (Lutuami)

1, 44, 45, 70, 77, 105, 137, 181, 195, 223, 254, 284, 447, 490, 524, 551, 568, 632, 675, 694, 765, 768, 829, 878, 931, 933, 960, 968, 975, 1017, 1029, 1101, 1111, 1116, 1117, 1118, 1121, 1124, 1153, 1158, 1187, 1200, 1237, 1254, 1287, 1294, 1329.

Mohave

454, 586, 694, 1021, 1121.

Monache (Western Mono)

13, 34, 110, 142, 256, 295, 296, 314, 380, 400, 636, 675, 860, 868, 971, 1017, 1018, 1043, 1054, 1121, 1135, 1174, 1291, 1335, 1337.

Mono Lake Paiute (see Paiute)

Nisenan (see Maidu)

Nomlaki (see Wintun)

Nongatl

33, 63, 64, 965, 1039, 1121, 1319, 1333.

Ohlone (see Costanoan)

Paiute

Mono Lake Paiute
15, 125, 128, 130, 137, 255, 306, 324, 446, 482, 524, 541, 607, 627, 675, 734, 739, 869, 906, 940, 1024, 1099, 1103, 1142, 1176, 1203, 1205, 1245, 1330.

Northern Paiute (Paviotso)
15, 50, 52, 58, 59, 66, 79, 102, 117, 177, 184, 227, 243, 298, 306, 310, 325, 332, 334, 335, 376, 378, 419, 448, 449, 468, 490, 551, 602, 626, 632, 656, 675, 691, 733, 734, 754, 756, 757, 758, 829, 834, 869, 874, 903, 920, 933, 972, 995, 1030, 1081, 1084, 1103, 1121, 1141, 1153, 1176, 1187, 1224, 1279, 1285, 1294, 1302, 1304, 1325, 1329.

Owens Valley Paiute
43, 85, 87, 99, 110, 136, 137, 148, 155, 188, 189, 263, 279, 296, 305, 306, 309, 317, 351, 395, 397, 400, 421, 444, 457, 459, 473, 482, 483, 524, 562, 573, 576, 603, 627, 684, 694, 720, 734, 771, 782, 804, 828, 829, 869, 884, 903, 928, 960, 972, 979, 998, 1017, 1040, 1061, 1063, 1082, 1099, 1103, 1114, 1121, 1123, 1135, 1140, 1142, 1153, 1158, 1176, 1203, 1214, 1219, 1220, 1222, 1271, 1272, 1286, 1287, 1294.

Panamint Shoshone (Koso)
99, 155, 194, 256, 273, 295, 327, 443, 458, 603, 695, 705, 782, 860, 869, 887, 960, 972, 984, 1018, 1021, 1043, 1078, 1098, 1121, 1125, 1158, 1212, 1213, 1220, 1244, 1251, 1337, 1338.

Patwin (see Wintun)

Pomo
13, 30, 32, 33, 34, 51, 60, 63, 64, 77, 88, 99, 105, 106, 110, 115, 122, 133, 135, 140, 145, 146, 165, 170, 194, 195, 196, 207, 210, 224, 229, 230, 231, 232, 235, 236, 254, 256, 260, 295, 317, 374, 380, 392, 399, 409, 412,

421, 425, 465, 502, 537, 556, 568, 573, 575, 578, 609, 624, 630, 631, 649, 675, 676, 681, 686, 693, 694, 727, 796, 798, 801, 821, 866, 871, 876, 878, 927, 931, 932, 939, 948, 958, 959, 966, 975, 977, 1001, 1010, 1039, 1043, 1052, 1053, 1056, 1057, 1058, 1061, 1062, 1071, 1074, 1078, 1120, 1121, 1123, 1125, 1129, 1132, 1135, 1141, 1144, 1147, 1150, 1153, 1158, 1160, 1173, 1226, 1228, 1250, 1251, 1254, 1257, 1262, 1264, 1268, 1274, 1279, 1282, 1287, 1300, 1319, 1325.

Pomo (Southeastern)
195, 229, 876, 931, 932, 971, 1010, 1060, 1062, 1281, 1291.

Pomo (Southern)
34, 115, 796, 1053.

Pomo (Southwestern)
13, 15, 21, 28, 33, 34, 35, 37, 53, 60, 63, 64, 67, 78, 91, 99, 101, 105, 106, 112, 113, 115, 122, 133, 140, 141, 148, 163, 165, 166, 194, 197, 201, 206, 225, 226, 230, 232, 236, 248, 254, 256, 262, 267, 268, 286, 295, 317, 322, 337, 352, 353, 363, 370, 374, 380, 392, 395, 421, 429, 447, 469, 471, 472, 502, 537, 544, 558, 564, 565, 566, 573, 587, 590, 592, 614, 619, 632, 633, 641, 649, 651, 653, 655, 661, 663, 676, 684, 709, 734, 735, 736, 752, 758, 766, 774, 786, 798, 801, 815, 824, 826, 844, 859, 867, 871, 876, 884, 902, 932, 941, 945, 959, 971, 974, 975, 977, 990, 1003, 1008, 1010, 1028, 1034, 1035, 1039, 1043, 1052, 1053, 1058, 1060, 1062, 1066, 1068, 1073, 1075, 1076, 1077, 1087, 1101, 1102, 1108, 1109, 1110, 1111 1112, 1123, 1128, 1132, 1141, 1143, 1145, 1152, 1153, 1160, 1173, 1186, 1210, 1227, 1235, 1246, 1250, 1251, 1254, 1259, 1260, 1265, 1272, 1273, 1276, 1277, 1283, 1285, 1287, 1290, 1291, 1293, 1294, 1297, 1300, 1301, 1303, 1311, 1319, 1323, 1325, 1326, 1333, 1343.

Pomo (Central)
34, 60, 115, 165, 194, 207, 229, 232, 256, 425, 465, 630, 798, 801, 931, 932, 1010, 1053, 1055, 1056, 1057, 1060, 1062, 1071, 1108, 1135, 1279, 1281, 1291.

Pomo (Eastern)
60, 194, 256, 575, 796, 1056, 1060, 1062, 1287, 1291.

Pomo (Northern)
194, 256, 878, 1056, 1058, 1173.

Salinan
31, 34, 106, 137, 163, 205, 339, 350, 433, 457, 564, 641, 649, 904, 1000, 1028, 1054, 1081,1106, 1123, 1135, 1140, 1259, 1318, 1336.

Serrano
135, 156, 647, 1337.

Shasta
1, 11, 14, 15, 30, 64, 71, 78, 96, 98, 106, 110, 115, 135, 144, 243, 317, 348, 374, 577, 579, 622, 628, 868, 876, 879, 966, 971, 975, 977, 1005, 1007, 1030, 1054, 1056, 1060, 1061, 1071, 1078, 1081, 1106, 1107, 1111, 1121, 1140, 1141, 1235, 1251, 1259, 1273, 1291, 1318, 1333.

Sinkyone
34, 64, 106, 110, 140, 242, 314, 375, 564, 655, 709, 797, 871, 876, 902, 959, 1054, 1106, 1110, 1140, 1173, 1235, 1259, 1323.

Shelter Cove Sinkyone
88, 205, 340, 678, 814, 876, 900.

Tolowa
13, 30, 33, 242, 374, 797, 965, 966, 1043, 1051, 1954, 1247, 1333.

Tubatulabal
1, 15, 18, 30, 23, 27, 34, 54, 55, 57, 64, 87, 98, 99, 110, 132, 142, 146, 167, 199, 203, 205, 214, 228, 231, 246, 247, 254, 263, 270, 283, 290, 293,

295, 301, 304, 308, 317, 319, 321, 324, 329, 332, 343, 375, 380, 383, 384, 386, 387, 399, 400, 401, 407, 411, 417, 427, 435, 444, 445, 453, 455, 459, 464, 467, 475, 478, 480, 481, 485, 486, 489, 490, 494, 497, 507, 508, 509, 510, 512, 514, 517, 523, 526, 527, 537, 538, 547, 548, 569, 573, 575, 576, 584, 599, 601, 629, 640, 662, 692, 695, 698, 707, 724, 726, 728, 751, 773, 775, 778, 779, 780, 781, 785, 789, 802, 807, 816, 820, 826, 832, 834, 837, 845, 846, 849, 851, 858, 860, 861, 869, 871, 873, 875, 883, 887, 888, 917, 923, 924, 938, 942, 946, 951, 960, 972, 977, 982, 988, 999, 1001, 1017, 1018, 1038, 1043, 1056, 1057, 1058, 1061, 1070, 1073, 1089, 1114, 1115, 1122, 1123, 1126, 1127, 1129, 1130, 1134, 1135, 1142, 1153, 1179, 1191, 1215, 1217, 1240, 1257, 1284, 1287, 1299, 1304, 1319, 1337, 1338, 1341.

Wailaki

13, 33, 46, 63, 64, 106, 254, 272, 295, 296, 317, 374, 403, 534, 535, 551, 577, 578, 839, 840, 846, 939, 948, 959, 965, 977, 1000, 1121, 1149, 1160, 1242, 1251, 1256, 1262, 1275, 1276, 1319, 1328.

Wappo

34, 62, 88, 98, 110, 130, 133, 142, 205, 256, 295, 314, 374, 641, 675, 868, 971, 977, 1054, 1121, 1140, 1158, 1259, 1291, 1318.

Washo

2, 14, 15, 89, 91, 125, 133, 135, 137, 142, 175, 176, 254, 295, 459, 472, 632, 674, 684, 694, 695, 757, 804, 830, 853, 858, 860, 899, 903, 912, 972, 1043, 1046, 1105, 1121, 1136, 1137, 1241, 1253, 1287, 1304, 1329.

Wilkut (Chilula)

33, 63, 64, 374, 965, 1039, 1121, 1319, 1333.

Wintun

30, 33, 34, 49, 64, 77, 98, 110, 142, 145, 195, 205, 223, 254, 295, 314, 317, 374, 380, 551, 675, 815, 854, 858, 948, 959, 971, 975, 977, 1061, 1078, 1121, 1153, 1190, 1251, 1324, 1333.

Wintun, Northern (Wintu)
1259, 1318.

Wintun, Central (Nomlaki)
34, 69, 77, 110, 163, 205, 797, 868, 966, 1054, 1106, 1140, 1259, 1318.

Wintun, River (Patwin)
868, 1054, 1106, 1121, 1140, 1158, 1318.

Wintun, Hill (Patwin)
34, 49, 110, 691, 868, 971, 977, 1318.

Wiyot
63, 123, 163, 206, 317, 374, 565, 586, 592, 604, 678, 869, 871, 932, 965, 966, 971, 1110, 1121, 1173, 1259, 1300, 1310, 1319, 1323, 1328.

Yana
13, 34, 91, 182, 187, 206, 208, 234, 295, 317, 577, 625, 675, 738, 802, 871, 931, 834, 1043, 1057, 1121, 1179, 1333.

Yokuts
87, 133, 195, 201, 205, 219, 254, 256, 295, 317, 339, 374, 380, 400, 427, 505, 532, 576, 675, 860, 915, 937, 951, 966, 1043, 1054, 1060, 1061, 1078, 1121, 1157, 1158, 1174, 1206, 1207, 1208, 1257, 1259, 1285, 1319.

Yokuts (Northern Hill)
34, 130, 142, 163, 276, 295, 314, 374, 400, 797, 966, 971, 972, 1081, 1106, 1140, 1153, 1318, 1333.

Yokuts (Southern Valley)
98, 110, 142, 314, 400, 868, 1054, 1120, 1140, 1158, 1173, 1292.

Yuki

1, 15, 16, 31, 34, 36, 51, 60, 74, 88, 99, 106, 108, 110, 115, 122, 135, 140, 145, 163, 184, 198, 207, 210, 219, 223, 229, 230, 243, 244, 254, 256, 279, 295, 314, 317, 337, 342, 346, 349, 352, 360, 368, 369, 374, 376, 392, 408, 421, 429, 449, 452, 470, 472, 502, 516, 530, 532, 537, 549, 551, 556, 566, 575, 586, 587, 632, 641, 658, 678, 686, 691, 709, 710, 716, 734, 748, 755, 762, 766, 767, 771, 772, 786, 797, 800, 801, 818, 820, 831, 858, 868, 876, 878, 902, 913, 932, 939, 948, 961, 971, 977, 980, 982, 988, 996, 1006, 1008, 1030, 1039, 1053, 1054, 1055, 1056, 1057, 1060, 1061, 1062, 1071, 1076, 1083, 1085, 1101, 1102, 1108, 1111, 1114, 1121, 1140, 1141, 1158, 1159, 1160, 1163, 1173, 1188, 1227, 1235, 1242, 1248, 1251, 1252, 1254, 1256, 1259, 1261, 1262, 1263, 1265, 1270, 1272, 1275, 1276, 1282, 1285, 1289, 1291, 1298, 1312, 1325, 1328.

Coast Yuki

31, 74, 123, 169, 218, 260, 271, 286, 337, 352, 470, 566, 587, 592, 678, 686, 771, 818, 859, 868, 902, 1002, 1010, 1045, 1051, 1108, 1109, 1111, 1141, 1159, 1173, 1289, 1291, 1294, 1300.

Cupeño

860, 1028, 1040, 1140.

Yurok

30, 33, 63, 64, 135, 186, 286, 374, 551, 568, 580, 586, 587, 653, 755, 768, 792, 868, 876, 948, 965, 966, 975, 1121, 1140, 1173, 1300, 1301, 1319, 1323, 1333.

Appendix B

Phonetics.

These are the phonetics as given in the literature cited. This section is not complete but only what the various authors described and utilized.

Goodrich, Lawson, and Lawson (1980).

From, *Kashaya Pomo Plants*.

The ′ indicates that the consonant is pronounced with an accompanying "clicking," "popping," or "sharp" sound. Long vowels are written with a rasied dot immediately following (a·).

letter	as in
a	*f*ather
a·	
b	*b*it
c	*j* asin "this job" (when said rapidly)
cʰ	*ch*urch
ċ	c (with accompanying popping sound)
d	d*u*d
e	b*ai*t (approximate example)
e·	
f	*f*oot
h	*h*ope
i	*b*eat (approximate example)
i·	
k	*sc*ald

k^h	*c*old
k´	k (with accompanying popping sound)
l	*l*ip
m	*m*all
n	*n*oo*n*
o	n*o* (approximate example)
o˙	
p	s*p*eak
p^h	*p*eek
p´	p (with accompanying popping sound)
q	like English k-sound but made further back in the mouth
q^h	with accompanying puff of air
q´	with accompanying popping sound
r	*r*ude
s	*s*ee
s´	like English ts-sound (in ca*ts*) with accompanying popping sound, or s (as in *s*ibo) with accompanying popping sound.
š	*sh*eet
t	blade of tongue is pressed against area between teeth and ridge between teeth.
t^h	with accompanying puff of air
t´	with accompanying popping sound
ʈ	s*t*one
$ʈ^h$	*t*one
ʈ´	with accompanying popping sound
u	so*u*p (approximate sound)
u˙	
w	*w*all, to*w*n
y	*y*awn, bo*y*
?	Uh-uh

Foster (1944).

From, *A Summary of Yuki Culture.*

A simplified phonetic recording is employed throughout. Symbols requiring special explanation are as follows:

ĉ	as in child
š	as in show
ŋ	as in sing
a^n	nasalized vowel
'	glottal stop
.	long vowel
á	accent

Gifford (1967)

From, *Ethnographic Notes on the Southwest Pomo.*

In general, the vowels in the Pomo words we have recorded have continental values. Three of the vowels have the following values, as indicated by the circumflex accent.

ê	as e̲ in shed
ô	as o̲u̲ in ought
û	as u̲ in but

We recognized two s̲ sounds, the normal English s̲ and the hissing s̲. The latter we have indicated by a capital S. We use c̲h̲ for Barrett's t̲c̲ and s̲h̲ for his c̲.

Kelly (1932)

From, *Ethnography of the Surprise Valley Paiute.*

Used Waterman (1911). From, *Phonetic Elements of the Northern Paiute Language.*

Loud (1918)

From, *Ethnography and Archaeology of the Wiyot Territory.*

In the various Wiyot names all the consanat sounds found in English are encountered, except f, v, and z. In addition to these there were several other sounds. One of these is similar to Welsh II, and has been written L in conformity with the usage of American anthropologists. A catch has been written '. Ch as in church; *x*, met with only three times has a similar sound to German ch in buch; g denotes the sound as in go; j has the English sound, written dj, never the French zj sound; *t* has a sound similar to th in thin. Where a syllable is strongly accented it has been marked this: '. The vowel sounds are as follows:

ä	as in father	ō as in note
a	as in hat	ū as oo in boot
ē	as in met	u as in put
e	as in met	ai as in aisle
ë	as in her	au as ou in loud
ī	as in machine	oi as in oil
I	as in pin.	

Steward (1933)

From, *Ethnography of the Owens Valley Paiute.*

The following phonetic symbols are used in native terms:

a	long as in far
ä	as in awe
c	always with the value of sh .
d	made with the tip of the tongue further back against the palate than in the English d, giving it a resemblance to r.
x	a voiceless fricative, resembling the German ch but farther forward.
ġ	the same voiced.
I	has its Continental value.
ö, ü	somewhat like the Germain in effect, but less clear.
ō, ū	nasalized.
ŋ	ng as in sing.
V	bilabial, often resembles b (which probably does not occur) when carelessly pronounced.

Small elevated letters, usually ending words, are whispered ' written elevated is a glottal stop or catch. Other words are pronounced as in English.

Waterman (1911)

From, *Phonetic Elements of the Northern Paiute Language.*

The sounds of the Paiute may be represented tabularly as follows:

Vowels	u, o, a, e, I.
Dipthongs	ai

[for more detail refer to the Waterman publication as the entire work is a discussion of the Northern Paiute language].

Timbrook (2007)

From, *Chumash Ethnobotany.*

John P. Harrington, upon whose field notes this work is largely based, was first and foremost a linguist.

Vowels

a	as in 'father'
e	as in 'they'
i	as in 'machine'
o	as in 'note'
u	as in 'rude'
ɨ	(barred I) is a distinct vowel midway between short 'I' in 'tick' and short 'u' as in 'tuck', sometimes written elsewhere as 'ə' (schwa).

Consonants are like those in English, with the following exceptions:

'	(glottal stop) is a catch in the throat, as in English 'uh-oh'
č	is pronounced 'ch' as in 'church'
ʰ	(superscript 'h') is a more aspirated sound
q	is like 'k' but further back in the throat
ł	(barred l) is a sort of whispered, shushing 'l'
š	is pronounced 'sh' as in 'ship'
x	is breathy and harsh, like 'ch' in the German pronunciation of 'Bach,' here rendered as 'kh' in the popularized spelling used in this book

Notes on Alphabetizing

'	(glottal stop) counts as a letter, comes before 'a' in the alphabet
ɨ	counts as a letter, comes after 'i' in the alphabet

Appendix C

Synonyms

The number on the left is the plants listing under which the synonym is listed.

3	*Abies concolor lowiana*
5	*Abies nobilis* var. *magnifica*
15	*Achillea borealis*
15	*Achillea Millefolium*
366	*Adenostegia spp.*
29	*Adenostoma sparsifolia*
32	*Adiantum emarginatum*
33	*Adiantum pedatum*
33	*Adiantum pedatum* var. *aleuticum.*
40	*Agoseris gracilens*
875	*Agoseris alpestris*
416	*Aira elongata*
52	*Allium acuminatum*
54	*Allium hayalinum*
59	*Allium pleianthum*
979	*Allocarpa* spp.
63	*Alnus oregana*
65	*Alnus tenuifoliu*
412	*Alsia abietina*

1078 *Rhus trilobata*

1087 *Ribes menziesii* var. *leptosum*

1083 *Ribes occidentale*

1092 *Ribes sanguineum variegatum*

1094 *Ribes velutinum* var. *glanduliferum*

1100 *Rorippa nasturtium-aquaticum*

1105 *Rosa Woodsii*

1111 *Rubus macropetalus*

1110 *Rubus menziesii*

1110 *Rubus spectabilis* var. *franciscanus*

1111 *Rubus vitifolius*

1164 *Salazaris mexicana*

138 *Salicornia subterminalia*

138 *Salicornia subterminalis*

1127 *Salix araquipa*

1123 *Salix argophylla*

1132 *Salix coulteri*

1123 *Salix fluviatilis argyrophylla*

1123 *Salix hindsiana*

1123 *Salix hindsiana* var. *leucodendroides*

1127 *Salix laevigata* var. *araquipa*

1125 *Salix nigra*

1123 *Salix sessilifolia*

1123 *Salix sessilifolia hindsiana*

1132 *Salix sitchensis coulteri*

1136 *Salvia carnosa*

1137 *Salvia dorrii* ssp. *carnosa*

1143 *Sambucus callicarpa*

1141 *Sambucus coerula*

Appendix D

Glossary

Ague
a fit of fever or shivering or shaking chills, accompanied by malaise, pains in the bones and joints, etc., chills.

Anthelmintic
a drug or medicine used in destroying or expelling intestinal worms.

Antispasmodic
a drug or medicine used in relieving or preventing spasms.

Atonic
lack of bodily tone or muscle tone.

Astringent
a substance which contracts the tissues or canals of the body, thereby diminishing discharges of mucus or blood.

Ataxia
loss of coordination of the muscles, especially of the extremities.

Cardiovascular
affecting the blood and heart vessels.

Carminative
causing gas to be expelled from the stomach and intestines.

Cathartic
a purgative; used for evacuating the bowels.

Colic
a severe attack or an increase in violence of a disease, usually recurring periodic pains in the abdomen or bowels.

Consumption
a wasting disease, especially tuberculosis,

of the lungs; often a progressive wasting of the body.

Corm
an enlarged, fleshy bulb-like base of a stem.

Cruciferous
plants belonging to the plant family *Cruciferae.*

Decoction
water in which a plant has been boiled and which therefore contains the constituents of the water soluble substances of the plant.

Dermatitis
an inflamation of the skin.

Discutient
an agent that causes the resolution of tumors, swellings, or the like.

Diuretic
a medicinal substance used to increase the volume of the urine excreted.

Dropsy
an excessive accumulation of fluid (serous) in the body cavities (serous), or in the subcutaneous cellular tissue .

Emetic
a medicinal substance which induces vomiting (emesis).

Emmenagogue
anything used to stimulate the menstrual flow.

Fomentation
the application of warm liquid, etc., to the surface of the body; the liquid, etc., so applied.

Galls
any abnormal vegetable growth or excrescence on plants caused by various agents, as insects, fungi, bacteria, mechanical injury, etc.

Gastroenteritus	inflamation of the stomach and intestines.
Grippe	influenza, an acute, extremely contagious, commonly epidemic disease, characterized by general prostration, occurring with inflamation of the mucus lining of the nasal passage, and bronchial inflammation; caused by a virus.
Hetchel	probably a mis-spelling of "hatchel," to cut roughly.
Hip	the ripe fruit of a rose, especially of a wild rose.
Infusion	the liquid prepared by the steeping or soaking of plant materials in water.
Labile	unstable; apt to change.
Mydriasis	excessive dilation of the pupil of the eye, as a result of disease, drugs, ore the like.
Pinole	a Spanish word applied both by Indians and whites to any dried and ground meal made from parched seeds.
Pleurisy	inflammation of the (serous) membrane around the lungs and thorax.
Poultice	a soft, moist mass applied as a medicament to the body.
Ramada	an open shelter, often having a thatched roof of branches.
Purgative	a medicine used to cause evacuation of the bowels.
Receptacle	in a flower that portion of the stem on which the sepals, petals, stamens, and pistils are borne.

Rubefacient	a substance which causes redness of the skin.
Stomachic	acting as a digestive tonic.
Tonic	producing or tending to produce good muscular tone or tension.
Tuber	a fleshy, usually oblong or rounded thickening or outgrowth of a subterranean stem or shoot of a plant.
Vasomotor	dealing with the diameter of blood vessels.
Vesication	to cause blisters.

About the Author

George R. Mead began to study anthropology in 1962 after being discharged (honorably) from the U. S. Army, Combat Engineers. He eventually received his degrees, a B.A., an M. A., and a Ph. D. in his chosen field, and many years later an M. S. W. in Clinical Social Work. He has worked in aerospace, taught at the college and university levels, worked in a community action agency, ran a restaurant, been unemployed, and worked for the U. S. Forest Service. He is now retired from the work-a-day world but does a certain amount of consulting, writing, and research. He lives seven miles outside of the small town of La Grande, Oregon, with his wife, and two cats.